VHDL Coding Styles and Methodologies

Second Edition

VHDL Coding Styles and Methodologies

Second Edition

Ben Cohen

KLUWER ACADEMIC PUBLISHERS
Boston / Dordrecht / London

Distributors for North, Central and South America:
Kluwer Academic Publishers
101 Philip Drive
Assinippi Park
Norwell, Massachusetts 02061 USA
Telephone (781) 871-6600
Fax (781) 871-6528
E-Mail <kluwer@wkap.com>

Distributors for all other countries:
Kluwer Academic Publishers Group
Post Office Box 322
3300 AH Dordrecht, THE NETHERLANDS
Telephone 31 78 6576 000
Fax 31 78 6576 474
E-Mail <orderdept@wkap.nl>

Electronic Services <http://www.wkap.nl>

Library of Congress Cataloging-in-Publication

Cohen, Ben, 1945 -
VHDL coding styles and methodologies / Ben Cohen. –2nd ed.
p. cm.
ISBN 0-7923-8474-1 (alk. Paper)
1. VHDL (Computer hardware description language) I. Title.
TK7885.7.C65 1999
621.39'2—dc21 99-17750

Printed on acid-free paper.

Printed in the United States of America

CONTENTS

PREFACE

VHDL Coding Styles and Methodologies, 2nd Edition is a follow up book to the first edition of same book and to *VHDL Answers to Frequently Asked Questions*, first and second editions. This book *was* originally written as a teaching tool for a VHDL training course. The author began writing the book because he could not find a practical and easy to read book that gave in depth coverage of both, the language and coding methodologies. This edition provides practical information on reusable software methodologies for the design of bus functional models for testbenches. It also provides guidelines in the use of VHDL for synthesis. All VHDL code described in the book is on a companion CD. The CD also includes the GNU toolsuite with EMACS language sensitive editor (with VHDL, Verilog, and other language templates), and TSHELL tools that emulate a Unix shell. Model Technology graciously included a timed evaluation version of *ModelSim*, a recognized industry standard VHDL/Verilog compiler and simulator that supports easy viewing of the models under analysis, along with many debug features. In addition, Synplicity included a timed version of *Synplify,* a very efficient, user friendly and easy to use FPGA synthesis tool. *Synplify* provides a user both the RTL and gate level views of the synthesized model, and a performance report of the design. Optimization mechanisms are provided in the tool.

This book is intended for:

1. **College students**. It is organized in thirteen chapters, each covering a separate aspect of the language, with complete examples. Students can compile and simulate the examples to get a greater understanding of the language. Each chapter includes a series of exercises to reinforce the concepts.

2. **Engineers.** It is written by an aerospace engineer who has many years of hardware, software, computer architecture and simulation experience. It covers practical applications of VHDL with coding styles and methodologies that represent what is current in the industry. VHDL synthesizable constructs are identified. Included are practical guidelines for the design of bus functional models used in testbenches, such as waveform generation, client/server control, text and binary file command methods, and binary file generation schemes. Also included is an elaboration of a project for the design of a synthesizable Universal Asynchronous Receiver Transmitter (UART), and a testbench to verify proper operation of the UART in a realistic environment, with CPU interfaces and transmission line jitter. An introduction to VHDL Initiative Toward ASIC Libraries (VITAL) is also provided. The book emphasizes VHDL 1987 standard but provides guidelines for features implemented in VHDL 1993.

This book differs from other VHDL books in the following respects:

1. Emphasizes VHDL core, *Ada* like sequential aspects and restrictions, along with the VHDL specific, concurrent aspects of the language.
2. Uses complete examples with good code, and code with common mistakes experienced by users to demonstrate the language restrictions and misunderstandings.
3. Provides a CD that includes all the book examples in addition to GNU EMACS language sensitive editor, other useful reference VHDL code material, and GNU TSHELL.
4. Uses an easy to remember symbology notation throughout the book to emphasize language rules, good and poor methodology and coding styles.
5. Identifies obsolete VHDL constructs to be avoided.
6. Identifies non-synthesizable structures.
7. Covers practical design of testbenches for modeling the environment and automatic verification of a unit under test.
8. Provides a complete design example that uses the guidelines presented in the book.
9. Provides an introduction to VITAL.
10. Provides guidelines for synthesis and identifies the VHDL constructs that are typically synthesizable.

This book is organized in four basic VHDL aspects:

1. **SEQUENTIAL LANGUAGE**. This is similar to the sequential aspects of other programming languages like C or *Ada*. Chapter 1 provides sufficient knowledge to compile and simulate a simple counter. Chapter 2 covers the basic language elements including the lexical elements, the syntax, and the types. Chapter 3 discusses the control structures.
2. **CONCURRENCY.** This differentiates VHDL from other sequential languages. Chapter 4 discusses drivers, chapter 5 covers the timing and chapter 6 emphasizes the concurrent statements.
3. **ADVANCED TOPICS**. This includes subprograms in chapter 7, packages in chapter 8, and attributes, specifications and configurations in chapter 9, and design for synthesis in chapter 10.
4. **APPLICATIONS.** This emphasizes reusable software methods to generate functional models, bus functional models, and testbench designs in chapter 11; a UART project with synthesizable transmitter and receiver in a testbench environment in chapter 12; VITAL coding style optional methodology in chapter 13.

The language rules, coding styles, and methodologies presented in this book support the structure necessary to create digital hardware designs and models that are readable, maintainable, predictable, and efficient.

About The CD

Table 1 summarizes the contents of the enclosed CD.

Table 1. Contents of Enclosed CD

Directory Name	Description
Src_dir	Contents of all source code described in this book, sorted by chapter. Standard packages, Std_Logic_1164 package, and Synopsys packages, Vital 2.0 packages
Usr	GNU toolset with Tshell and emacs editor with VHDL, Verilog, and other programming language modes
man	GNU help files in Windows Help format. Root file is *ManPagesDir*
Etc	Csh.cshrc and my.cshrc startup files for TSHEL
VHDL_Syntax	VHDL'87 and VHDL'93 syntax in HTML format
PDF_FIles	ModelSIm5_2 reference guide, VHDL and Verilog reference cards, Std_Logic_1164 reference card, and European Space Agency Modeling guidelines
NotGnu	GNU Fast EMACS editor with minimal disk space requirements. Editor does not read Lisp, and thus does not run in VHDL or Verilog modes.
ModelSim	Model Technology's VHDL *Modelsim* compiler/simulator
Synplify	Synplicity's *Synplify* synthesizer

NOTATION CONVENTIONS

The following symbols and syntactic description are used to facilitate the learning of VHDL.

SYMBOLS

M	Methodology and guideline.
👍👍	Two thumbs up. Good methodology or approach.
👎👎	Two thumbs down. Poor methodology or approach.
👍👎	Disagreement in community on methodology or approach.
☺	Legal or OK code
💣	Coding Error
	Synthesizable
	Non-Synthesizable
...	Ellipsis points in code: Source code not relevant to discussion.
[1]	Quotations reprinted from IEEE Std 1076-1993 IEEE Standard VHDL Language Reference Manual (LRM). Quotations printed in *"italic and in this font"*. Syntax reprinted from the LRM "in this font", but without the prefix [1].
Boldface	Boldface in text: Emphasizes important points. Boldface in syntax and sample code: Emphasizes VHDL reserved words.

SYNTACTIC DESCRIPTION

left_hand_side ::= right_hand side
left_hand_side is the syntactic category
right_hand_side is a replacement rule
::= (read as "can be replaced by")

Vertical bar separates alternative items
Example: letter_or_digit ::= letter | digit

Square brackets [] enclose optional items
*Example: return_statement ::= **return** [expression]*

Braces {} enclose a repeated item (zero or more times).
Example: index_constraint ::= (discrete_range, {discrete_rang})

Underlined identifies that the notation is applicable for VHDL'93 ONLY
*Example: ... **end** [**configuration**] [configuration_simple_name]*

Acknowledgments

VHDL Coding Styles and Methodologies, 2nd *Edition* evolved from the previous edition of this book, and from *VHDL Answers to Frequently Asked Questions*, first and second editions. It also evolved from several documents and discussions with several individuals, along with personal experiences and frustration of students in using VHDL.

I thank Model Technology for allowing me access to *ModelSim,* an excellent and easy to use VHDL/Verilog compiler/simulator, and for their excellent product support. I thank Synplicity for allowing me access to *Synplify,* a very efficient, user friendly and easy to use FPGA synthesis tool. I also thank these two companies for providing evaluation copies of their tools in this book.

I thank Peter Sinander from the European Space Agency for publishing on the Internet the document VHDL *Modelling Guidelines*[2]. I thank Janick Bergeron from Qualis Design Corp for publishing on the Internet the document *Guidelines for Writing VHDL Models in a Team Environment*[3]. Those documents contributed to many of the coding styles presented in this book. I thank Richard Hall from Cadence Design Systems, Inc. who reviewed the original version of this book and provided many suggestions. I thank Larry Saunders, Steve Schoessow, Johan Sandstrom, and John Coffin for various VHDL discussions we had over the years on the use of VHDL. I thank Synopsys, Inc. for the release of their VHDL packages.

I thank Geoff Voelker, Andrew Innes and Reto Zimmermann for their effort in providing GNU Emacs for Windows NT and Windows 95/98. I thank James Fulcomer and Drew Davidoff for their inquisitive challenges in the use of the language, in addition to compiling the GNU software into an easy to install package.

I also thank my publisher Carl Harris for supporting in these endeavors of publishing books.

I acknowledge my daughter Lori Hillary, and my son Michael Lloyd for inspiring me to teach.

I especially thank my wife, Gloria Jean, for her patience and support in these projects

---

2 The *VHDL Modelling Guidelines* document is available through anonymous ftp from ftp.estec.esa.nl in the "/pub/vhdl" directory.

3 The *Guidelines for Writing VHDL Models in a Team Environment* is available via ftp from vhdl.org as /pub/misc/guidelines.paper.ps.

About the Author

Ben Cohen has an MSEE from USC and is a Scientist engineer at Raytheon Systems Company. He has technical experience in digital and analog hardware design, computer architecture, ASIC design, synthesis, and use of hardware description languages for modeling of statistical simulations, instruction set descriptions, and hardware models. He applied VHDL since 1990 to model various bus functional models of computer interfaces. He authored *VHDL Coding Styles and Methodologies, 1st Edition,* and *VHDL Answers to Frequently Asked Questions*, first and second editions. He was one of the pilot team members of the VHDL Synthesis Interoperability Working Group of the Design Automation Standards Committee who authored the *IEEE P1076.6 Standard For VHDL Register Transfer Level Synthesis.* He has taught several VHDL training classes, and has provided VHDL consulting services on several tasks.

email: **VhdlCohen@aol.com**
Web page: **http://members.aol.com/vhdlcohen/vhdl**

VHDL Coding Styles and Methodologies

Second Edition

1. VHDL OVERVIEW AND CONCEPTS

This chapter presents an overview of VHDL design units and provides guidelines and definitions where applicable. Enough concepts and features of VHDL are introduced to allow the user to compile and simulate the exercises, thus getting the VHDL "feel".

1.1 WHAT IS VHDL

VHDL[4] is all of the following:

1. **Non-proprietary language.** VHDL is defined in IEEE-1076 standard 1987, and IEEE-1076 standard 1993.
2. **Widely supported Hardware Description Language (HDL).** Several vendors have adopted the standard and are supplying VHDL compilers, simulators, and synthesis tools.
3. **Programming language**. Sections of VHDL are similar to *Ada* and include data types, packages, sequential statements, procedures, functions, control structures, and file I/O.
4. **Simulation language**. VHDL includes structures to define and simulate events, timing, and concurrency.
5. **Documentation language.** VHDL is capable of documenting instruction set architectures, state machines, structures, and hardware design hierarchies.
6. **Usable for logic synthesis.** VHDL provides constructs that imply hardware. The language is technology independent, but allows user defined attributes to tailor the synthesis process into a user-defined direction. Several vendors are supplying synthesis tools that read and convert VHDL code into a gate level

---

[4] VHDL is an abbreviation for VHSIC Hardware Description Language. VHSIC is an abbreviation of Very High Speed Integrated Circuit.

description targeted toward specific technologies. The *IEEE P1076.6 Standard For VHDL Register Transfer Level Synthesis* is a document prepared by the *VHDL Synthesis Interoperability Working Group of the Design Automation Standards Committee.* "The purpose of this standard is to define a syntax and semantics that can be used in common by all compliant RTL synthesis tools to achieve uniformity of results in a similar manner to which simulation tools use the IEEE 1076 standard". The document is available from the Institute of Electrical and Electronics Engineers, Inc. (*http://stdsbbs.ieee.org/*).

7. **Functional verification language.** :VHDL can be used to define the environment necessary to verify the units under test (UUTs). This environment include software or hardware models of the UUTs' interfaces. It can also include a verifier model to verify the correct functionality of the UUTs when exposed to the test stimuli.

1.2 LEVEL OF DESCRIPTIONS

VHDL is a hardware description language with a vocabulary rich enough to span a very wide range of design descriptions. VHDL inherits from *Ada* the **typing definitions** (e.g. enumerations, arrays, records, pointers), the strong *Ada* **type checking**, and the **overloading of operators and subprograms** (e.g., "+" for integer and "+" for Std_Logic_Vector). In addition, to separate the supporting programming structures from the problem domain (i.e., information hiding), VHDL inherits from *Ada* the concepts of **packages**. Packages enable the reusability of common routines and definitions (e.g., constants, types, functions and procedures). While *Ada* uses one construct (the "task") to describe concurrency, **VHDL provides several concurrent constructs** that relate more closely to real hardware design, including the description of hierarchy. VHDL provides specific constructs to define the structures or inter-connectivity of hierarchical hardware designs. With the use of configuration declarations, a user can configure a design with different alternate architectures for its subelements, thus allowing the analysis of various design alternatives.

As a result of its flexibility, VHDL is used at several stages of the design processes to describe and verify designs. These stages are generally classified as "levels of representation" of the design aspect. There are many different interpretations, definitions, and opinions of the modeling levels, some of which are overlapping and typically include the following:

1. **System Level** (SL). :There are several interpretations as to what a "system" is. One such interpretation is a **statistical** model. This represents tokens transmitted through a Petri net to emulate transactions that make demands on system resources. The purpose of this type of simulation is to assess the efficiency of a design, in terms of tasks or jobs, imposed on resources. These resources could be busses, processors, memories, FIFO's, etc. Other interpretations of system level modeling include the **algorithmic** level that defines the algorithms for a particular implementation (e.g. image processing algorithms), and the **protocol** level that defines the communication protocol between units.
2. **Board Level** (BL). :This is often referred to as "system level" because a board typically represents a subsystem function. Board level simulation is typically performed using VHDL components, modeled at various levels, that simulate the

system environment and component interfaces. Bus Functional Models (BFM, see chapter 11) are often used to represent the bus interfaces of components because of their efficiency. Large gate level designs are often simulated with gate level accelerators or cycle-based simulators in mixed mode simulation (i.e., accelerated gate level simulation in a non-accelerated environment).

3. **Instruction Set Level** (ISA). :The ISA level is typically used to define and simulate the instruction operations of a processor. Instructions are fetched and decoded based on the instruction format. The executions of the instructions are typically performed using the algebraic and logical operators.
4. **Register Transfer Level** (RTL). :This level describes the registers and the Boolean combinatorial equations (or logic clouds) between the registers. It is often used as an input to a logic synthesizer. It is also used to define state machines for controller designs where the state registers may be either explicitly or implicitly defined depending on the declarations and coding style (see chapter 10).
5. **Structural Level** (SL). :This level represents the structure and interconnections of a design. This level could be generated using either a manual text editor or automatically from a tool that converts a schematic to a VHDL structure.
6. **Gate level** (GL). :This level describes the structural interconnections of low level elements (gates and flip-flops) of a design. It is generally a VHDL output of a synthesizer or layout tool and is used either as documentation of the netlist, or as a VHDL model of the structural definition of a design for VHDL simulation.

1.3 METHODOLOGY AND CODING STYLE REQUIREMENTS

Methodology is an art that represents an orderly approach in performing a task. Its beauty lies in the eye of the beholder. Not everyone agrees with a methodology because what is orderly and consistent to one individual may be viewed as inconsistent, cumbersome or uneconomical to another. It is possible to build anything with almost any methodology; however, a good methodology usually would create a unit of higher quality or less effort. It is also important to note that not all projects need the same process or methodology.

This book presents a coding style and methodology that abides by the VHDL rules. The methodology presented in this book stresses the following requirements:

1. **Code must abide by the VHDL language rules**. The language is explained in terms of its capabilities and legal constructs. Legal and illegal constructs are also identified and highlighted with examples.
2. **Code should have a common look and feel** to enhance code familiarity between different models.
3. **Code should be easily readable and maintainable** not only by the author, but also by others.
4. **Code must yield expected results**, whether the description is a behavioral level or synthesizable description. Guidelines are provided with regards to synthesis and coding style.
5. **Obsolete or outdated VHDL should be avoided**. Those constructs are defined, and alternate constructs are explained.

6. **Simulatable (versus synthesizable) code should be efficient from a simulation viewpoint**. Guidelines to enhance simulation speed are provided.
7. **Code should be cohesive**. Thus, common functions should be lumped in common packages, partitions, or architectures.
8. **Synthesizable code must abide to vendor's synthesis rules.** Compliance to the *IEEE P1076.6 Standard for VHDL Register Transfer Level Synthesis* is recommended for enhanced portability to other synthesizers. In any case, synthesizable VHDL structures must be selected to match vendor's synthesis requirements.

1.4 VHDL TYPES

Types and subtypes (see section 2.3) represent the *[1] set of values and a set operations*, structure, composition, and storage requirement that an object, such as a variable or a constant or a signal, can hold. VHDL has a set of predefined types in package Standard. A package is a design unit that allows the specification of groups of logically related declarations. Table 1.4 presents a summary of the type declarations defined in package Standard of the LRM. Appendix B presents the full package.

Table 1.4 Type Declarations Summary of Package Standard for VHDL'93

```
type boolean is (false, true);
type bit is ('0', '1');
type character is (nul, ..., '0', ..,'9', ':', ';' ... '>', '?', '@', A', .., 'Z', 'a', .., 'z', .., DEL, -- VHDL'87
    c128, ..., 'À', ..., 'à', ...'ÿ' ); -- VHDL'93 only
type severity_level is (note, warning, error, failure);
type integer is range implementation defined;
-- type universal_integer must include range -2147483647 to +2147483647;
type real is range implementation defined;
-- type universal_real must at least be of range -1.0E38 to +1.0E38 and
-- must include a minimum of six decimal digits of precision
type time is range implementation_defined units ... end units;
  -- Guaranteed to include the range -2147483647 to 2147483647 (-2^31 to 2^31)
subtype delay_length is time range 0 fs to time'high;
subtype natural is integer range 0 to integer'high;
subtype positive is integer range 1 to integer'high;
type string is array (positive range < >) of character; -- array size is not constrained
type bit_vector is array (natural range < >) of bit; -- -- size of the array is not constrained
type file_open_kind is (read_mode, write_mode, append_mode); -- VHDL'93 only

type file_open_status is (open_ok, status_error, name_error, mde_error); -- VHDL'93 only
```

1.5 VHDL OBJECT CLASSES

VHDL categorizes objects into four classes (LRM 4.3). *[1] An object is a named item that has a value of a given type* that belongs to a class. These classes include:

1. **Constant** *[1] An object whose value may not be changed.*
2. **Signal** *[1] An object with a past history.* The simulator must maintain the necessary data structures to maintain this time history. Components ports are signals. During simulation signals require a lot of overhead storage and overhead performance. In synthesis, signals represent either wires, or the outputs of combinational logic, or latches, or registers. Care must be taken in using signals to prevent the unwanted creation of latches or registers. .
3. **Variable** *[1] An object with a single current value.* In simulation, variables have a simpler data structure because they do not have a history (or time) associated with them, and thus require less storage. As a result they are also more efficient than signals. In synthesis, variables in processes represent either temporary combinational logic, or latches, or registers. Care must be taken in using variables in processes to prevent the unwanted creation of latches or registers. . .
4. **File** An object used to represent file in the host environment. . .

A class and a type represent different concepts.

1. The **type** of an object represents its **structure, and storage** requirement (e.g. type integer, real).
2. A **class** is relevant to the nature of the object, and represents **HOW** the object is used in the model. Thus, a signal's value can be modified but has "time" element associated with it. A variable can also be modified, but has no "time" association, whereas a constant is like a variable, but its value cannot be modified. A file cannot be modified, but interacts with the host environment. Constants, signals, and variables can however be of any type, but with some class restrictions. The objects defined in a file are of a user-defined type, with some restrictions. Thus, it is possible to define a file of character strings (text), integers, reals, or records.

The following suffixes (or prefixes) are recommended to denote the class of an object:

Suffix	Prefix	Comments	Example
_c	c	Represent constants.	Width_c, cWidth
_s _r	s r	Represent feedback signals (see section 1.6.1.2) Represent signals implemented as registers. *NOTE: No prefix or suffixes used to represent Ports or signals that represent Wires or combinational logic. This approach enhances readability*	Addr_s sAddr Addr_r rAddr
_v	v	Represent variables	Data_v, vData
_f	f	Represent files.	Data_f, fData

Chapter 2, section 2.1.1.2 discusses the subject of naming convention and suffixes.
Rationale: *Readability is enhanced if object class labels are attached to the object instance names.*

1.5.1 Constant

[1] (LRM 4.3.1.1) A constant is an object whose value may not change. The syntax for a constant declaration is: .

```
constant_declaration ::=
    constant identifier_list : subtype_indication [ := expression ];
```

Use constants to define data parameters and table lookups. Table lookups can be used in a manner similar to function calls to either translate a type or lookup a data value with reference to an index (see arrays). For example, use the "ToTypeName_c" as the style for the constant identifier if the constant is used for type conversion.
Rationale: *Constants play a very important role in VHDL because they create more readable and maintainable code (see rationale for generics in section 1.6.1.1.3). The use of constants as type translators or data value lookup is very efficient from a simulation viewpoint. The suffix or prefix in the identifier enhances readability.*

Figure 1.5.1 represents an example of type and constant declarations declared in an entity (see section 1.6). Arrays are discussed in chapter 2.

```
...
subtype  TwoBits_Typ   is integer range 0 to 1;
type     ArrayBits_Typ is array(TwoBits_Typ) of bit;
type     ArrayInt_Typ  is array(bit) of TwoBits_Typ;
--natural is subtype of integer
constant WordWidth_c : natural := 16; -- width of a word.
-- These constant tables converts a natural number to a bit
-- and a bit to an integer.
-- example: if Int_v is an integer variable, and
--             Bit_v is a bit variable then the following is true
--   Int_v := 1;
--   Bit_v := ToBit_c(Int_v); -- Bit_v becomes '1'. Table lookup through array
--   Bit_v := '0';
--   Int_v := ToInt_c(Bit_v); -- Int_v becomes 0.   Table lookup through array

constant ToBit_c     : ArrayBits_Typ := (0 => '0',
                                         1 => '1');
constant ToInt_c     : ArrayInt_Typ  := ('0' => 0,
                                         '1' => 1);
```

Figure 1.5.1 Example of Type and Constant Declarations (ch1_dir\Cdemo_p.vhd)

1.5.2 Signal and Variable

In synthesis, signals and variables can represent the outputs of combinational logic or can represent registers.

A **SIGNAL** can be declared in various declarative sections:

1. **Port of an entity**. A port is a signal.
2. **Architecture declarative part of an architecture (see 1.6.2).** Such a signal represents wires or registers, or latches internal to the architecture.
3. **Block declarative part of a block.** Such a signal is local to the block.
4. **Package declaration** (see chapter 8). Such a signal is considered "global" because it may accessed by design units that uses the specified package
5. **Formal parameters of a subprogram (i.e. function and procedure).** During a subprogram call, the actual signal being passed to the subprogram becomes associated with the formal signal parameter.

A signal has three properties attached to it, including:

1. **Type** and **type attributes**. The type insures consistency in operations on objects. Attributes defines characteristics of objects (e.g. S'high, see chapters 2 and 5 for the definitions of attributes).
2. **Value.** This includes current, future, and past value (e.g. S'last_value).
3. **Time.** This represents a time associated with each value.

A **VARIABLE** can be declared in various declarative sections:

1. **Architecture declarative part of an architecture (see 1.6.2).** This is only allowed as a shared variable (for VHDL'93 only). Shared variables are not synthesizable.
2. **Block declarative part of a block.** This is only allowed as a shared variable (for VHDL'93 only).
3. **Package declaration** (see chapter 8). This is only allowed as a shared variable (for VHDL'93 only). Such a variable is considered "global" because it may accessed by design units that uses the specified package
4. **Formal declaration section of a process.** In synthesis, variables in processes represent either temporary combinational logic, or latches, or registers. Care must be taken in using variables in processes to prevent the unwanted creation of latches or registers.
5. **Formal parameters of a subprogram (i.e. function and procedure).** During a subprogram call, the actual variable being passed to the subprogram becomes associated with the formal variable parameter.

A variable has two properties attached to it including:

1. **Type** and **type attributes** just like the signal properties, but with no attributes associated with time.
2. **Value** with no time history.

M Use **signals** as **channels of communication** between concurrent statements (e.g. components, processes). In non-synthesizable models, avoid using signals to describe **storage elements** (e.g. memories). Use **variables** instead.
Rationale: *Signals occupy about* ***two orders*** *of magnitude more storage than variables during simulation. Signals also cost a performance penalty due to the simulation overhead necessary to maintain the data structures representing signals.*

Table 1.5.2 compares the amount of storage for various objects declared as signals and as variables. Those Figures are based upon Model Technology's memory requirements. Those Figures are simulator dependent and will vary among vendors. Section 10.1.1 presents the model of a memory that uses a variable for storage element.

Table 1.5.2 Storage for Elements Declared as Signals and Variables

ELEMENT TYPE	VARIABLE	SIGNAL
Enumeration type < 256 states	**1 byte** -- 8 bits	**1 + ~ 100 byte**:
Enumeration type > 256 states	**4 bytes** -- 32 bits	**4 + ~ 100 byte**:
Array(1 to n) of EnumType (<256 states)	**n * 1 byte**	**n * (1 + ~ 100 byte)** :
integer	**4 bytes** -- 32 bits	**4 + ~ 100 bytes**
real	**8 bytes** -- 64 bits	**8 + ~ 100 bytes**
time	**8 bytes** -- 64 bits	**8 + ~ 100 bytes**
array 1 to 64,000 of Std_LogicVector(31 downto 0) -- 64K X 32 bits Memory	64K * 1 * 32 = 2,048 Kbytes **~ 2 MBytes**	64K * 101 * 32 = 206,848 Kbytes **~ 207 Mbytes**
array 1 to 64,000 of integer -- 64K integer Memory	64K * 4 = 256 K **~ 1/4 Mbyte**	64K * 104 = 6,656 Kbytes **~ 7 Mbytes**

1.5.3 File

A file represents objects stored in files in the host environment. Section 8.2 expands the use of files that relates to the TextIO package.

1.6 VHDL DESIGN UNITS

VHDL contains several *[1](LRM 11.1) design units constructs that can be independently analyzed and stored in a design library.* A library is a collection of compiled VHDL design units. Design units can be stored in separate files or grouped in common files. The VHDL design units are shown in Table 1.6:.:::

Table 1.6 VHDL Design Units

Design Unit	Comments
1. **ENTITY**	Represents the interface specification (I/O)
2. **ARCHITECTURE**	Represents the function or composition of an entity. Together the entity/architecture pair represents a component.
3. **PACKAGE DECLARATION**	Provides a collection of declarations (types, constants, deferred constants (see chapter 8), signals, components) or subprograms (procedures, functions). Shared variables (VHDL'93) can also be declared. The subprogram bodies are NOT described.
4. **PACKAGE BODY**	Provides a complete definition (i.e. the algorithm or body) of the subprograms. Values for deferred constants are also declared.
5. **CONFIGURATION DECLARATION**	Binds a particular architecture to an entity or binds an entity/architecture pair to a component.

M 👍👎 It is recommended to use the design unit name as the file name.

DESIGN FILE	NAME
Entity	EntityName. vhd
Architecture	ArchitectureName. vhd
entity/architecture pair 👍👎	EntityName._ArchitectureName. vhd
Package Declaration	PackageName_p.vhd
Package Body	PackageName_b.vhd
Package declaration and body 👍👎	PackageName_pb.vhd
Configuration	ConfigName.vhd

Rationale: *The file name should relate to the design name. Using the design unit name as the file name simplifies the location of design units within the file storage system. Separate design files are easier to maintain, particularly on a large project. Combining an entity and an architecture or a package body and package declaration in one file is more difficult to maintain and can cause unnecessary recompilation of other design units because of the design units dependencies. Thus, if a package declaration is recompiled, then all design units that make use of that package must also be recompiled. However, if a package body is recompiled, no design units requires recompilation. Similarly, if an entity is recompiled then all the architectures of that entity must be recompiled, and all the architectures that instantiate components of those entity/architectures must also be recompiled. If however, only the architecture of an entity is recompiled, no other recompilation is necessary (see chapter 6 and 8).*

1.6.1 ENTITY

(LRM 1.1) An entity defines and represents the interface specification of a design and defines a component from the external viewpoint. Thus, an entity defines the inputs/outputs or "pins" of a component. It assigns the component name and the port names. See chapter 6 for an expanded discussion of entities. Figure 1.6.1 represents an entity for a counter. An entity may include port declarations, which define the names, data types and directions of the port signals. An entity may have NO port declaration (e.g., testbench or system). Note that ports are of the class "signal".

1.6.1.1 Style

This example demonstrates several concepts in methodology including:

1. Comments
2. Header
3. Generics
4. Indentation
5. Line lengths
6. Statements per line
7. Declarations per line
8. Alignment of declarations

The enclosed EMACS editor (on CD) with the Language Sensitive VHDL templates facilitates the use of headers, indentation, and alignment of declarations.

```
--------------------------------------------------------------------------
--
--   Project               : ATEP CLASS
--   File name             : Adder.vhd
--   Title                 : Adder
--   Description           : Adds two Integer numbers (up to 32 bits)
--                         : Maximum range is defined in a generic
--   Design Libray         : Atep_Lib
--   Analysis Dependency: none, does not require any library
--   Simulator(s)          : Model Technology on PC
--                         : Cadence Leapfrog on Sun workstation
--   Initialization        : none
--   Notes                 : This model is designed for synthesis.
--                         : Compile with VHDL'93
--------------------------------------------------------------------------
--    Revisions    :
--          Date              Author  Revision          Comments
--        7/02/98  cohen     Rev A             Creation
--                            VhdlCohen@aol.com

--------------------------------------------------------------------------
    entity Adder is
      generic          -- Interface constants, can be modified by configuration
        (MaxValue_g    : Integer := 255);

      port
        (A      : in  Integer range 0 to MaxValue_g; -- constrained range
         B      : in  Integer range 0 to MaxValue_g;
         Y      : out Integer range 0 to MaxValue_g);
    end entity Adder;
```

Header

Generics enhance design

Figure 1.6.1 Counter Entity (ch1_dir\adder.vhd)

1.6.1.1.1 Comment

[1] A comment starts with two adjacent hyphens and extends up to the end of the line. A comment can appear on any line of a VHDL description.

M The purpose of **comments** (any characters following "--") is to allow the function of a design to be understood by a designer not involved in the development of the code. Thus, a comment helps in understanding the code. Comments can be on the same line with the code if they are short, otherwise comment lines should be lines on their own. In this case, the block comments should immediately precede the code they describe. Example:

```
   Subtype Int6_Typ is integer range 0 to 5;  -- Used for counter

     -- This type defines the states used in XYZ processor. The default
     -- state is the idle state.  The state machine is described
     -- in detailed in document ABC_XYZ.
   type  State_Typ is (Idle, Fetch, Decode, Execute,
                       Interrupt, DMA);
```

Rationale: *Trailing comments after the code are cumbersome. The comment should help understand the code that is about to be read, not the other way around.*

1.6.1.1.2 Header

M 👍👍 Each file should have a descriptive **header** that is composed of a set of comments containing the following information (Note: The project should decide which header information is appropriate) .

1. Project name.
2. File name.
3. Title of the design unit.
4. Description of the design unit including its purpose and model limitations.
5. Design library where the code is intended to be compiled in.
6. List of analysis dependencies, if any (e.g. packages, components (for simulation)).
7. Initialization of model (e.g. hardware RESET, port and signal initialization).
8. Notes or other items (e.g. synthesis aspect of design unit).
9. Author(s) and full address (email, phone number, etc.).
10. Simulator(s), simulator version(s), and platform(s) used.
11. Revision list containing version number, author(s), the dates, and a description of all changes performed.

Rationale: *Comments and headers are important to maintain proper documentation.*

Note: Throughout this book, the headers will include only a minimum set of documentation to conserve space. The files on disk contain full headers.

1.6.1.1.3 Generics

An entity also may define **generics** values as component parameters. *[1] Generics provide a channel for static information to be communicated to a block* (e.g., architecture) *from its environment. Unlike constants, however, the value of a generic can be supplied externally, either in a component instantiation* (i.e., plugging in of a component, see chapter 6*) or in a configuration specification or declaration* (see chapter 9). Generics can be used to control the model size (e.g. array width and depth sizes), component instantiations, timing parameters, or any other user defined parameter. Chapter 3 discusses limitations in the use of objects that are based on the values of generics.

M 👍👎 For non-VITAL compliant models use suffix "_g" to denote a generic. For VITAL compliant models (see chapter 13), use the recommendations defined in the VITAL specification.

Rationale: *A generic is like a constant but is declared in an entity. User readability is enhanced when the object class is known at a glance. VITAL models must conform to the VITAL specification.*

M 👍👍 **Avoid** using "**Hard-Coded**" numbers for characteristics that may change throughout the lifetime of the model. **Use generics or constants.**

Rationale: *Generics or constants promote code reusability and increase the usefulness and lifetime of a design because the model can adapt to a variety of environments by postponing or modifying those parameters late in the design cycle (see chapter 8 on deferred constants, and chapter 9 on configurations). Another benefit of using constants and generics is the increased readability.*

1.6.1.1.4 Indentation

M 👍👍 All declarative regions and block statements should be **indented by at least 2 spaces**. A block statement can be the conconcurrent statement part of an architecture, the **else** clause of an **if** statement, the body of a subprogram, etc. If at all possible, indentation levels in sequential statements should not exceed 4. Use subprograms to break the code into manageable parts (see chapter 2, and 3 for specific examples).

Rationale: *Indentation enhances readability. Too many indentation spaces quickly occupy a line. A large number of indentation levels is often an indication of bad programming style.*

1.6.1.1.5 Line length

M 👍👍 Lines should not exceed 80 characters in length.

Rationale: *Restricting line lengths to less than 80 characters avoids confusing wrap-arounds when viewing the source on a regular text terminal, or when printed on a standard-sized printout.*

1.6.1.1.6 Statements per line

M 👍👍 Each statement should start on a new line. Example:

```
Statement;
if condition then
  statement;
else
  statement;
end if;
```

Rationale: *More than one statement on a line makes the code difficult to read or scan quickly.*

M 👍👍 Long lines should be broken where there are white spaces and continued on the next line. Align the continuation line two or more spaces from the current level of indentation. Example:

```
Variable SomeVeryLongName_v :
    Atep_lib.SomePackageName_Pkg.SomeLongTypeName_Typ :=
      "1010101010110011100001100111001";
Variable X_v                    : Integer;
```

Rationale: *Indenting line continuations makes it possible to quickly distinguish the continuation of a line from new statement.*

M 👍👍 Use blank lines to group logically related text of code.

Rationale: *Blank lines enhance readability and modular organization.*

1.6.1.1.7 Declarations per line

M 👍👍 Each declaration should start a new line.

Rationale: *It is easier to identify individual ports, generics, signals, variables, and change their order or types when their declarations are on separate lines .*

1.6.1.1.8 Alignment of declarations

M 👍👍 Elements in interface declarations should be vertically aligned. Example:

```
procedure  Test
  (signal     Data_s          :  out    A_Typ;
   constant   Adder_c         :  in     B_Typ;
   variable   VeryLongName_v  :  inout  C_Typ);
```

Rationale: *Vertical alignment of interface declarations allows quick identification of the various kind of interface declarations, their names, directions, and types.*

1.6.1.2 Entity Ports

Entities use *[1] ports that provide channels of communication between the component and its environment.* A port is a SIGNAL (or a wire) with a specified data flow direction. The following port types are allowed:

in -- input. A variable or a signal can READ a reference value of an **in** port, but cannot assign a value to it. ::

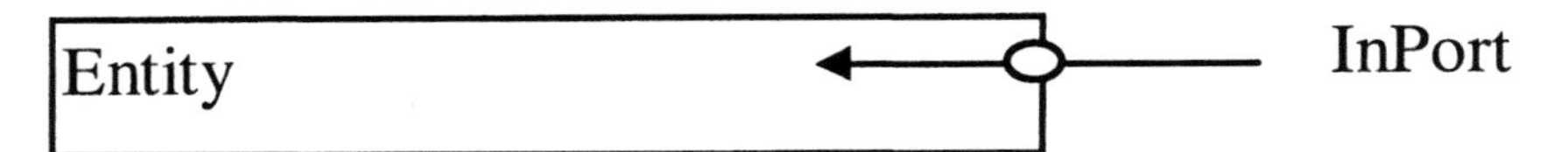

-- Multiple reads of **in** ports are allowed.

Var := InPort; -- legal (reading an **in** port named InPort)
Data_s <= InPort; -- legal

InPort <= 5; -- Illegal (writing to an **in** port named InPort) 

out -- Output. Signal assignments can be made to an **out** port, but data from an **out** port cannot be read . ::

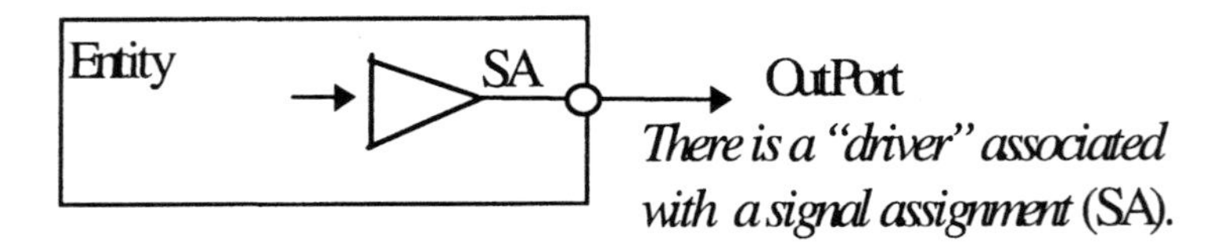

-- Multiple signal assignments are allowed (see chapter 4)

OutPort <= 7; -- legal (writing to an OUT port named OutPort) 

Var := OutPort; -- ILLEGAL (reading from an OUT port named OutPort) 

inout -- Bi-directional. Signal assignments can be made to an **inout** port, and data can be read from an **inout** port. ::

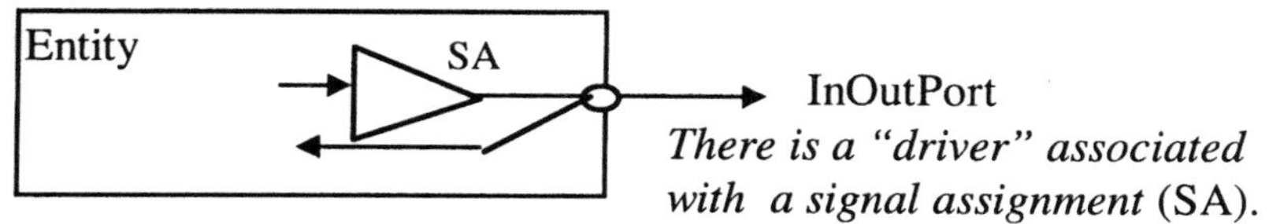

-- Multiple signal assignments are allowed (see chapter 4)

InOutPort <= 7; -- legal (writing into an INOUT port named OutPort) 

-- Multiple reads of **in** ports are allowed.

Var := InOutPort; -- legal (variable reading from an INOUT port) 

---

buffer -- **Out port** with read capability. A buffer port may have **at most ONE signal assignment** within the architecture : (i.e., a buffer port can be the only driver on a net) :

-- Concurrent statements

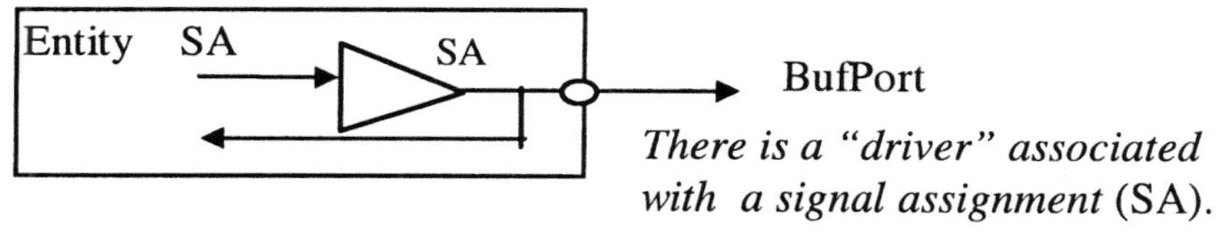

BufPort <= 7; -- legal (writing into a Buffer port from one source) 

-- Another concurrent statement

BufPort <= 8; -- ILLEGAL (writing to a Buffer port from a second source) 

-- Multiple reads of **buffer** ports are allowed.

Data_s <= BufPort; -- LEGAL (reading from a buffer port)

> Buffer ports should NOT be used.
>
> ***Rationale:*** *Buffer ports have no correspondence in actual hardware and they impose restrictions on what can be connected to them. If internal feedback is required, use an* ***out*** *port with an internal signal and a concurrent signal assignment.*

It is important to note that in synthesis ports of direction OUT, INOUT and BUFFER has NO correlation with the kind of hardware drivers that is implemented (e.g. push-pull, open collector, or tri-state driver). This decision is determined by the values of the signals driven onto the ports and by the directed technology. Thus, if a 'Z' is assigned onto a port, the synthesizer will implement a tri-state or open collector hardware driver. However, if only forcing values (e.g. '1' and '0') are assigned with no 'Z', then a push-pull type of hardware driver will most likely be implemented. The technology library and the VHDL architectural code determine the kind of output design. For example, a specific type of port interface device (e.g., open drain) can be instantiated as a component within the design. VHDL drivers and sources are discussed in chapter 4.

1.6.2 ARCHITECTURE

Architectures *[1]* (LRM 1.2) ***describe the internal organization or operation of a design entity. An architecture body is used to describe the behavior, data flow, or structure of a design entity.*** Language rules for architectures include:

1. A single entity can have **several architectures** or implementations (e.g., behavioral descriptions, structural interconnects, ASIC revisions).
2. There can be no architecture without an entity.
3. Libraries, use statements, ports, generics, and all other declarations (e.g., types, attributes, subprograms) defined within an entity are **fully visible and accessible** by the architectures of that entity.
4. Architectures can contain zero or more **concurrent statements** (i.e., pieces of code that operate concurrently with other pieces of codes). The concurrent statements are shown in Table 1.6.2 in order of importance, and graphically in Figure 1.6.2:.

Table 1.6.2 Concurrent elements within an architecture

Order of Importance *	Concurrent Statement	Comment
1	process_statement	Sequential statements
2	concurrent_signal_assignment_statement	Data flow
3	component_instantiation_statement	Hardware blackbox
4	generate_statement	Build elements
5	block_statement	Hierarchy
6	concurrent_procedure_call	Software blackbox
7	concurrent_assertion_statement	User defined violations

* This order of importance is subjective, and is based on the author's experience.

A hardware blackbox is a hardware view of a concurrent statement and represents a component with port interfaces that are equivalent to pins of a component. A component is a blackbox because the internal operation is not necessarily known, and is not directly visible to the instantiating architecture. What is visible are the port interfaces. A component can be instantiated multiple times in an architecture, just like real hardware devices of the same design (e.g. SN54L00 gate) can be instantiated multiple times in a hardware design.

A software blackbox is a software view of a concurrent statement and is represented by a concurrent procedure with a parameter list representing the interfaces. Unlike a component, a concurrent procedure does not have an entity declaration since it does not have ports. However, like a component, it functions as a blackbox because it can be instantiated multiple times and its internal operations are not necessarily known and visible to the instantiating architecture. A concurrent procedure interfaces to the architectural signals through the interface list. An example of a concurrent procedure is a timing check procedure instantiated multiple times, once for each signal being verified. Concurrent procedures represent combinational logic in synthesis.

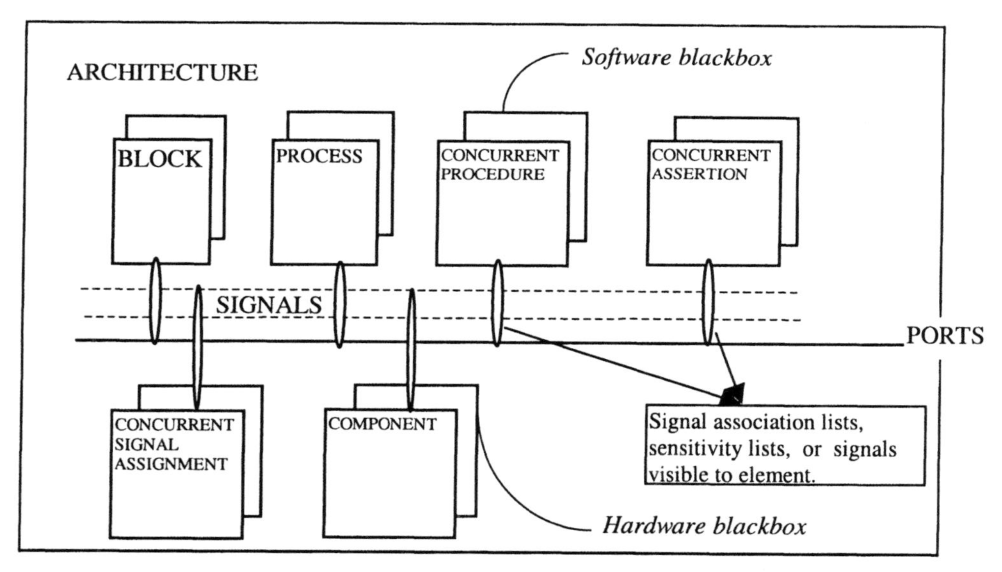

Figure 1.6.2 Concurrent Elements within an Architecture

1.6.2.1 Process

(LRM 9.2) A process is a concurrent statement of an architecture. Thus, an architecture can have several processes to describe the concurrent operation of the various pieces of the architecture (e.g. state machine, counter, ALU, multiplier). In RTL synthesis, a process is the only concurrent statement that can be used to represent registers. This process is called a "clocked process". A process may also represent combinational logic. The process is discussed in greater details in chapter 6.

A process executes the **sequential statement in sequence UNTIL it gets suspended with a wait statement**. A suspended process may resume after the time-out of a **wait** statement or as a result of an event (change in value) occurring on any signal in the sensitivity list of the process or the wait statement. A **wait** statement with no parameters cause the process to stop indefinitely. The **wait** statement is fully described in chapter 5. Examples of wait statements:

```
wait for 100 ns;          -- Process gets suspended for 100 ns
 wait until Clk = '1';    -- process gets suspended until
                          -- CLK changes to a '1' value
 wait;                    -- Process is suspended forever
```

[1] The execution of a process statement consists of the ***repetitive*** *execution of its sequence of statements. After the last statement in the sequence of statements of a process is executed, execution will* ***immediately continue with the first statement*** *in the sequence of statements* (i.e., automatic looping back to the start). A sensitivity clause implies a **wait** statement at the end of a process (see chapter 6). Processes interface among each other through either internal signals or through global signals declared in packages (see chapter 8 and 10). They also communicate with the environment (i.e. the I/O) through the ports of the entity. Figure 1.6.2.1-1 represents the VHDL coding structure of an architecture. Figure 1.6.2.1-2 represents an architecture of the counter entity described in Figure 1.6.1. A variable is used for demonstration. This process could have been written without a variable.

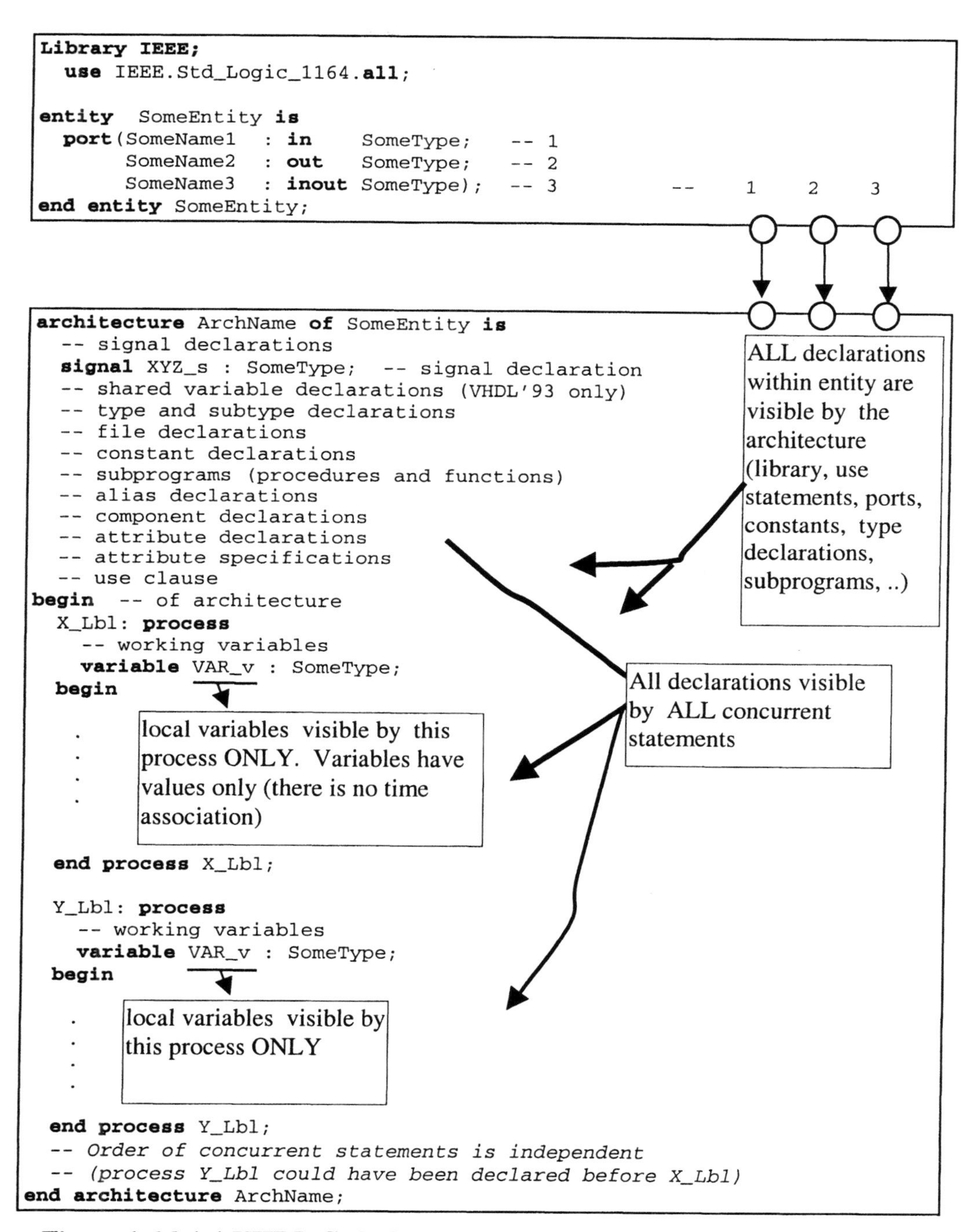

Figure 1.6.2.1-1 VHDL Code Structure of an Architecture Using Processes

```
----------------------------------------------------------------
--
--    Project      :  ATEP CLASS
--    File name    :  Adder_a.vhd
--    Title        :  Adder
--    Description  :  Architecture Description
--                 : VHDL'93
----------------------------------------------------------------
--    Revisions    : Date    Author   Revision  Comments
--    Creation     7/02/98   cohen    Rev A
----------------------------------------------------------------
architecture Adder_a of Adder is
  -- Declarations go here
  -- (signals, constants, types, subprograms)
  -- All declarations defined here
  -- are visible by the architecture
begin
  Y <= (A + B);
end architecture Adder_a;
```

Figure 1.6.2.1-2 Counter Architecture (ch1_dir\adder_a.vhd)

1.7 COMPILATION, ELABORATION, SIMULATION

Figure 1.7 represents the compilation and elaboration process prior to simulation. All VHDL analyzers/simulators store analyzed designs in **design libraries**. The VHDL'87 and VHDL'93 are provided with library STD that includes two packages: **Standard** and **TextIO.** A package represents a program unit that can specify groups of logically related declarations (see chapter 8). The Standard package includes various data type and functions definitions; it is reproduced in appendix B and a copy of Model Technology's[5] version is supplied on CD. *[1] Every design unit except package Standard is assumed to contain the following implicit context items as part of its context clause:*

library Std, Work;

use Std.Standard.all;

[5] Model Technology Incorporated, 8905 SW Nimbus Avenue, Suite 150, Beaverton, OR 97005-7159

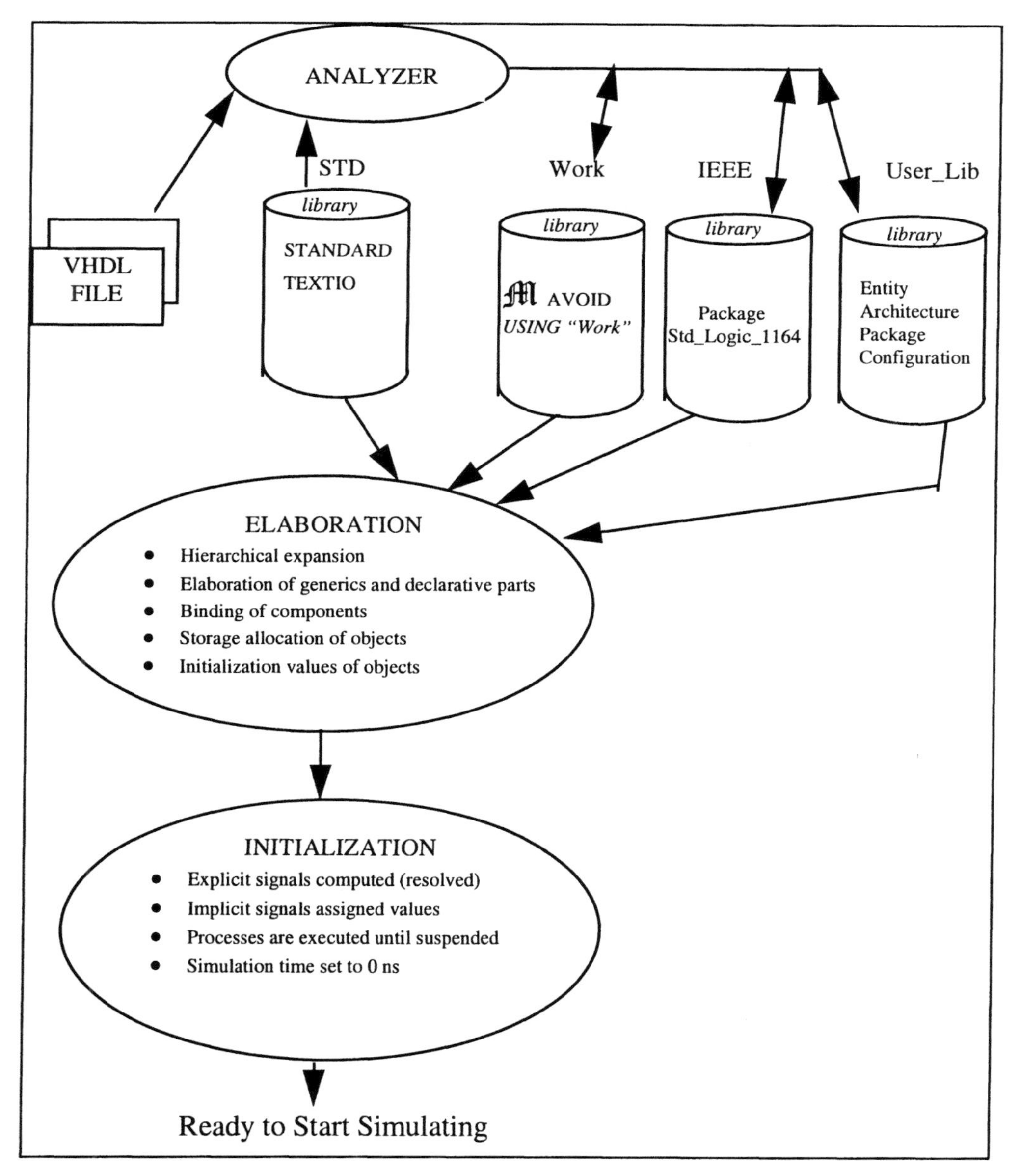

Figure 1.7 Compilation and Elaboration Prior to Simulation

Since the analyzer always reads package Standard, every type and function defined in this package is accessible to the user. Package TextIO provides type definitions and routines to read and write data from/to text files. This package is reproduced in appendix C and a copy of Model Technology's version is also supplied on CD. TextIO package is discussed in chapter 8.

Files ready for compilation are stored in files in the host environment. The VHDL analyzer (sometimes referred to as compiler) reads the package Standard from the STD

library. It then reads the VHDL files and starts the compilation. If the VHDL code includes the statement "*library Name_Lib*;" then the compiler verifies that this library exists and provides an error message if it does not. A user must create a library that is intended to be used. Most VHDL vendors supply tools to create libraries, to compile and simulate VHDL code. These tools provide a comprehensive set of error messages that occur during compilation and simulation. They supply a source language debugger that allows a user to set breakpoints, to single step through the code, and to perform many other debug activities (e.g., viewing or tracing variables, signals, and drivers). They also supply a user interface to display simulation waveforms and a listing of signals and variables as they change in values during the simulation.
During compilation the compiler reads any package declared in the user file. If the analyzer finds no error in the analyzed file, it stores the compiled design into the designated library. The default library is "work". Once a design with an entity and an architecture is compiled, it is ready to simulate. When the simulator is called, a user supplies the design unit to be simulated from the designated library and initiates the simulation process. Prior to simulation, the design is first elaborated. **Elaboration** consists of the following tasks: :

1. **Hierarchical expansion**. This is an expansion of the design hierarchy (e.g., components within components)
2. **Elaboration of generics and declarative parts**. This step creates a generic constant of the indicated subtype for each of the generics. It elaborates the declarative parts including the type declarations and constraints of types and subtypes. Thus if a range constraints (e.g., width of a memory bus) is defined in terms of a generic, that constraint is expanded during this step. This enables the user to fix the dimensions or parameters of the model late in the design process as a function of generics
3. **Binding of components.** :This step binds components to architectures.
4. **Storage allocation of objects.** This step allocates memory storage for the various objects in preparation for simulation.
5. **Initialization values of objects.** All signals and variables are initialized either to their default values or to the user defined initialization values.

After elaboration, the **initialization** process takes place. Specifically,

1. **Explicit signals computed**. The explicit signals, including ports of entities and signals of architectures, are initialized, resolved and assigned values (see chapter 2, 4 and 6 for resolved signals).
2. **Implicit signals assigned values.** Signals declared implicitly with attributes (see chapter 5) are assigned values.
3. **Processes are executed until suspended.** Every concurrent statement is executed until it is suspended as a result of an explicit or implicit **wait** statement (see chapter 5).
4. **Simulation time is reset to 0.**

Chapter 4 discusses the issue of initializations. Note that the elaboration and simulations are steps accomplished internal to the simulator and may not necessarily be directly controlled by the user.

M👍👍 Design units should be placed in design libraries **other** than *work*. The library name should reflect the family of devices or of functions the library holds. Thus F54_Lib can hold components of the 5400 library, Design_Lib can hold the actual design.

Rationale: *Some tools do not allow the use of library* ***work****. In addition, Every design except package* Standard *contains the implicit context item as part of its context clause:*

```
library STD, Work;
use STD.Standard.all;
```

If library ***work*** *is used then no library reference statement is needed. When large designs are implemented by different organizations, the use of one library can create ambiguities and errors. For example assume that one organization defines a package called "Utilities_Pkg" in file UtilX_pb.vhd, and another organization defines a different package called "Utilities_Pkg" in file UtilY_Pb.vhd. In that case, the last compiled design unit in the library will be the design unit in effect (even though the packages are defined in separate files). If both design units are compiled in library* ***work,*** *then incorrect operation will result for the models that attempt to access elements from the compiled design unit which were intended for other designs. This is because those elements may either not exist or may be defined differently in the packages.*

Separating the packages in different libraries, and referencing the libraries needed for each model prevents this design unit association problem.

1.7.1 Compilation Example

The compilation and simulation of the counter example was performed using V-Systems from Model Technology Design Environment. The compilation results are presented in Figure 1.7.1-1. The library used throughout this book is called "ATEP_Lib". Other VHDL tool vendors have similar compilation procedures.

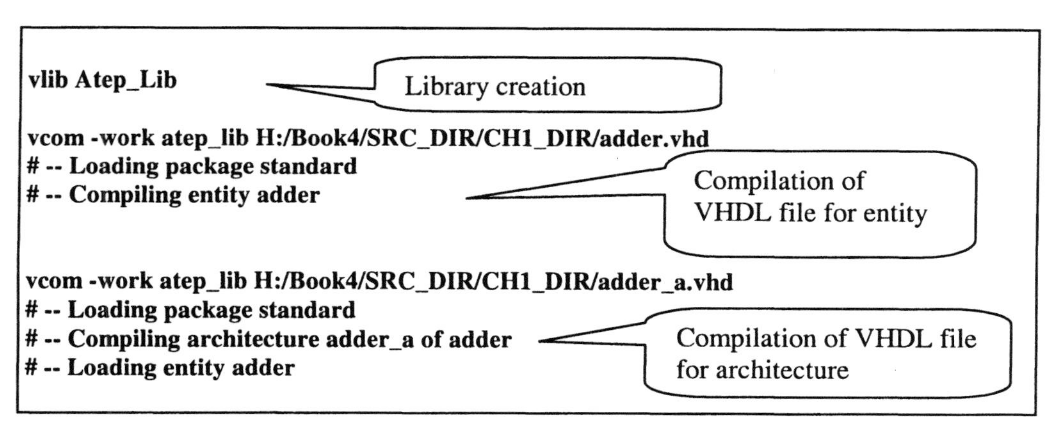

Figure 1.7.1-1 Compilation of Counter Entity and Architecture with Model Technology ModelSim

1.7.2 Simulation Example

Figure 1.7.2 shows the simulation script for Model Technology's simulator, the simulation output waveforms, and the output list file for the adder model. Force vectors using the simulation command language (SCL) is used in this example. It is generally recommended NOT to use SCL for the generation of test vectors because they are simulator dependent and are not portable. Instead, it is recommended to use VHDL code, as described in chapter 10 (testbench designs).
Note that Model Technology provides the following views during simulation:

1. **Waveform display.** Any signal (or port) and variable can be viewed by name and path in hex, octal, binary, or symbolic notation (for enumeration types).
2. **List view.** Displays a listing of times and values of signals as they change in values.
3. **Variable view.** Displays variables, constants, and generics for the active process.
4. **Signal view.** Displays signals within a level of hierarchy.
5. **Structure view.** Displays levels of hierarchy for each component instantiation.
6. **Source view.** Displays source code within a structure. It also enables a user to insert breakpoints.
7. **Transcript window.** This represents the user interface to insert simulation commands and to display the commands, and all the assertions and writes to Outputs.
8. **Process windows.** Displays all the processes in the current region.

Many simulators, including *ModelSim EE* and like *Leapfrog*[6], provide the capability to display variables in the waveform display.

[6] *Leapfrog* is a product of Cadence Design Systems, Inc

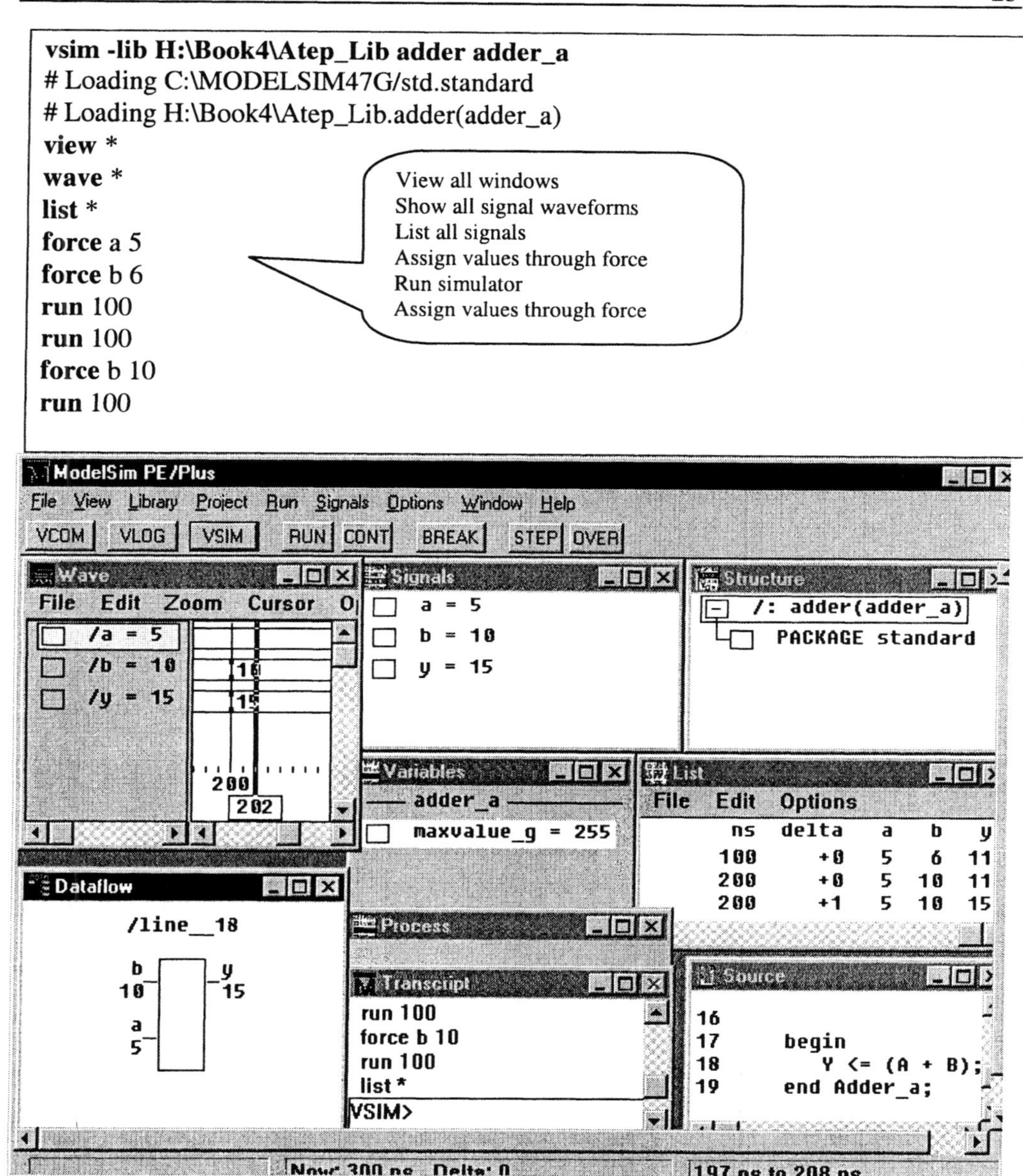

Figure 1.7.2 Adder Model Simulation

1.7.3 Synthesis Example

Synthesis tools can be used to automatically synthesize the HDL code into the target technology. Figure 1.7.3-1 provides a sample output of the RTL schematic view of such a tool . Figure 1.7.3-2 represents a portion of the technology view for the *Quicklogic PL3012* component.

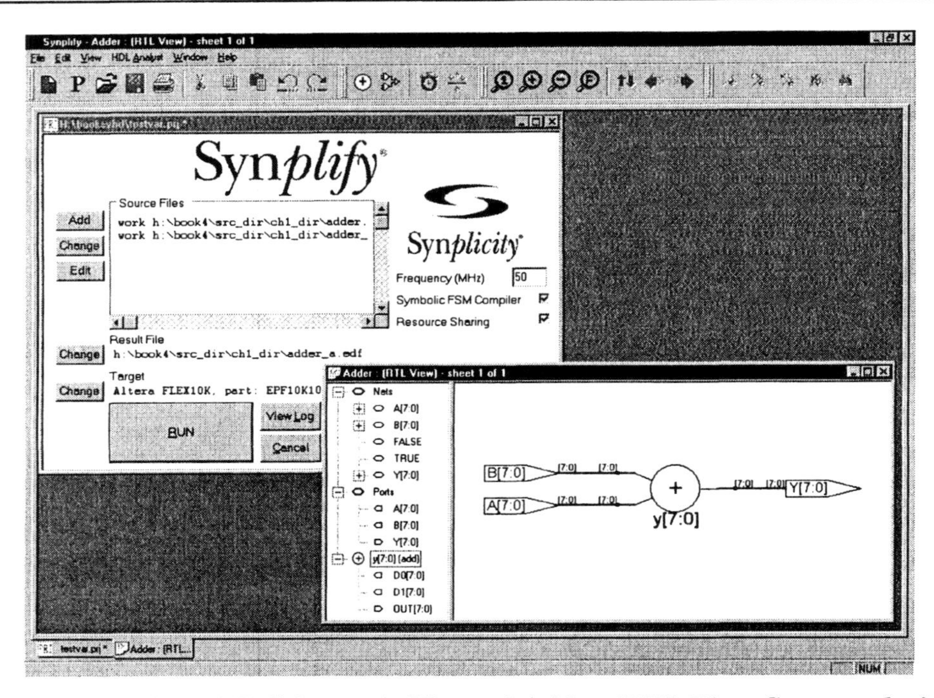

Figure 1.7.3-1 RTL Schematic View of Adder ***(RTL View Generated with Synplify from Synplicity)***

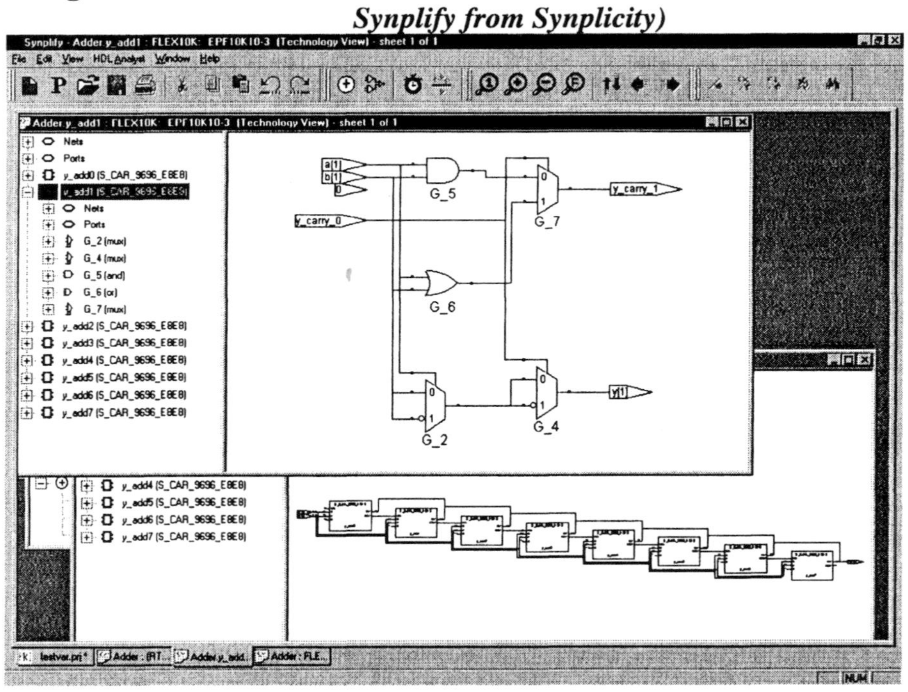

Figure 1.7.3-2 Portion of the Technology View for the Quicklogic PL3012 ***(RTL View Generated with Synplify from Synplicity)***

EXERCISES

1. Modify the adder architecture from an adder to a substractor.

2. Compile and simulate the substractor.

3. Write and compile an entity that includes the following pins:

Port Name	Port type	Port Direction
In1	Integer	input
In2	Integer	input
Out1	Integer	output

4. Provide specific examples for the various levels of modeling descriptions.

2. BASIC LANGUAGE ELEMENTS

This chapter covers the fundamental structures of the language including the identifiers, delimiters, literals, types, subtypes, predefined attributes and aliases. Coding styles and guidelines with regards to these concepts are also presented with examples. A reader familiar with VHDL should study these coding rules and may use this chapter as reference. An instructor may wish to cover, on an as needed basis, the sections of this chapter necessary to progress to other concepts.

2.1 LEXICAL ELEMENTS

[1] Lexical element are identifiers, delimiters, literal, and comments.

2.1.1 Identifiers

Identifiers *(LRM 13.3)* are names that identify various VHDL objects, procedures, functions, processes, etc. The syntax for a basic identifier is:

```
basic_identifier     ::= letter {[underline] letter_or_digit}
letter_or_digit      ::= letter | digit
letter               ::= upper_case_letter | lower_case_letter
```

There are eight rules to follow when constructing VHDL identifiers.

1. Identifiers can only contain letters, numbers, and underscores[8].
2. Identifiers can be of any length as long as the entire identifier appears within a line[8].
3. VHDL does not distinguish between upper and lower case letters[8].
4. Identifiers cannot have the same name as a keyword[8].
5. Identifiers must begin with a letter[8].
6. Identifiers cannot have underscores at the beginning, end, or side by side[8].
7. Underscores in identifiers are significant[7].
8. No space is allowed within a basic identifier since a space is a separator.

Note: VHDL'93 also allows extended identifiers, braced between two back-slash characters. The extended identifiers can be used to integrate VHDL code with other tools that use extended identifiers. Its syntax is:

```
extended_identifier ::= \graphic_character {graphic_character}\
```

 Avoid the use of extended identifiers.

Rationale: *Extended identifiers are not compatible with VHDL'87. The use of spaces and operators reduce readability. Extended identifiers are case sensitive and can cause errors. Most synthesis tools do not yet support extended identifiers.*

EXAMPLES:

Identifier	Comment	
INTGR9	-- legal	☺
Intgr1_5	-- legal	☺
Intgr1-5	-- the "-" is illegal	💣
Acquire_	-- "_" at end is illegal	💣
_On_State	-- "_" as 1st character is illegal	💣
Zero_To_3	-- legal	☺
0to3	-- First character must be a letter	💣
Abc@deF	-- @ is a special character	💣
\1 2bcx@#- +Name_\	-- VHDL'93	☺
\1 2bcx@#- +Name_\	-- VHDL'87	💣

[7] *RendezVous with Ada, A Programmer's Introduction*, David J. Naiditch, John Wiley & Sons, Inc. Copyright ©1989 by David Naiditch.

2.1.1.1 Port Identifiers

The philosophy of self-documenting code used by experienced modelers depends heavily on the descriptive names for user defined identifiers. There are several objectives in the naming of VHDL objects, some of which are conflicting in nature. Table 2.1.1-1 provides a list of port identifiers objectives and requirements.

Table 2.1.1-1 Port Identifiers Objectives / Requirements and Comments.

ITEM #	OBJECTIVE/REQUIREMENT	RATIONALE/COMMENT
1	Port names should not exceed 16 characters, including size of array.	Size limitation on some tools (e.g. synthesis tools). This requirement is very tool specific.
2	Names of ports should be same as name of connecting signal in next level architecture, and must be same as port names in other connecting entities.	Use of automated drawing tools to draw signal names from port names. Readability is also enhanced.
3	For large designs, names of ports should identify the origin of the design partition.	Enhances readability because it identifies source of signal (e.g. Microprocessor, memory).
4	Names of ports should identify polarity of signals	Enhances readability.
5	Names of ports may identify a registered or delayed version of an original, non-delayed signal.	Enhances readability.

2.1.1.2 Identifier Naming Convension

The following rule was recommended by Michael Keating and Pierre Bricaud in the book *Reuse Methodology Manual*, Kluwer Academic Publishers, 1998, ISBN 0-7923-8175-0:

> M "Develop a naming convention for the design. Document it and use it consistently throughout the design".
>
> ***Rationale:*** *Code consistency*

A general naming convention for identifiers is provided below as a set of guidelines. These guidelines can be modified as required.

M IdentifierName ::=

[prefix]**Identifier_Name**[Polarity][_][Suffix]

PREFIXES are used in the naming of timing parameters in a manner compliant with the VITAL (VHDL Initiative Toward ASIC Libraries) specification that allows back-annotation of timing parameters. Table 13.3 provides a list of prefixes that are VITAL compliant. A prefix may also be used to identify the originating partition or source of a signal. Some users have also used prefixes instead of suffixes for the identity of the class of the objects classes (i.e., *constant, signal, variable, file*).

IDENTIFER_NAME represents the name of the object in mixed or upper case, with the first character in upper case. The name may also use underscores for separators. Common nouns should be used when naming non-Boolean objects. They generally should be full descriptive words. When naming Boolean objects, predicate clauses or adjectives should be used. Boolean objects should easily be read as if bracketed by the clause "*if .. then*". Acronyms should only be used if their meaning is obvious to the design community.

POLARITY represents a logical polarity for signals of type standard logic or standard logic vectors, where "T" represents active TRUE, and "F" or "N" represents active FALSE. If the identifier name ends with a "T" and the polarity is active TRUE, then the polarity is omitted. Conversely if the identifier name ends with an "F" and the polarity is false, then the polarity is omitted. Examples:

```
ExtracT    -- active TRUE        nBuffT  -- active True
nExtractF  -- active false       nBufF   -- active false
```

SUFFIX represents additional information about the identifiers. A design contains several design units and objects of various types that belong to classes (i.e., *constant, signal, variable, file*). Identifying these items with an easily recognizable notation enhances readability. The suffix notation defined in table 2.1.1.2 was found very helpful. It represents a variation to the Hungarian notation often used in C^{++} applications. In addition, many synthesizers or timing analysis tools provide information about implied registers. The use of this notation facilitates the user's recognition of those registers. For example, if a signal with no *_r* or *_vr* suffix notation is synthesized as a register, then the possibility of an error may exist because the suffix *_r* is reserved for intended registers.

COMMENTS: For simple loops, users have experienced the characters *'I'*, *'J'*, *'K'* as more practical for loop indices. However, for more complex loops, where additional meaning is attached to the loop index, the use of the *_i* suffix is recommended. In addition, the differentiation between locally static versus a globally static (see chapter 3) is maintained in the notation for constants only (with *_c* or *_gc* suffixes). That differentiation is not carried to types, signals, or variables because it was felt to be excessively burdensome. A user who wants to keep track of the globally static objects may however carry that notation further with *_gs, _gr, _gv, _gvc*, or *_gtyp*.

Table 2.1.1.2 Recommended Suffix Notation

Suffix	Explanation	Example
_Pkg	Package identifier	**package** Design_Pkg **is**
_Lib	Library identifier	**Library** ProjectX_Lib;
_Cfg	Configuration identifier	**configuration** EEPROM_Cfg **of**
_Typ	Type or subtype identifier	**subtype** Read_Typ **is** Integer **range** 20 **downto** 18
_a, _Beh, _Fnc _Rtl, _Str,	architecture (general case) Behavior level Functional (e.g. ISA level) Register transfer level Structural level	**architecture** EntityName_a **of** EntityName **is** **architecture** EntityName_Beh **of** EntityName **is** **architecture** EntityName_Fnc **of** EntityName **is** **architecture** EntityName_Rtl **of** EntityName **is** **architecture** EntityName_Str **of** EntityName **is**
_Blk	Block label	ALU_Blk: **Block**
_Proc	Label for Process	Alu_Proc: **process**
_Lp	Loop Labels	Count_Lp: **for** I **in 1 to 10 loop**
_Lbl	Other labels	TestActive_Lbl: **if** T **then**
_v	Variable implemented as combinatorial logic	**variable** Data_v : Std_logic_Vector(31 **downto** 0)
_vr	Variable implemented as register	**variable** Dat_vr : Std_logic_Vector(Read_Typ)
_s	Signal implemented as a bus	**signal** Data_s : Std_logic_Vector(31 **downto** 0)
_r	Signal implemented as register	**signal** Data_r : Std_logic_Vector(31 **downto** 0)
_c	Constant, locally static	**constant** Read_c : Std_Logic_Vector(Read_Typ) := "001";
_gc	Constant, globally static	**constant** Size_gc := 2 ** NumBits_g;
_g	Generic identifier	**generic**(DataWidth_g : Integer := 32);
_I	Index for a loop	**for** Bit_i **in** Count_v'range **loop**
_f	File identifier	**file** Data_f : Std.TextIO.Text **is**

Port names should reflect manufacturer's port name if they are available because it enhances readability. Thus, **DO NOT use suffixes for signals that connect to component ports. However, use the suffix for signals that are not ports, or for formal subprograms parameters that are of class signal.** When naming signals connected to ports, use the port name of the driving signal. Figure 2.1.1.2 provides an example of signals interconnecting ports of components.

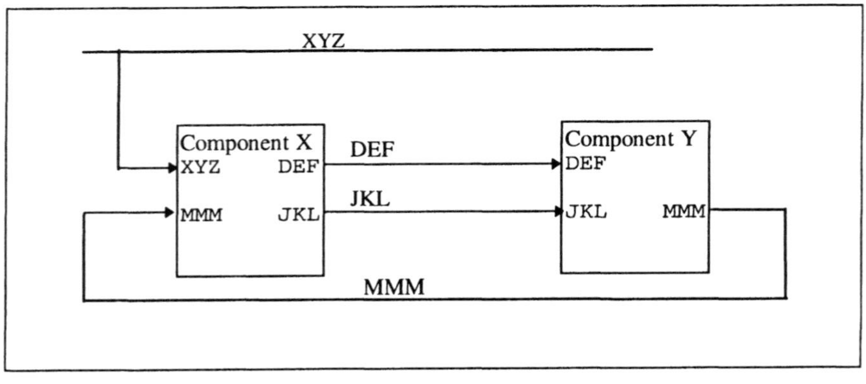

Figure 2.1.1.2 Naming Signals Interconnecting Ports of Components.

Identifiers for **library** and **package** design units should be allocated from a single source (project management). Library and package names should consist of noun phrases in **mixed or upper case**. Library names should depict an appropriate program, project, design tool, or logical name.

Rationale: *Maintains consistency and avoid confusion across the project.*

Identifiers for **entities** should consist of noun phrases depicting the appropriate hardware or logical name in *upper or mixed case* (e.g., StateMachine).

Rationale: *Enhances readability.*

The VHDL name of predefined identifiers, including the identifiers in the *Std* and *IEEE* design libraries shall never be used for other identifiers. Examples:

type DoNotUse_Typ **is** (FF, Time, Min, Ns, Ms, ACK, Real, Std, On);

Rationale: *Use of predefined identifiers causes confusion. In addition, some VHDL compilers and simulators have difficulty dealing with such identifiers. This reduces code portability.*

Identifiers for ports or signals should consider down stream design process tools for limitations on character set, length, case and bus naming conventions that would supersede any general guides. As an example, if schematic capture is used to create VHDL, the underscore may be avoided because a wire will mask the underscore if the signal name is placed over the wire.

Rationale: *Prevent confusion and enhancing readability.*

Examples of port or signal names:

```
AddressT        -- address True, port name
AddressT_s      -- address True, internal signal
AddressT_v      -- address True, variable
pRWF            -- Processor  Read Write Active False.
pRwRF           -- Processor Read Write, Registered  Active False
IsDone          -- Boolean, , no delay,  no polarity
POwnBus         -- Boolean, Processor owns bus state, no partition
```

𝔐 👍 👍 Identifiers for **subprograms** (i.e., procedures and functions) should be verb phrases descriptive of their actions such that their functionality can be interpreted from the context of usage. Examples:

```
Upper_To_Lower    UpperToLower
BIT_TO_INTEGER   Initialize_Memory   Convert
```

Rationale: *The designer should not have to go to the subprogram source code to understand the code from which it's called.*

𝔐 👍 👍 Identifiers for **types** and **subtypes** should be noun phrases more general to the scope of the design unit containing it. Examples:

```
subtype Int8_Typ is Integer range 1 to 8;
subtype RLV31_0_Typ  is Std_Logic_Vector (31 downto 0);
type MachineState_Typ is (Standby, Transmit, Receive);
```

Rationale: *Enhances readability*

2.1.1.3 Accessing Identifiers Defined in Packages

𝔐 👍 👎 For non synthesizable code, if an identifier is declared in a package, use either of the following methods to explicitly document the identifier:

1. Include the following two statements *"use Library_Name.Package_Name.all;" "use Library_Name.Package_Name;"*. This allows the definition of the path without the library name. When declaring or using the identifier, include the package name (e.g., *Package_Name.Identifier*). For Example:

```
use Std.TextIO.all;
use Std.TextIO;
...
variable Line_v : TextIO.Line;
      TextIO.Write(Line_v, string'("Done test");
```

Exception: If the user community is very familiar with the contents of a package, then the path may be excluded from being connected with the identifier. An example of such package is IEEE.Std_Logic_1164. Another exception may be TextIO package. It is recommended that the path be included for the TextIO procedures "READ" and "WRITE" because those procedures are often overloaded with user defined "READ" and "WRITE" procedures for the transfer of data.

Rationale: *It is essential that the code be maintainable. Relying on the compiler to find all the identifiers declared in packages is adequate for computers, but not humans. Providing the package name as the path, without the library name, is adequate for because the reader is provided full knowledge about the source of the identifier (the library name because of the "use" statement, and the package name attached to the identifier).*

2.1.1.4 Capitalization

M 👍 👎 Lower case should be used for all VHDL reserved words and VHDL attribute definitions. Mixed case should be used for all other identifiers, with consistent casing in all the code. Some users recommend that subprograms (i.e., functions and procedures) be written in upper case to distinguish them from identifiers. Other users insist on using upper case for the reserved words. What is important is consistency throughout the design.

Rationale: *User defined keywords should not depend on capitalization to be readable or meaningful. When user identifiers are in mixed or upper case, they stand out.*

Table 2.1.1.3 provides a list of reserved words.

Table 2.1.1.3 [1] List of Reserved Words.

Reserved Words	Std 1076-1987			*Std 1076-1993*
abs	else	nand	select	*group*
access	elsif	new	severity	*impure*
after	end	next	signal	*inertial*
alias	entity	nor	subtype	*literal*
all	exit	not		*postponed*
and		null	then	*pure*
architecture	file		to	*reject*
array	for	of	transport	*rol*
assert	function	on	type	*ror*
attribute		open		*shared*
	generate	or	units	*sla*
begin	generic	others	until	*sll*
block	guarded	out	use	*sra*
body				*srl*
buffer	if	package	variable	*unaffected*
bus	in	port		*xnor*
	inout	procedure	wait	
case	is	process	when	
component			while	
configuration	label	range	with	
constant	library	record		
	linkage	register	xor	
disconnect	loop	rem		
downto		report		
	map	return		
	mod			
Predefined Attributes				~~*'structure*~~
				~~*'behavior*~~
'pos(x)	'val(x)	'base		*'ascending*
'succ(x)	'pred(x)	'leftof(x)	'rightof(x)	*'image*
'left	'right	'high	'low	*'value*
'left[(n)]	'right[(n)]	'high[(n)]	'low[(n)]	*'asending(n)*
'range[(n)]	'reverse_range[(n)]		'length[(n)]	*'driving*
'event	'active	'last_event	'last_active	*'driving_value*
'last_value	'behavior	'structure		*'simple_name*
				'instance_name
'delayed[(t)]	'stable[(t)]	'quiet[(t)]		*'pathname*
'transaction[(n)]				

Note: Reserved words that are deleted from the language are shown with a strike-through line.

2.2 SYNTAX

A syntax is the pattern or structure of the word order in a phrase. When this explanation is applied to VHDL, it refers to rules for statements spelled out in the LRM and verified by an VHDL analyzer (or compiler). However, latitude in the rules allow a programmer the choice of syntax for particular program statements that reflects his style. This area of latitude can cause significant variations in style, and can affect the overall readability of a program.

2.2.1 Delimiters

Delimiters are symbols that have special meaning within VHDL. These symbols are used, for instance, as operators and statement terminators. There are single delimiters that consist of only one symbol. There are double delimiters that consists of two symbols. Table 2.2.1 represents the VHDL delimiters.

Table 2.2.1 VHDL Delimiters.

<table>
<tr><th>Delimiter</th><th>Name</th><th>Example</th></tr>
<tr><td>+</td><td>addition or positive</td><td>Count_v := +5 + 9;</td></tr>
<tr><td>-</td><td>subtraction or negative</td><td>Count_s <= -7 - 19;</td></tr>
<tr><td>/</td><td>division</td><td>Sum_v := 7 / 5; -- = 1</td></tr>
<tr><td>=</td><td>equality</td><td>if Sum_v = 9 then</td></tr>
<tr><td><</td><td>less than</td><td>if Count_s < Max_g - 1 then</td></tr>
<tr><td>></td><td>greater than</td><td>if Count_s > Max_g then</td></tr>
<tr><td>&</td><td>concatenator</td><td>FourBits_v := Twobits_v & "10"</td></tr>
<tr><td>|</td><td>verical bar</td><td>when 'Y' | 'y' =></td></tr>
<tr><td>;</td><td>terminator</td><td>end HA_Structure;</td></tr>
<tr><td>#</td><td>enclosing based literals</td><td>Total_v := 16#2A4F#</td></tr>
<tr><td>(</td><td>left parenthesis</td><td>type OnOff_Typ is (On1, Off);</td></tr>
<tr><td>)</td><td>right parenthesis</td><td>Sum_v := (5 + Add_s) * 7;</td></tr>
<tr><td>.</td><td>dot notation</td><td>OnOff_v := Message_v.Switch_v;</td></tr>
<tr><td>:</td><td>separates data object from type</td><td>variable Switch_v : OnOff_Typ;</td></tr>
<tr><td>"</td><td>double quote</td><td>report "A message"</td></tr>
<tr><td>'</td><td>single quote or tick</td><td>X_v := OnOff_Typ'left & Off;</td></tr>
<tr><td>**</td><td>exponentiation</td><td>Sum_v := 2 ** 5; -- =32</td></tr>
<tr><td>=></td><td>arrow (read as "then")</td><td>when On1 => Sum_v := 6;</td></tr>
<tr><td>=></td><td>arrow (read as "gets)
(for named notation)</td><td>Array_v := (Element1 => 5,
others => 100);</td></tr>
<tr><td>:=</td><td>variable assignment</td><td>Sum_v := Numb1_s + 7;</td></tr>
<tr><td>/=</td><td>inequality (not equal)</td><td>if Sum_s /= 7 then</td></tr>
<tr><td>>=</td><td>greater than or equal</td><td>if Sum_v >= 2 then</td></tr>
<tr><td><=</td><td>less than or equal</td><td>if Sum_v <= 4 then</td></tr>
<tr><td><=</td><td>signal assignment</td><td>Count_s <= 5 after 10 ns;</td></tr>
<tr><td><></td><td>box</td><td>type Stack_Typ is array
(Integer range <>) of B16_Typ;</td></tr>
<tr><td>--</td><td>comment</td><td>-- comment line
Y := -- comment till end of line
5; -- Y := 5; on two lines</td></tr>
</table>

No spacing is allowed before or after the following delimiters

1. The sharp # -- 16#AB10#
2. The period . -- Data.Enable
3. The single quote ' -- Bus'transaction

Bit literals must be bracketed with double quotes with no spaces inside the quotes

1. The double quote " -- "1011"

Avoid using spacing before the following delimiters:

1. The semicolon ;
2. The colon : -- after a label only, use spacing elsewhere
3. The parentheses ( ) -- Parentheses may be started on a continuation line

For ALL other delimiters use one or more spacing before and after using the delimiter.
Examples:

```
signal Addr_s   :  Std_Logic_Vector(31 downto 0);
SendCharacter
  (Data_c   =>  ToSendData_v,
   Data_s   =>  DataBus);
If Data_v > Data_Typ'high then ...
```

Rationale: *These rules enhance code readability. The semicolon is a terminator and thus needs no separator. The colon after a label can be thought of as part of the label. The parentheses embrace parameters that are part of the object's definition. Long lines are more readable when elements embraced in parentheses are started on a new continuation line.*

2.2.2 Literals

[1] A literal is a value that is directly specified in the description of a design. A literal can be a bit string literal, enumeration literal, numeric literal, or the literal null.

Use underscores for numerical literals.
Rationale: *Enhance readability*

2.2.2.1 Decimal literals

The syntax for decimal literals is:

```
decimal_literal ::= integer [.integer][exponent]
integer         ::= digit [underline] digit
exponent        ::= E [+] integer | E [-] integer
```

```
Examples: LRM:13.4.1
  12   0   1E6   123_456             -- integer literals
  12.0   0.0 0.456   3.14159_26      -- real literals
  1.34E-12   1.0E+6   6.023E24       -- real literals with exponents
```

2.2.2.2 Based literals

```
based_literal (LRM: 13.4.2)::= Literals:
  base#based_integer [.based_integer]#exponent
  Examples:
  2#1111_1111#   16#FF#   016#0FF#   integer literal of value 255
  16#E#E1     2#1110_0000#   integer literal of value 224
  16#F.FF#E+2    2#1.1111_1111_111#E11   - real literal of value 4095.0
```

2.2.2.3 Character literals

[1] A character literal is a literal of the character type (Std.character from the package STANDARD). *Character literals are formed by enclosing one of the graphics characters (including the space and non-breaking space characters) between two apostrophe(') characters.* Examples include 'A' '*' ''' for the A, *, and the single quote.

2.2.2.4 String literals

[1] A string literal is a sequence of graphics characters, or possibly none, enclosed between two quotation marks ("). The type of a string literal is determined from the context.

```
"This is a string"
""                    -- an empty string literal
" " "A" """"          -- three strings of length 1
"1011"                -- a string of four characters.
```

A string *(LRM 13.6)* must fit on one line. Longer sequences can be obtained by concatenation of string literals.

```
"First part of a sequence of characters " &
"that continues on the next line"
-- one string literal in all.
```

2.2.2.5 Bit string literals

[1] A bit string literal (LRM: 13.7) is formed by a sequence of extended digits enclosed between two quotations and preceded by a base specifier. The type of a bit string literal is determined from the context. The length of a bit string is the length of its string literal value. Figure 2.2.2.5 represents and example that utilizes bit string literals. The syntax for bit string literals is:

bit_string_literal ::=
  base_specifier "[bit_value]"
base_specifier ::=
  B -- for Binary notation (0, 1)
  O -- for Octal notation (0 ..7)
  X -- for heXadecimal notation (0 ..9, A, B, C, D, E, F)

An underline character inserted between adjacent character digits of a bit string does not affect the value of this literal.

VHDL'87 bit string literals can only be defined for Bit_Vector type. VHDL'93 is more flexible, and bit string literals can be used for any enumeration type that uses '0' and '1' in its enumeration (e.g., type IEEE.Std_Logic_1164.Std_Logic_Vector, and String type). Conversion functions can be used to convert Based Literal (integer) or string literals to Std_Logic_Vector. Specifically, the *To_StdLogicVector* in the Std_Logic_1164 package converts a bit vector into Std_Logic_Vector. Thus, the following is legal:

```
S_Std <= To_StdLogicVector(X"A67ABCD9"); -- VHDL'87 only
```

The value in parenthesis is interpreted as a Bit_Vector as defined in package Standard. However, for portability between VHDL'87 and VHDL'93, the following is recommended:

```
S_Std  <= To_StdLogicVector(Bit_Vector'(X"A67ABCD9"));
```

In addition, the Synopsys arithmetic package (Std_Logic_Arith) can convert an integer number into an Std_Logic_Vector, as shown below:

```
S_Std <= CONV_STD_LOGIC_VECTOR(ARG => 16#01BCDEF8",
                               SIZE => 32);
```

```
architecture String_Beh of String is
begin
  -- Process (conconcurrent statement)
  Test_Lbl: process
     -- type string is defined in Standard Pkg.
    variable S1_v           : string(1 to 12);
    -- type BitVector is defined in Standard Pkg
    variable BitVect_v      : Bit_Vector(12 downto 1);
  begin
    S1_v       := "111111111111";          -- string •
    -- Incompatible types for assignment.• VHDL'87 ONLY
    -- B"1111_1111_1111" is not a string! • for VHDL'93••
    -- S1_v        := B"1111_1111_1111";
    -- Incompatible types for assignment.•VHDL'87 ONLY•
    -- S1_v        := X"FFF"; • for VHDL'93 (12 bits of ones)

    BitVect_v := "111111111111";
    -- B"1111_1111_1111" is a bit string literal •
    BitVect_v := B"1111_1111_1111";

    -- X"FFF" is a bit string literal •
    BitVect_v := X"FFF";
    wait;  -- This statement suspends the process
  end process Test_Lbl;
end String_Beh;
```

Figure 2.2.2.5 Bit String Literals (ch2_dir\strng_ea.vhd)

2.2.3 Operators and Operator Precedence

The operators that may be used in expressions are shown in Table 2.2.3. *[1] (LRM 7.2) Each operator belongs to a class of operators, all of which have the same precedence levels.* The precedence of operators is maintained even if the operators are overloaded (see chapter 7).

Table 2.2.3 VHDL Operators

Precedence	Operator Class	Operators					
LOW	Logical operator	**and**	**or**	**nand**	**nor**	**xor**	**xnor**[8]
	relational operator	=	/=	<	<=	>	>=
	shift operator[6]	**sll**	**srl**	**sla**	**sra**	**rol**	**ror**
	adding operator	+	-	**&**			
	sign	+	-				
↓	Multiplying operator	*	/	**mod**	**rem**		
HI	Miscellaneous operator	**	**abs**	**not**			

[8] **xnor** and shift operators are for VHDL'93 only

2.2.3.1 Logical operators

[1] (LRM: 7.3.1) The logical operators ***and, or, nand, nor, xor*** *(also* ***xnor*** *for VHDL'93), and* ***not*** *are defined for ONLY for the predefined types Bit and Boolean* (i.e., the right and left hand operands for those operators can bit either a Bit or a Boolean) and the result of the operation is of type Boolean (see 2.3.1.2.1). *[1] For short-circuit operands* ***and, or, nand,*** *and* ***nor*** *on types Bit and Boolean, the right operand is evaluated only if the value of the left hand operand is not sufficient to determine the result of the operation. For operations* ***and*** *and* ***nand,*** *the right hand is evaluated only if the value of the left operand is True; for* ***or*** *and* ***nor****, the right operand is evaluated only if the left operand is False.*

Note that the *Std_Logic_1164* package defines overloaded operators **and, or, nand, nor, xor, xnor,** and **not** for operands of type *std_ulogic_vector* and *std_logic_vector* (where the size of the vector is 1 to "n" bits). These overloaded operators return results that are of type *std_ulogic_vector* or *std_logic_vector*, and are different than the predefined operators that operate on *bits* and *Boolean* and return a *Boolean* type.

2.2.3.2 Relational Operators

[1] (LRM:7.2.2) Relational Operators include tests for equality, inequality, and ordering of operands. The operands of each relational operator must be the same type unless the relational operators are overloaded (see chapter 7). *[1] The result type of each relational operator is the predefined type Boolean.* Figure 2.2.3.2 represents an example that applies Boolean types, logical and relational operators, and demonstrate the short-circuit operation.

```
architecture Bool_Beh of Bool is
  type States_Typ is (Idle, InUse, Waiting, Suspended);
begin  --  Bool_Beh
  Test_lbl:  process
    variable BusActive_v          : boolean := true;
    variable BusState_v           : States_Typ := Idle;
    variable UnitAllocated_v      : boolean := true;
    variable I_v         : integer := 0;
    variable K_v         : integer := 9;
  begin
    BusActive_v := BusState_v  = InUse    -- yields boolean
               and UnitAllocated_v;
    if BusActive_v and K_v = 10 then
      UnitAllocated_v := False;
    else
      UnitAllocated_v := True;
    end if;

    -- division by ZERO avoided because of short circuit operator "and"
    if  I_v /= 0 and K_v / I_v > 4 then
          UnitAllocated_v :=  BusState_v  /= Idle or I_v > 10;
    end if;
    wait;
  end process Test_lbl;
end Bool_Beh;
```

Note that "and" operator operates on boolean types

Figure 2.2.3.2 Logical and Short-Circuit Operation (ch2_dir\bool_ea.vhd)

2.2.3.3 Shift Operators

Shift operators *(LRM 7.2.3)* are new for VHDL'93 and are *[1] defined for any one-dimensional array* (see section 2.3.2.1.1) *whose element type is either of the predefined types Bit or Boolean.* Table 2.2.3.3 summarizes the shift operators. The left operand is any one-dimensional array type whose element type is Bit or Boolean. The right operand type is an integer. The result type is the same as the left operand. *[1] If the right operand of the shift operation is 0 or if the left operand is a null array, then the result of all these operations is the value of the left operand. In addition, if the right operand (the integer) is negative, the direction of the shift is reversed.* The right operand indicates the amount of shifts or rotate to perform. Figure 2.2.3.3 demonstrates all these shift operations.

Note that the predefined shift operators for VHDL'93 operate on types *Bits* or *Boolean* ONLY. However, numeric packages[9] perform most of these shift operations on vectors of type *SIGNED* and *UNSIGNED*, which are unconstrained arrays of *Std_Logic_Vector* and *Bit_Vector*. The definition of those types is provided below:

```
type UNSIGNED is array (natural range <> ) of Std_Logic;
type SIGNED   is array (natural range <> ) of Std_Logic;
type UNSIGNED is array (natural range <> ) of Bit;
type SIGNED   is array (natural range <> ) of Bit;
```

The advantages of using those packages is that the shift operators can be used with VHDL'87. In addition, some synthesis tools do not yet support VHDL'93.

[9] Numeric packages are available via ftp from "vhdl.org /vi/vhdlsynth/numeric_std.vhd". Other numeric packages are included on CD in the Synopsys subdirectory.

Table 2.2.3.3 Shift Operators

Operator	Operation	Explanation	Example L :Bit_Vector(7 downto 0) := "11001001"	Result
sll	Shift left logical	T'left (T is the array Type)	Q := L **sll** 3; Q := L **sll** - 3;	01001000 00011001
srl	Shift right logical	T'left (T is the array Type)	Q := L **srl** 3; Q := L **srl** - 3;	00011001 01001000
sla	Shift left arithmetic		Q := L **sla 3**; Q := L **sla - 3**;	01001111 11111001
sra	Shift right arithmetic		Q := L **sra 3**; Q := L **sra - 3**;	11111001 01001111
rol	Rotate left logical		Q := L **rol** 3; Q := L **rol** - 3;	01001110 00111001
ror	Rotate right logical		Q := L **ror** 3; Q := L **ror** - 3;	00111001 01001110

```
architecture Shift_a of Shift is
begin  --  Shift_a
  Shift_Lbl : process
    variable    Q_v      : Bit_Vector(7 downto 0) := "11001001";
    variable    R_v      : Bit_Vector(7 downto 0);
  begin  --  process Shift_Lbl
    R_v := Q_v sll 3;         -- 01001000
    R_v := Q_v sll - 3;       -- 00011001
    R_v := Q_v sll 10;        -- 00000000

    R_v := Q_v srl 3;         -- 00011001
    R_v := Q_v srl - 3;       -- 01001000
    R_v := Q_v srl 10;        -- 00000000

    R_v := Q_v sla 3;         -- 01001111
    R_v := Q_v sla - 3;       -- 11111001
    R_v := Q_v sla 10;        -- 11111111

    R_v := Q_v sra 3;         -- 11111001
    R_v := Q_v sra - 3;       -- 01001111
    R_v := Q_v sra 10;        -- 11111111

    R_v := Q_v rol 3;         -- 01001110
    R_v := Q_v rol - 3;       -- 00111001
    R_v := Q_v rol 10;        -- 00100111

    R_v := Q_v ror 3;         -- 00111001
    R_v := Q_v ror - 3;       -- 01001110
    R_v := Q_v ror 10;        -- 01110010

    wait;
  end process Shift_Lbl;
end Shift_a;
```

Figure 2.2.3.3 Shift Operations, VHDL'93 (ch2_dir\shift_ea.vhd)

2.2.3.4 The Concatenation "&" Operator

The concatenation operator is predefined for any one- dimensional array type. *[1] If both operands are one-dimensional arrays of the same type, the result of the concatenation is a one-dimensional array of the same type whose length is the sum of the lengths of its operands, and whose elements consist of the elements of the left operand (in left to right order) followed by the elements of the right operand (in left to right order). The direction of the result is the direction of the left operand, unless the left operand is a null array, in which case the direction of the result is that of the right operand.*

If the range of an array is "L to R" (e.g., *Bit_Vector* (0 **to** 10)) then this *[1] range is called an ascending range; if L > R, then the range is a null range.* Conversely, if the range of an array is L downto R (e.g., *Bit_Vector*(10 **downto** 0)) then this *[1] range is called a descending range; if L < R, then the range is a null range.*

[1] If both operands are null arrays, then the result of the concatenation is the right operand. Otherwise, the direction and bounds of the result are determined as follows: Let S be the index subtype of the base type of the result. The direction of the result of the concatenation is the direction of S, and the left bound of the result is S'left. (See section 2.3.2.1.6 on the definition of a null array). Figure 2.2.3.4 provides some examples that demonstrate this rule. See section 2.3.2.1.2 for recommendations on slices declaration.

```
subtype Byte_Typ is Bit_Vector(7 downto 0);
-- Base type of result is Byte_TYp with direction 7 downto 0.
-- Thus, the left bound of the result is Byte_Typ'left, or 7.
constant N1234_c  : Byte_Typ := "1001" & "0100"; -- 7 downto 0
--                                        1         2         3
--                              1234567890123456789    01234567890
constant ErrorMsg_c  : string := "Error Number A8765 " & "Bad Access ";
constant BadAccess_c : string := ErrorMsg_c(20 to 30);  -- "Bad Access"

-- Base type of result is Bit_Vector with direction 0 to integer'high.
-- Thus, the left bound of the result is Bit_Vector'left, or 0
constant C1_c : Bit_Vector := N1234_c & N1234_c; -- range 0 to 15,
                                                  -- value "1001010010010100"
-- Slice direction does not match subtype direction
-- N1234_c(1 to 2) is a null array because it is of type Byte_Typ that
-- is descending in order, but the range (1 to 2) is ascending.
-- Thus result range is C2_c(7 downto 4), and its value is 0100

constant C2_c : Byte_Typ   := N1234_c(1 to 2) & N1234_c(3 downto 0);
```

Figure 2.2.3.4 Concatenation Examples (ch2_dir\concat.vhd)

When several bits are concatenated together to create a vector there are situation when the compiler may not know the type of the resulting vector because it can be one of several types. For example, assume that *S1, S2* are of type *Std_Logic*, and *S2bits* is of type *Std_Logic_Vector(1 downto 0).* Also assume that the package *IEEE.Std_Logic_Arith* is in visible.

```
S2bits <= S1 & S2; -- OK because type of concatenation is defined by the target
case (S1 & S2) is  --ERROR. Type of result is undefined.
                   -- Could be Std_Logic_Vector, Signed, or Unsigned.
case Std_Logic_Vector'(S1 & S2) is – OK, type is qualified
```

2.2.3.5 Remainder and Modulus

The **rem** and **mod** operators *(LRM: 7.2.6)* are similar operators. For positive numbers, they both yield the remainder of integer division. Thus, 5/3 yields 1 with a remainder of 2.

Reminder hint:

Sign of **rem** operation is sign of the *"left"* operand because *"remainder"* is what's *"left"!.*

[1]The integer division and remainder are defined by the following relation:

*A = (A / B) * B + (A rem B)*

where (A rem B) has the sign of A and an absolute value less than the absolute value of B. Integer division satisfies the following identify:

(-A) / B = - (A / B) = A / (-B)

The result of the modulus operation is such that (A mod B) has the sign of B and an absolute value less than the absolute value of B; in addition, for some integer value N, this result must satisfy the relation:

*A = B * N + (A mod B)*

Mod is useful when defining counters (up or down) that roll over when computing pseudorandom number within a desired range. Figure 2.2.3.5-1 provides an example of the **mod** and **rem** operators.

J	K	J mod K	J rem K	J	K	J mod K	J rem K
0	5	0	0	0	-5	0	0
1	5	1	1	1	-5	-1	1
2	5	2	2	2	-5	-2	2
3	5	3	3	3	-5	-3	3
4	5	4	4	4	-5	-4	4
5	5	0	0	5	-5	-0	0
0	5	0	0	0	-5	0	0
-1	5	4	-1	-1	-5	-1	-1
-2	5	3	-2	-2	-5	-2	-2
-3	5	2	-3	-3	-5	-3	-3
-4	5	1	-4	-4	-5	-4	-4
-5	5	0	0	-5	-5	-0	-0

Figure 2.2.3.5-1 Example of the mod and rem operators

A pseudo-random number generator making use of the **mod operator** is provided in Figure 2.2.3.5-2.

```
architecture Random_a of Random is
  signal Random_s : integer;
begin  --  Random_a

  PseudoRandom_Lbl: process
    variable Seed_v         : INTEGER := 17654;
    constant Multiplier_c   : INTEGER := 25173;
    constant Increment_c    : INTEGER := 13849;
    constant Modulus_c      : INTEGER := 65536;   -- rollover value
```

```
  begin
    SEED_v :=  (Multiplier_c * Seed_v + INCREMENT_c) mod Modulus_c;
    Random_s <= Seed_v;
    wait for 10 ns;  --repeat process every 10 ns
  end process PseudoRandom_lbl;
end Random_a;
```

Figure 2.2.3.5-2 Pseudo-random Number Generator (ch2_dir\randm_ea.vhd)

The addition, sign, multiplying and miscellaneous operators are discussed in the next subsection because they relate to scalar types for the predefined operators.

2.3 TYPES AND SUBTYPES

In VHDL, every object belongs to a type *(LRM 4.1)*. The classification of types is shown in Figure 2.3. The syntax for a type declarations is as follows:

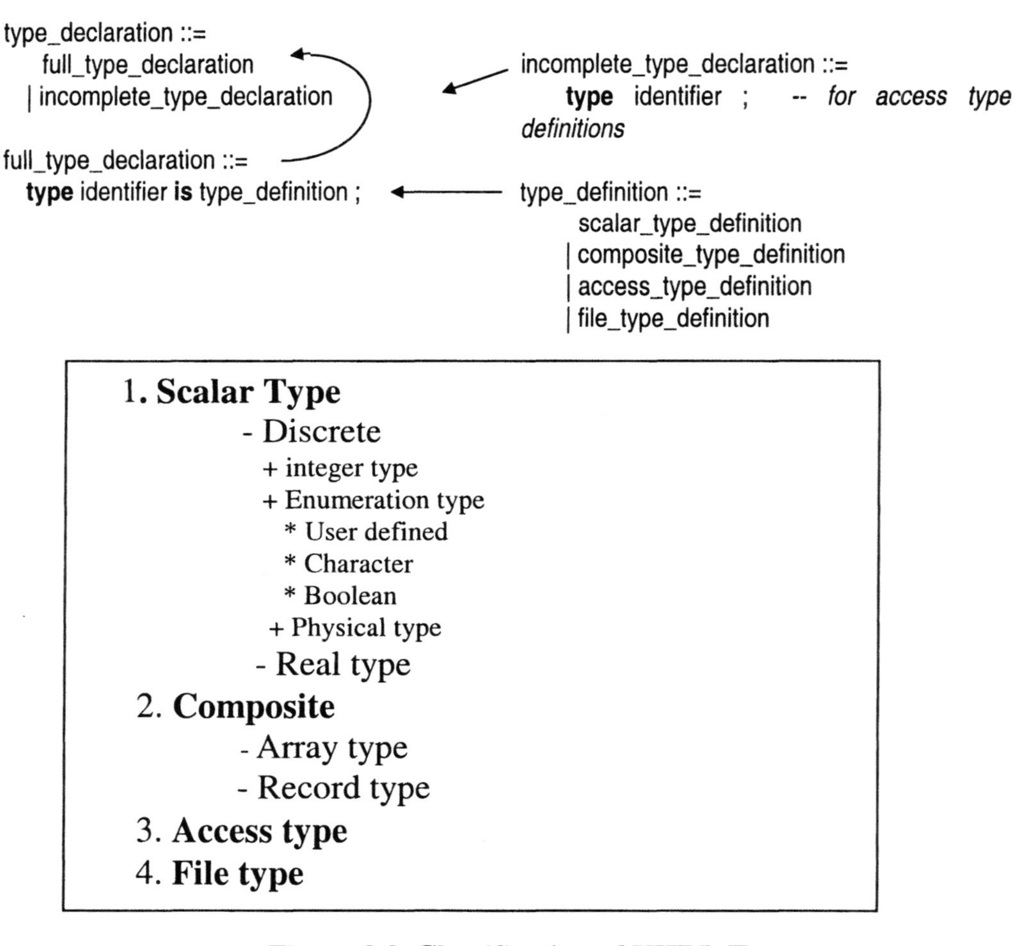

Figure 2.3 Classification of VHDL Types

[1](LRM 4.2) A subtype is a type with a constraint which specifies the subset of values for the type. *The type is called the base type of the subtype.* ***Subtypes are compatible with the base type****, provided the constraints are not violated.* The syntax for subtype declarations is as follows:

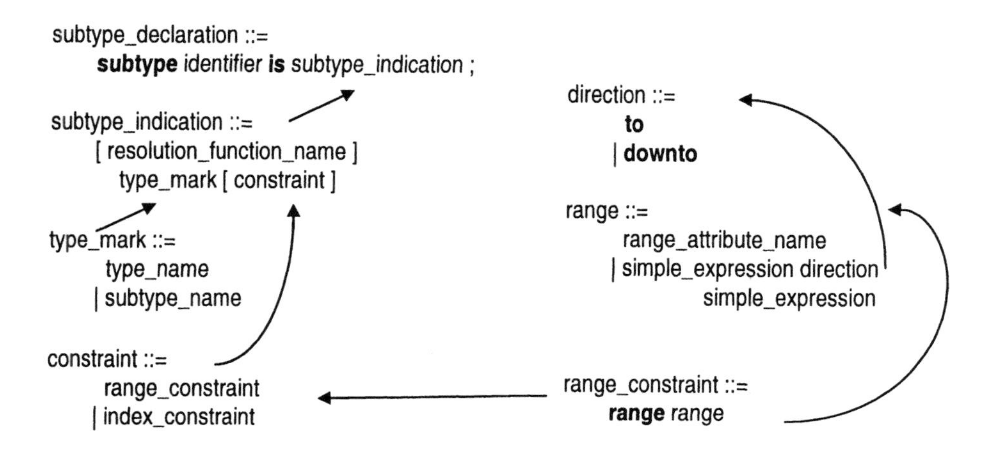

MI ✍ ✍ Use subtypes to constraint the range of a type.

Rationale: *Constraint errors can be detected at compile and/or run time. Subtypes are compatible with the base type, and do not create new types. In synthesis, subtypes are used to constrain the size of any implied registers.*

Examples of types and subtypes are presented in the following subsections.

2.3.1 Scalar Type

[1] (LRM:3.1) Scalar types are types whose values have no elements and do not have any internal component or structure (unlike composite types). *All scalar types are ordered; that is, all relational operators are predefined for their values. Enumeration types and integer types are called discrete types and are composed of objects that have an immediate successor and predecessors. A scalar real type is composed of numbers that form a continuum and have no immediate predecessors or successors.*

2.3.1.1 Integer Type and Subtypes

Maximum range for the integer type is implementation dependent. The minimum range from -2,147,483,647 to +2,147,483,647 (i.e. -2**31 to 2**31 - 1). Package standard defines the following:

```
-- type integer is range implementation_defined; -- per LRM
```

```
-- Model Technology's Standard package, version 4.2f
--  uses the following definition:
type integer is range -2147483648 to 2147483647;
subtype natural is integer range 0 to integer'high;
subtype positive is integer range 1 to integer'high;
```

Most synthesizers and simulator vendors have adopted this extended integer range to allow the synthesis of 32 bit signed numbers. When using integer types, a user must insure that the range is not extended (such as multiplying the equivalent unsigned values of two 16-bit numbers). If the range exceed the integer limits, then signed or unsigned types must be used instead.

Operations on type integers are defined in Table 2.3.1.1-1. Table 2.3.1.1-2 provides language warnings on integer operations.

Table 2.3.1.1-1 Operations on Type Integers

Operator	Operation	Left_Oprd	Right_Oprd	Result
+	Addition	integer	integer	integer
-	Subtraction	integer	integer	integer
*	Multiplication	integer	integer	integer
/	Division	integer	integer	integer
mod	modulus	integer	integer	integer
rem	Remainder	integer	integer	integer
=	equal	integer	integer	boolean
/=	unequal	integer	integer	boolean
>=	greater than or equal	integer	integer	boolean
<=	less than or equal	integer	integer	boolean
>	greater than	integer	integer	boolean
<	less than	integer	integer	boolean

Table 2.3.1.1-2 Language Warnings on Integer Operations

1. Fractional part of integer division is discarded. Thus, 5/4 yields 1.
2. It is illegal to place 2 arithmetic operators side by side. Thus,
   Var := 2 * -4; -- illegal 💣
   Var := 2 * (- 4) ; -- legal ☺
3. Cannot use negative exponents on integers. Thus,
   Var := 2 ** (-3); -- compile error 💣
4. Bits of an integer type value are not directly accessible. No assumption is made on the location of these bits. Must use specific functions translating an integer into a bit vectors.

Figure 2.3.1.1 represents an example using the predefined type **integer**. This example demonstrates operations on integer types and subtypes. It also demonstrates the declaration of subtypes, and range constrained errors on subtypes. The example makes use of a process with a **wait** statement to suspend the process execution at the end of the process, thus emulating a sequential programming language like "C".

```
architecture Integer_a of Integer_N is
  -- subtypes ensure that the range of the variable is not violated
  subtype Int0_5_Typ is integer range 0 to 5;    -- 👍👍

  -- Constant declarations
  constant MaxCount_c   : Int0_5_Typ := 4;
  ...

begin
  Test_Lbl: process
    variable MaxCount_v   : Int0_5_Typ := 3;
  begin
    Sum1_s <= MaxCount_c + 7;   -- signal assignment
    Sum0_s <= ((3 + 8) * 9) / 4 after 10 ns; -- 99/4 = 24

    --      MaxCount_v MUST be between 0 and 5.
    MaxCount_v  := ((3 + 8) * 9) / 4; -- 99/4 = 24 💣
    MaxCount_v := 20;      -- compile error 💣
    Count_s <= abs (-36) mod 7;     -- 1
    wait;                      -- wait forever (i.e. suspend process)
  end process Test_Lbl;
end Integer_a;
```

Figure 2.3.1.1 Operations on Integers (ch2_dir\intgr_ea.vhd)

2.3.1.2 Enumeration Types

The enumeration types include the user defined and the predefined enumeration types. **Loop indices of enumeration data types may not necessarily be synthesizable**. Some synthesis tools require that the parent type for the loop index be of integer type (see chapter 3 for control structures).

2.3.1.2.1 User Defined Enumeration Types

An enumeration type *(LRM: 3.1.1)* defines a type that has a set of user defined values consisting of identifiers and character literals. The order in which enumeration literals are listed is significant. The compiler assumes that enumeration literals are listed in ascending order, from left to right. In synthesis it is possible to enforce a numeric encoding that is different than the default encoding. Objects of enumeration types are synthesizable.

> M 👍👍 Attempt to set the order of the literals so that the first literal shall be the default state.

Rationale: *If the value of any object is not initialized, then the default value for that object is the leftmost value for that type, and is referred to as ObjectType'left (pronounced ObjectType "tic" left). In addition many synthesizers implement the order of enumeration type starting with "0" for the leftmost value.*

M 👍👍 Use of enumerated type in modeling is **STRONGLY**_recommended to model objects that can take discrete enumeration values (such as states of a state machine)

Rationale: *Use of enumeration represents one way of remaining in an abstract level and increases readability. .*

Figure 2.3.1.2.1-1 demonstrates the use of user defined enumeration type.

```
architecture Enum_Beh of Enum is
  type CpuOp_Typ is (Load, Store, Add, Sub, Mult, Div,
                                    And2, Or2, Xnor2); -- can't use reserved words
  type StateMachine_Typ is (Idle, Attention, Ready, Fire, Shutdown);
  signal   CpuOp_s  : CpuOp_Typ := And2;
begin
   Demo_Lbl: process
     subtype ArithOp_Typ is CpuOp_Typ range Add to Div;
     variable CpuOp1_v : CpuOp_Typ;
     variable CpuOp2_v : CpuOp_Typ;
     variable States_v : StateMachine_Typ;

   begin
     CpuOp_s <= Store after 10 ns,
                Add   after 100 ns,
                Mult  after 150 ns; -- signal assignment (waveform)
     CpuOp1_v := Add;
     CpuOp2_v := Or2;
     if CpuOp1_v = CpuOp2_v then
       States_v := Attention;
     else
        States_v := Fire;
     end if;
     wait;
   end process Demo_Lbl;
end Enum_Beh;
```

Figure 2.3.1.2.1-1 User-Defined Enumeration Type (ch2_dir\ enum_ea.vhd)

Two type definitions always define two distinct types, even if they are lexically identical.

M 👍👍 Avoid using different types for lexically identical elements unless there IS a good reason.

Rationale: *The two types are distinct, incompatible, and can cause confusion to the reader.* **Separate types** *for the same subset of values* ***are not compatible*** *and need type conversion. Synthesis tools have restrictions on types that they can accept.*

M Avoid declaring (and using) separate subtypes (2 or more) for the same subset of values of the parent type. For example:

subtype INT8_Typ **is** Integer **range** 1 **to** 8;

subtype I1to8_Typ **is** Integer **range** 1 **to** 8; -- This is redundant

Rationale: *This is redundant information and worsens readability.*

Figure 2.3.1.2.1-2 demonstrates the declarations and use of objects of two distinct types that are lexically identical.

```
architecture TtlEcl_Beh of TtlEcl is
    subtype   TTL_Typ is bit;       -- '0', '1' from package Std.Standard
    type      ECL_Typ is ('0', '1');
begin  --  TtlEcl_Beh

 Test_Lbl : process
    variable   TTL1_v : TTL_Typ := '0';
    variable   TTL2_v : TTL_Typ := '1';
    variable   ECL1_v : ECL_Typ := '0';
    variable   ECL2_v : ECL_Typ := '1';

  begin  --  process Test_Lbl
    TTL1_v := TTL2_v;   --  '1'
    TTL2_v := '0';
    TTL1_v := ECL1_v;          --  Illegal, different types (#40) M
     --  Must use conversion functions to change types
    ECL2_v := TTL1_v;          --  Illegal  (#42) M
    ECL1_v := ECL2_v;
    wait;
  end process Test_Lbl;
end TtlEcl_Beh;
```

Separate type definitions prevents operations on incompatible types

Figure 2.3.1.2.1-2 Objects of Lexically Identical Types (Ch2_dir\ttlec_ea.vhd)

When an enumeration type is defined, the relational operators are implicitly defined to allow comparison operations on objects of that type. These operators include:

```
  =   /=                          -- equality and inequality operators
 <  <=  >  >=    -- Ordering operators (less than, less than or equal, ...
```

Thus, for the above example *enum_ea.vhd* the following can be asserted for the variables CpuOp1_v and CpuOp2_v that are of type CpuOp_Typ, where type CpuOp_Typ is (Load, Store, Add, Sub, Mult, Div,And2, Or2, Xnor2):

if CpuOp1_v = CpuOp2_v **then** -- legal ☺
if CpuOp1_v > CpuOp2_v **then** -- legal ☺
  -- Load < Store < add < Xnor2
  -- Xnor2 > Sub > Load
if CpuOp1_v /= CpuOp2_v **then** -- legal ☺

if CpuOp1_v > CpuOp2_v **then** -- legal ☺
if CpuOp1_v < CpuOp2_v **then** -- legal ☺
if CpuOp1_v + CpuOp2_v **then**
  -- illegal, "+" is not defined 💣
if CpuOp1_v **and** CpuOp2_v **then**
  -- illegal, "**and**" is not defined 💣

It is possible for a user to overload (or *define*) the "+", "-",..., "and", "*" (arithmetic and logical) operator for any specific enumerated data type through the use overloaded functions (chapter 6). The "=" and "/=" and ordering operators are automatically instantiated once the enumerated type is defined, and should not be redefined.

An enumeration literal can be overloaded *(LRM 10.5)* by including the definition of two or more enumeration types. Compiler does its best to determine literal's type. In some circumstances, this is not possible, and it must be qualified by explicitly stating its type. Figure 2.3.1.2.1-3 provides an example of overloaded enumeration literals.

```
architecture Ovnum_Beh of OvNum is
 type Color_Typ is (Yellow, Orange, Red, Green, Blue);
 type PrimaryColor_Typ is (Red, Yellow, Blue);
 signal BrushPrim_s   : PrimaryColor_Typ;
 signal Brush_s       : Color_Typ;

begin
  Demo2_Lbl:   process
  begin
    BrushPrim_s <= Orange; -- not a literal of that type! 💣
    BrushPrim_s <= Red;  -- ** which type Color_Typ? PrimaryColor_Typ?
                         -- Compiler picks PrimaryColor_Typ
                         -- because of context.
    BrushPrim_s <= PrimaryColor_Typ'(Red);  -- qualified expression
    Brush_s     <= Color_Typ'(Red);  --
    BrushPrim_s <= Color_Typ'(Red); -- error.
    wait;
  end process Demo2_lbl;
end Ovnum_Beh;
```

Enumeration literal "Red" is overloaded because it is a literal of two different types. The 2 "Red" literals are incompatible.

Type conflict in qualified expression. Type color_typ versus primarycolor_typ. 💣

Figure 2.3.1.2.1-3 Overloaded Enumeration Literals (ch2_dir\ ovnum_ea.vhd)

2.3.1.2.2 Predefined Enumeration Types

The predefined enumerated types *(LRM 3.1.1.1)* in package Standard include:

type Boolean **is** (false, true); -- false < true
type Bit **is** ('0', '1'); -- '0' < '1'
type Character **is** (--VHDL'93 has a greater character set than VHDL'87)
  nul, soh, stx, etx, eot, enq, ack, bel,
  ... '0', '1', '2', '3', '4', '5', '6', '7',
  ... '@', 'A', 'B', 'C', 'D', 'E', 'F', 'G',
  ... '`', 'a', 'b', 'c', 'd', 'e', 'f', 'g',
  ... 'ø', 'ù', 'ú', 'û', 'ü', 'ý', 'þ', 'ÿ'); --'93
  (-- see standard package for full definition)
type Severity_level **is** (note, warning, error, failure);
type File_open_kind **is** (read_mode, write_mode, append_mode);

type file_open_status **is** (open_ok, status_error, name_error, mode_error);

Enumeration literals are listed in ascending order, from left to right. Thus, 'a' < 'b'; 'a' > 'A'; false < true;

Type *Std_uLogic* is defined in *Std_Logic_1164* package (see appendix D). The *Std_uLogic* type is an enumerated data type with special meaning by the user and simulator (when displaying values). The *Std_Logic_1164* package is discussed in chapter 7. Table 2.3.1.2.2 provides an explanation of the Std_Logic type states and the common interpretation of the logic states.

Table 2.3.1.2.2 Explanation of the Logic States for Std_ULogic Type

Std_ULogic Enumerated State	IEEE Std 1164-1993 definition	Draft VHDL Synthesis Package Interpretation	Comments
type Std_Ulogic is			
('U',	-- Uninitialized	Model behavior	* No value assigned.
'X',	-- Forcing unknown	Model behavior	* Forcing No distinguishable state.
'0',	-- Forcing 0	Logic level	Strong 0, or driving 0.
'1',	-- Forcing 1	Logic level	Strong 1 , or driving 1.
'Z',	-- High Impedance	Disconnect	Signal source makes no contribution.
'W',	-- Weak Unknown	Model behavior	* Weak No distinguishable state.
'L',	-- Weak 0	Logic level	Resistive pulldown equivalent.
'H',	-- Weak 1	Logic level	Resistive pull-up equivalent.
'-');	-- Don't care	match all	* Match ALL with Std_Match function in Numeric_Std package[10]

* The values 'U', 'X', 'W', and '-' are *metalogical* values; they define the behavior of the model itself rather than the behavior of the hardware being synthesized.

The value "-" is a literal of the enumeration, and is therefore a value. In strict VHDL, if a standard logic bit vector is compared against a constant that has "-" in it, it does NOT mean that the comparison should ignore those bits. Instead, it means an exclusive comparison against the value "-". For example:

if InPort(3 **downto** 0) = "1--0" **then** ...

implies that the values of InPort(3 downto 0) are exactly "1--0", and not that just bit 3 is a '1' and bit 0 is a '0'. For synthesis though, many synthesizers have adopted the convention that the "-" or the "X" is truly a "don't care" or "wild card" when used in an assignment statement (and not in the expression used for comparison). Thus, *S <= "1--1";* means that the center bits of the assignment can take any value.

To overcome this problem between the synthesis interpretation of VHDL and actual definition and simulation of the enumeration literals, the IEEE DASC Synthesis Working Group have included in the Standard VHDL Synthesis numeric package a function called *STD_MATCH* that provides "don't care" or "wild card" testing of values based on the *STD_uLogic* type.

---

[10] The numeric package is available via ftp from "vhdl.org /vi/vhdlsynth/numeric_std.vhd"

This package also includes the overloaded operators for arithmetic ("+", "-"), multiplication ("*"), division ("/"), modulus (mod, rem), comparison ("<", ">", "<=", ">=", "=", "/=") and logical (not, and, or, nand, nor, xor, xnor) for types SIGNED and UNSIGNED. Those types represent unconstrained arrays of Std_Logic_Vector.

Vendors of synthesis tools also provide similar packages. See the Synopsys subdirectory provided on disk for a set of public domain packages that provide these functions.

Signed and unsigned arithmetic is a computer science subject, rather than a VHDL subject and is not covered in this book.

2.3.1.2.3 Boolean Type

The Boolean type is predefined in package standard as an enumerated data type as:

type Boolean is (false, true);

Boolean constants, variables, and signals can be declared as shown below:

variable IsActive_v : Boolean; -- defaulted to boolean'left (i.e., False)
signal Data_Rcvd_v : Boolean := True;

Boolean operators may use the following relational operators:

<, >, <=, >=, =, /=

No matter what type of objects these relational operators are comparing, the result is a boolean value of either TRUE or FALSE.

> When testing a boolean object for TRUE or FALSE, use the object directly. For example:
>
> **variable** InIdle_v : **boolean := False**;
> ...
> **if** InIdle_v **then** --
>
> **if** InIdle_v = **TRUE then** --
>
> ***Rationale:*** *Testing an object against true or false is redundant and decreases readability.*

2.3.1.3 Physical types 

Physical types *(LRM:3.1.3)* are used to represent physical quantities such as distance, current, time, etc. A physical type provides a base unit, and successive units are defined in terms of this unit. *[1] Each value of a physical type has a position number that is an integer value.* Objects of physical types are not synthesizable. Delays of type *time* are ignored by synthesizers. For example assigning a value to a signal after a delay of 15 ns does not cause the synthesizer to add delay gates to approach the 15 ns delay. The definition of time is in package Standard and is defined as shown below. With this definition, the maximum time that can be described is 2^{31} femtoseconds, or 2.147 microseconds.

```
type time is range -2_147_483_647 to 2_147_483_647
    units
     fs;
     ps = 1000 fs;        ns = 1000 ps;        us = 1000 ns;
     ms = 1000 us;        sec = 1000 ms;       min = 60 sec;     hr = 60 min;
    end units;
```

This, of course, is a very limited time range. Simulation vendors have chosen to built the time limit into the simulator, so the range in the package is not related to the actual implementation. Some vendors interpret the maximum simulation time as 2^{31} times the unit resolution of time (e.g. ps, ns), defined as an option prior to simulation. Thus, if the simulation resolution is nanoseconds, then the maximum simulation time is 2,147,483,647 ns or 2.147 seconds. In addition, any time value less then 1 ns (e.g. 575 ps) is considered as 0 ns. Again, the 2 second maximum simulation time with a resolution of 1 nanosecond was too limiting for long simulation runs. As a result, many vendors have either extended the time range definition to 2^{63}, or have extended the simulation time to 2^{63} times the unit resolution of time defined prior to simulation (while maintaining the time range definition to 2^{31}). A literal of a physical type can be expressed as shown in the following examples:

101 ns   10.7 ns

The predefined function "**now**" in package STANDARD returns the current simulation time. This is a very useful function to either measure time between events, or to verify time against an absolute simulation time.

To represent an integer multiplicand of a physical type use the following relationship:
(integer * physical_type) yields a physical type -- "*" represents multiplication

To represent an real multiplicand of a physical type use the following relationship:
(real * physical_type) yields a physical type -- "*" represents multiplication

To convert physical type to integer use the following relationship:
(physical_type / physical_type) yields a universal integer, where "/" represents division.

Figure 2.3.1.3-1 demonstrates operations on physical type time. Figure 2.3.1.3-2 defines other physical types.

```
Test_Lbl: process
    variable Delay_v        : time := 10 ns;
    variable Int_v          : integer;
    constant PropDelay_c    : time := 25 ns;
    constant Margin_c       : positive := 2;
  begin
    Delay_v := now;  -- current time
    Delay_v := PropDelay_c * Margin_c;  -- 25 ns * 2 = 50 ns
    Int_v  := (4 * PropDelay_c) / Delay_v;  -- 100 ns / 50 ns  = 2
    wait;
  end process Test_Lbl;
```

Figure 2.3.1.3-1 Operations on Physical Type Time (ch2_dir\time_ea.vhd)

```
architecture Physical_Beh of Physical is
    type Current_Typ is range -2147483647 to 2147483647
       units
         nA;                 -- base unit, nano Amp
         uA  = 1000 nA;      -- micro Amp
         mA  = 1000 uA;      -- milli Amp
         Amp = 1000 mA;      -- Ampere.  MAX Current = 2.147 Amp
       end units;
    type Volt_Typ is range -2147483647 to 2147483647
       units
         uV;                 -- base unit, micro volt
         mV  = 1000 uV;      -- milli volt
         V   = 1000 mV;      -- Volt.  MAX voltage = 2147 volts
       end units;
    type DegreeC_Typ is range -1E3 to 1E3
       units
         C;                  -- base unit, Celsius
       end units;
  begin
    Test_Lbl: process
      variable I_v           : Current_Typ := 10 mA;
      variable OutVolt_v     : Volt_Typ    := 5 mV;
      variable Temperature_v : DegreeC_Typ := 100 C;
    ...
    end process Test_Lbl;
  end Physical_Beh;
```

Figure 2.3.1.3-2 Physical Type Examples (ch2_dir\phys_ea.vhd)

2.3.1.4 Distinct Types and Type Conversion

M Use subtypes to constrain objects of base type so that objects declared with these subtypes remain compatible. Do not use separate type definitions.

Rationale: *Two type definitions always define two distinct types, even if they are lexically identical. Objects of a common subtype are compatible.*

However, when working with closely related numeric or array types, type conversion *(LRM 7.3.5)* allows the conversion to be performed without the use of a function. The format for type conversion is:

```
type_conversion ::=                    type_mark ::=
    type_mark (expression)                 type_name
                                         | subtype_name
```

For example, to convert a variable of type real to type integer the following can be used:
SomeInteger <= Integer(VarReal) **after** 1 ns;
Figure 2.3.1.4 provides other examples of type conversions.

```
architecture TypeTest_Beh of TypeTest is
begin
  Test_Lbl: process
    type A_Typ is range 1 to 10;  -- Type declarations
    type B_Typ is range 1 to 10;
    subtype C_Typ is integer range 1 to 10;
    variable A1_v    : A_Typ := 1;  -- variable declarations
    variable A2_v    : A_Typ := 1;
    variable B_v     : B_Typ := 2;
    constant C_c     : C_Typ := 3;
   begin
    A1_v  := 1 + A1_v;       -- OK, addition to a universal integer
      -- ()
    A2_v  := A2_v + A1_v;    -- OK, same type addition
    B_v  := 1 + B_v;         -- OK
    C_c  := 3;               -- ERROR, cannot modify constant

    A1_v := A2_v + B_v;      -- ERROR, not same type
                             No feasible entries for infix op: "+"
                             Bad right hand side in assignment.
                             UNLESS "+" is overloaded (chptr

    B_v  := 2 + C_c;         -- ERROR, not same type
                             Incompatible types for assignment.
                             UNLESS "+" is overloaded (chptr
                             7)

    A1_v := A2_v + A_Typ(B_v);  -- OK, Type conversion, allowed between
                                -- closely related numeric or array type

    B_v  := 2 + B_Typ(C_c);   -- Type conversion
    wait;                     -- suspend process
  end process Test_Lbl;

  Enum_Lbl: process            -- a second concurrent process
    type Enum1_Typ is (R, G, B);
    type Enum2_Typ is (R, G, B);
    variable Enum1_v : Enum1_Typ;
    variable Enum2_v : Enum2_Typ;
  begin                                   Incompatible types for assignment.
       Enum1_v := B;          -- OK
       Enum2_v := R;          -- OK
       Enum1_v := Enum2_v;    -- ERROR, not same type
       Enum1_v := Enum1_Typ(Enum2_v); -- ERROR, not closely related
                                      -- numeric or array type.
                                      -- Illegal type conversion.  Must
                                      -- use a conversion function (chapter 6)
       wait;                          -- suspend process
  end process Enum_Lbl;
end TypeTest_Beh;
```

Figure 2.3.1.4 Type Conversions Examples (ch2_dir\typet_ea.vhd)

2.3.1.5 Real type

Range of real *(LRM: 3.1.4.1)* is defined in package STANDARD, and is implementation dependent. It must at least cover the range of -1.0E38 to +1.0E38. It must allow for at least six decimal digit precision. Operations on real types is shown in Table 2.3.1.5. Figure 2.3.1.5 demonstrates the use and restrictions of real types.

Table 2.3.1.5. Operations on Real Type

EXPRESSION	OPERATION	EXAMPLE **variable** Var : **real;** **variable** Int : **integer;**	COMMENT
+ -	real + real ☺ real + universal real ☺ real + integer 💣	Var := Var + Var; ☺ Var := Var + 3.2; ☺ Var := Var + Int; 💣 Var := 2.34 + real (5); ☺	type conversion needed
* /	real * | / real ☺ real * | / universal real ☺ real * | / universal integer 💣 real * | / integer 💣 universal integer * | / universal real ☺ universal real * | / integer 💣	Var := Var * Var; ☺ Var := Var * 3.1; ☺ Var := Var * 2; 💣 Var := Var * Int; 💣 Var := Var * real(Int); ☺ Var := 2 * 3.5; ☺ Var := 2.0 * Int; 💣 Var := 2.0 * real(Int); ☺	type conversion needed
abs	absolute function	Var := **abs**(-3.4);	-- = 3.4
**	exponentiation	1.5**2	-- = 2.25

```
architecture Real_Beh of Real is
  subtype Real5to10_Typ is real range 5.0 to 10.0;
begin
Demo2_Lbl:
   process
     variable Real_v  : real := 30.0;
     variable Real2_v : Real5to10_Typ;
     variable Int_v   : integer := 3;
   begin
     Real_v  :=  3.0 * Real_v;
     Real_v  :=  Real_v * real(Int_v);  --   type conversion
     Real_v  := 2 * 3.0;     --   universal integer * universal real -> OK ☺
     Real2_v := 2 * 3.5;     --   universal integer * universal real -> OK ☺
     -- Real_v  := 2.0 * Int_v;         --   variable must be real 💣
     --   No feasible entries for infix op: "*"    UNLESS "+" is overloaded (chptr 7)

     -- Real_v  := 2 * Real_v;  -- 💣
     -- Universal integer * real variable is illegal
     wait;
   end process Demo2_lbl;
end Real_Beh;
```

Figure 2.3.1.5 Use and Restrictions of Real Types (ch2_dir\real_ea.vhd)

2.3.2 Composite

Compositess consist of arrays and records.

2.3.2.1 Arrays

[1] (LRM 3.2.1) An array type is a type, the value of which consists of elements that are all of the same subtype (and hence, of the same type). Each element is uniquely distinguished by an index (for a one-dimensional array) or by a sequence of indexes (for a multidimensional array). Each index must be a value of a discrete type and must lie in the correct index range. An entire array is referenced with a single identifier. An individual component of the array is referenced with the array identifier, followed by an index value placed in parentheses. Arrays are categorized as constrained or unconstrained. A **constrained** array is an array whose size is constrained (e.g., 10 elements). An **unconstrained** array defines an array type, and a name denoting that type. See section 2.3.2.1.2 for definition of unconstrained array. The syntax for the array definitions is shown below.

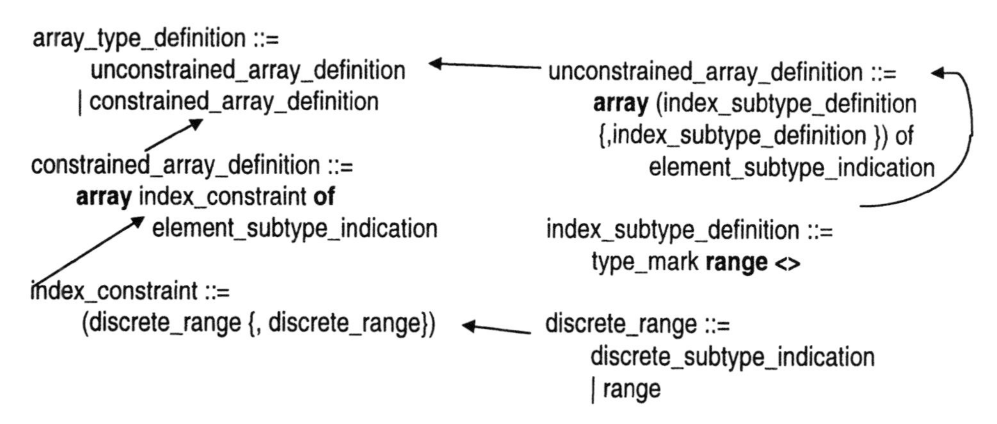

2.3.2.1.1 One Dimensional Arrays

One dimensional arrays define vectors. For example:

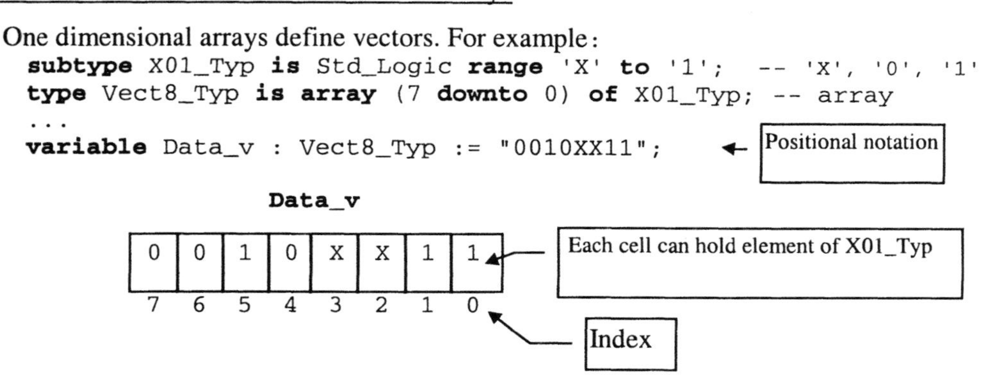

```
type ID_TYP  is
  (ID0, ID1, ID2, ID3, ID4, ID5, ID6, ID7);
subtype ID03_TYP  is ID_TYP range ID0 to ID3;
type CountID03_typ is array(ID03_TYP) of integer;
...
variable CountId03_v : CountID03_typ :=
   (ID0  => 5,
    ID1  => 45,          <-- Named notation
    ID2  => 16,
    ID3  => 8);
```

CountId03_v

Index	CountId03_v
ID0	5
ID1	45
ID2	16
ID3	8

Each cell can hold an integer element

[1] (LRM 7.3.2) An aggregate is an expression denoting a value of a composite type. The value is specified by giving the value of each of the elements of the composite type.

The syntax for aggregates is as follows:

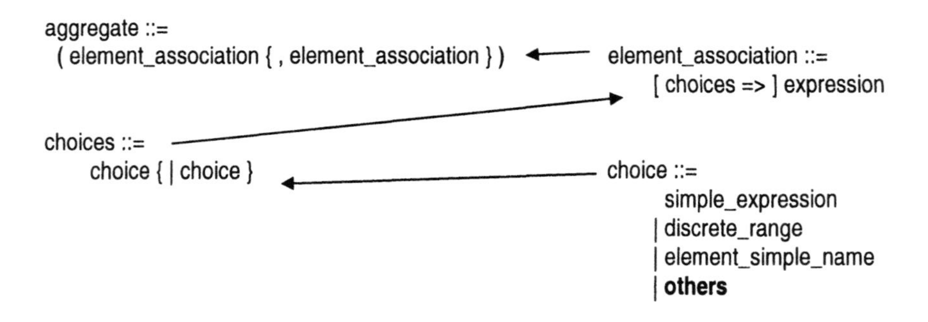

*[1] Either Positional association or a named association may be used to indicate which value is associated with which element. When the "others" is used, it represents ALL of the other elements not listed in the named association. The "others" statement MUST be the **LAST** statement in the associated list.*

Figure 2.3.2.1 represents operations on a 1 dimensional array of Boolean type and demonstrates the use of aggregates.

```
architecture Array_a of Array_N is
begin
  Simple_Lbl : process
    subtype  R10to20_Typ is integer range 10 to 20;
    subtype NumbStudents_Typ is integer range 0 to 35;
    type WantArray_Typ is array (NumbStudents_Typ) of boolean;
    variable WantBook_v  : WantArray_Typ; -- initialized to boolean'left
  begin
    -- set WantBook_v aggregate, All elements assigned true
    WantBook_v:= (WantBook_v'range => true);
    -- All elements assigned false
    WantBook_v:= (WantBook_v'left to WantBook_v'right => false);
    WantBook_v := (others => false);      --   the whole aggregate
    --   Filling a portion of the whole aggregate.
    WantBook_v(10 to 20) := (10     => true, -- 10, 12, 15 get true
                             12     => true, -- 11, 13, 14, 16 to 20 get false
                             15     => true,
                             others => false);
    WantBook_v(R10to20_Typ)  := (others => false);

    -- filling up array slices in aggregates
    WantBook_v := (1          => true,   -- need complete aggregate
                   3 to 5     => true,   -- in this format because the whole
                   10         => true,   -- object (WantBook_v) is updated
                   25 to 35   => true,
                   others     => false);
    -- an update of a slice.
    WantBook_v(25)  := false;
    WantBook_v(25 to 30)  := (others => false);
    WantBook_v(25 to 30)  := (25        => true,
                             26 to 27  => false,
                             others    => true);  --  index 28, 29, 30
    wait;
  end process Simple_Lbl;
end Array_a;
```

See section 2.5 for definition of attributes

Aggregate with named association

Figure 2.3.2.1 Operations on a 1 Dimensional Array of Boolean Type (ch2_dir\moreary.vhd)

If it is desired to use an aggregate of length one, then the use positional notation is necessary. Otherwise, the compiler interprets the value as a single element (e.g., Bit) rather than an aggregate. For example:

signal Clock7 : Bit_Vector(1 **to** 1) := ('1'); -- 💣

signal Clock9 : Bit_Vector(1 **to** 1) := '1'; -- 💣

signal clock7 : Bit_Vector(1 **to** 1) := (**others** => '1'); -- ☺

signal clock7 : Bit_Vector(1 **to** 1) := (1 => '1'); -- ☺

A string literal notation can also be used as follows:

signal clock9 : Bit_Vector(1 **to** 1) := "1"; -- note the double quote. ☺

2.3.2.1.2 Unconstrained Array Types

An unconstrained array is an array that defines the type of elements, but not its size. This allows users to declare arrays that differ only in size (the number of index value in a given dimension) to be of the same type. An unconstrained array does not include information about the size of the array. Thus the array indexes are not constrained to a particular range of values. Determination of the array size is deferred until an array is declared to belong to this unconstrained type[11]. An example of an unconstrained array is:

```
type IntArray_Typ is array (integer range <>) of integer; -- <> called "box"
  -- to indicate no constraint is imposed on the range of integer index values.
```

To use the array two methods are available:

1. **Constrained subtype:** Create a subtype of the unconstrained type, and define the size constraints. When declaring the object, define the type as the declared subtype.

```
subtype RegSize_Typ is Integer range 1 to 4;          -- Size subtype
subtype R4_Typ is Bit_Vector(RegSize_Typ);  -- Constrained subtype
subtype Inflexible_Typ is Bit_Vector (1 to 4); -- range is hard-coded
variable R4_v    : R4_Typ := "1010";
variable R4b_v   : Inflexible_Typ := "1010";
```

2. **Direct object definition**: Declare the object as the unconstrained subtype, but define the size constraints.

```
variable OtherR4_v   : Bit_Vector (1 to 4) := "1010";
```

Use a subtype for the range definition instead of hard coded range unless the range is known not to ever change. Declare subtypes to constrain unconstrained array sizes to enable use of the subtype attributes.

Rationale: *Attributes (section 2.5) represent a significant role in the definition and readability of VHDL code. Several attributes can only use a type or subtype for a prefix. Other attributes can use the object or the type as a prefix. Hard coded range definitions are difficult to maintain when the size of the arrays are changed.*

Figure 2.3.2.1.2-1 represents assignments on a one-dimensional array of bit type (i.e., the Std.Bit_Vector type). It represents the same example as shown in Figure 2.3.2.1.1 except for the type definition for the elements of the array. This example makes use of the predefined unconstrained Bit_Vector type defined in package *Standard* as:

```
type bit_vector is array (natural range <>) of bit;
```

---

[11] *Rendez Vous with Ada,* A Programmer's Introduction, David J. Naiditch, John Wiley & Sons, Inc. Copyright ©1989 by David Naiditch

For arrays that represent bits, use **positional notation** when defining the value of the array grouped as a word (e.g., address or data word). When the bit array represents individual control elements, use **named notation**. Example:

```
Address_s <= "1011000010101111";
RelaySelect_s <= (2  =>  '1',     -- Transmitter
                  1  =>  '0',     -- Receiver
                  0  =>  '1');    -- Monitor
```

Rationale: *Enhanced readability.*

...

```
architecture Array2_a of Array_n is
begin
  Simple_Lbl : process
    subtype  R10to20_Typ is Integer range 10 to 20;
    subtype NumbHits_Typ is Integer range 0 to 35;
    subtype HitArray_Typ is Bit_Vector(NumbHits_Typ);
    variable Hit_v   : HitArray_Typ; -- initalized to boolean'left
                                     --  by default (i.e. false)
  begin
    -- set Hit_v aggregate, All elements assigned true
    Hit_v:= (Hit_v'range => '1');

    -- All elements assigned '0'
    Hit_v:= (Hit_v'left to Hit_v'right => '0');
    Hit_v := (others => '0');      --   the whole aggregate

    --   Filling a portion of the whole aggregate.  The next aggregate
    --   is from 10 to 20

    Hit_v(10 to 20) := "10100100000";

                          -- (10     => '1', -- 10, 12, 15 get '1'  --
                          --  12     => '1', -- 11, 13, 14, 16 to 20 get '0'
                          --  15     => '1',
                          --  others => '0');
    Hit_v(R10to20_Typ) := (others => '0');

    -- filling up array slices in aggregates
    Hit_v := (1         => '1',    -- need complete aggregate
              3 to 5    => '1', -- in this format.
              10        => '1',
              25 to 35  => '1',
              others    => '0');

    -- an update of a slice.
    Hit_v(25) := '0';
    Hit_v(25 to 30) := "000000"; -- (others => '0');
    Hit_v(25 to 30) := "100111";
                          -- (25        => '1',
                          --  26 to 27 => '0',
                          --  others   => '1');
    wait;
  end process Simple_Lbl;
end Array2_a;
```

All bits (25 to 35) get the same value ('1')

All non-mentioned bits (2, 6 .. 9, 11 .. 24) get the same value ('0')

Figure 2.3.2.1.2-1 Operations on Array using Bit_Vector type (ch2_dir\array2.vhd)

Figure 2.3.2.1.2-2 represents another example of array utilization with aggregates.

Array size can be redefined through the generic in a configuration.

```
entity UArray is
  generic (ArraySize_g : Positive := 5);
end UArray;

architecture UArray_Beh of UArray is
  subtype RegSize_Typ  is Integer range 1 to ArraySize_g;
  type    IntArray_Typ is array(natural range <>) of Integer;
  subtype IntA_Typ    is IntArray_Typ(RegSize_Typ);

begin  -- UArray_Beh
  ---------------------------------------------------------------------------
  -- Process:   Demo_Lbl
  -- Purpose:   Demonstrates uses of unconstrained arrays and aggregates
  ---------------------------------------------------------------------------
  Demo_Lbl : process
    variable IntA1_v    : IntA_Typ;
    variable IntA2_v    : IntArray_Typ(1 to 5); -- Use IntA_Typ

  begin  -- process Demo_Lbl
    IntA1_v := (1, 2, 3, 4, 5);   --   positional notation 1, 2, ,3 ,4, 5
    IntA1_v := (others => 0);     --   0, 0, 0, 0, 0.  others must be last
    IntA1_v := (3 | 5 | 1 => 555,
                others => 999);   --   2 and 4
    IntA1_v(3) := 3;              --   0, 0, 3, 0, 0
    IntA1_v := (1 => 2,           -- Good approach for control logic
                2 => 3,
                3 => 7,
                4 => 9,
                5 => 11);
    IntA1_v := (1 |2 |5 => 100,
                3 | 4   => 50);
    IntA1_v := (1 to 5 => 1000);


    IntA2_v := (1, 2, 3, 4, 5);   --   positional notation 1, 2, ,3 ,4, 5
    IntA2_v := IntA1_v;           -- all elements copied
    IntA2_v := (others => 0);     --   0, 0, 0, 0, 0
    IntA2_v(3) := 3;              --   0, 0, 3, 0, 0
    -- IntA2_v := (1 => 2,   -- Aggregate length is 2. Expected length is 5.
    --             2 => 3);    --
    -- IntA2_v := (1 |2 |5 => 100, -- Aggregate length is 4. Expected length is 5.
    --              3       => 50);  --
    IntA2_v := (1 to 5 => 1000);

    if IntA1_v = IntA2_v then
      assert false
        report "arrays are equal";
    end if;
    wait;
  end process Demo_Lbl;
end UArray_Beh;
```

Figure 2.3.2.1.2-2 Array Utilization with Aggregates (ch2_dir\uarry_ea.vhd)

Avoid constraining constants of array types with the aggregate value of that constant. Instead, declare the type of the object as either a constrained array subtype or as an unconstrained array subtype with a constraint. Examples:

```
constant  UpWord_c   : Bit_Vector := "1000";    -- range is 0 to 3 most likely
-- range is 0 to 3 most likely
constant  Mix_c      : Bit_Vector := UpWord_c(2 to 3) & UpWord_c(0 to 1); --

constant  DnWord_c   : Bit_Vector(7 downto 0) := X"84"; --Direction is bound

constant  Mix2_c     : Bit_Vector  := -- type is range is upward direction
                                DnWord_c(3 downto 0) & DnWord_c(7 downto 4);
-- Mix2_c'range   = 3 downto -4 ?   ** VHDL'87 spec **
-- Mix2_c'range   = 0 to 7          ** VHDL'93 spec **
constant  Mix2_c     : Bit_Vector(7 downto 0)    -- Direction is bound
     := DnWord_c(3 downto 0) & DnWord_c(7 downto 4); -- range 7 downto 0
```

Rationale: *Do NOT rely on the language to automatically pick the range because this is currently an unsettled issue when concatenation operators are used, and may cause the code to be non-portable.*

Always MATCH slice direction with subtype direction. In addition, stick to ONE convention on the direction of the arrays (ascending or descending).

Rationale: *The design is not cohesive if different directions are used. If the direction of a slice is incompatible with the subtype direction, then the compiler will interpret the array as a "null" array (see 2.3.2.1.6). In some situations, the compiler provides a warning. This produces incompatible code.*

The leftmost bit of an array shall be the most significant, regardless of the bit ordering.

Rationale: *This guidelines provides consistency in design.*

Figure 2.3.2.1.2-3 represents an example using the concatenation operator with the rules described above.

```
subtype R70_Typ is integer range 7 downto 0;
  subtype DnWord70_Typ is Std_logic_Vector(R70_Typ); --   constrained array


  constant DnWordOK_c   : DnWord70_Typ := "10000011";     --
   -- array constrained by the definition


  constant DnWordBad_c : Std_Logic_Vector     :=          --
    DnWordOK_c(3 downto 0) & DnWordOK_c(7 downto 4);
  -- 0 to 7 ?   Most likely

  constant DnWordOK2_c : Std_Logic_Vector(8 downto 0) := --
      DnWordOK_c(3 downto 0) & DnWordOK_c(7 downto 4)& '1';  -- 8 downto 0
```

Figure 2.3.2.1.2-3 Using the Concatenation Operator (ch2_dir\conct_ea.vhd)

2.3.2.1.3 Multi-dimensional Array types

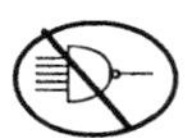

Mutli-dimensional arrays are arrays with more than one index. Thus, a memory can be declared as a two-dimensional array with one index representing the depth, and the other representing the width.

subtype Depth_Typ **is** Integer **range** 0 **to** 1024; -- address range
subtype Width_Typ **is** Integer **range** 7 **downto** 0; -- data range
type Mem_Typ **is array**(Depth_Typ, Width_Typ) **of** bit;

Generally, multi-dimensional arrays are not allowed for synthesis. One way around this is to declare two one-dimensional array types. This approach is easier to use and more representative of actual hardware.

subtype Data_Typ **is** Bit_Vector(Width_Typ);
type Memory_Typ **is array**(Depth_Typ) **of** Data_Typ;

Figure 2.3.2.1.3-1 demonstrates the declaration, initialization, and assignment of values to a memory whose size is defined using generics.

```
library IEEE;                                  --   used because Std_Logic
  use IEEE.Std_Logic_1164.all;                 --   signals are resolved

entity Memory is
  generic (WordSize_g : Natural := 7;
           Depth_g    : Natural := 10);
end Memory;

architecture Memory_a of Memory is
begin  --  Memory_a
  Memory_Lbl : process
    subtype Depth_Typ  is Integer range 0 to Depth_g;           -- address range
    subtype Width_Typ  is Integer range WordSize_g downto 0;    -- data range
    subtype Data_Typ   is Std_Logic_Vector(Width_Typ);
    type    Memory_Typ is array(Depth_Typ) of Data_Typ;

    -- Declare memory and initialize ALL to '0'
    variable Memory_v  : Memory_Typ := (others => (others => '0'));
```

Initialization using aggregates

```
begin  --  process Memory_Lbl
   -- Initialize word 0 to "11111111"
   --                 1 to "10101010"
   --                 2 to "00000010"
   --             3 to end "01LHZXWU";
   Memory_v(0) := "11111111";
   Memory_v(1) := "10101010";
   Memory_v(2) := "00000010";
   Memory_V(3 to Depth_Typ'high) := (others => "01LHZXWU");
   wait;
  end process Memory_Lbl;
end Memory_a;
```

Range defined using attributes of address range

Figure 2.3.2.1.3-1 Memory Declaration, Initialization, and Assignment of values (ch2_dir\mem_ea.vhd)

Other applications of multi-dimensional arrays is the specification of delay parameters for a design. Figure 2.3.2.1.3-2 demonstrates the use of two-dimensional arrays to specify the time parameters of a design.

```
entity TimeSpec is
  port (Address    : out    Integer;
        Data       : inout  Integer;
        Control    : out    Integer);
end TimeSpec;

architecture TimeSpec_a of TimeSpec is
  type TimeParam_Typ is
    (tpd_Address,     tpd_Data,     tpd_Control,
     tSetUp_Address, tSetUp_Data, tSetUp_Control,
     tHold_Address,   tHold_Data,   tHold_Control);
  type Cond_Typ is (tMin, tTyp, tMax);
  type tSpec_Typ is array(TimeParam_Typ, Cond_Typ) of time;
  constant Spec_c : tSpec_Typ :=
    (tpd_Address     => (tMin => 20 ns, tTyp => 25 ns, tMax => 30 ns),
     tpd_Data        => (tMin => 21 ns, tTyp => 26 ns, tMax => 30 ns),
     tpd_Control     => (tMin => 22 ns, tTyp => 27 ns, tMax => 30 ns),

     tSetUp_Address => (tMin => 23 ns, tTyp => 25 ns, tMax => 30 ns),
     tSetUp_Data    => (tMin => 24 ns, tTyp => 25 ns, tMax => 30 ns),
     tSetUp_Control => (tMin => 25 ns, tTyp => 25 ns, tMax => 30 ns),

     tHold_Address  => (tMin => 1 ns,  tTyp => 2 ns,  tMax => 3 ns),
     tHold_Data     => (tMin => 1 ns,  tTyp => 2 ns,  tMax => 3 ns),
     tHold_Control  => (tMin => 1 ns,  tTyp => 2 ns,  tMax => 3 ns));

begin  --  TimeSpec_a
  --  Process:   Test_Lbl
  --  Purpose:   Demonstrate use of timing parameters
  Test_Lbl : process
  begin  --  process Test_Lbl
    Address  <= 100 after Spec_c(tpd_Address, tTyp);
    Control  <= 2   after Spec_c(tpd_Control, tTyp);
    Data     <= 25  after Spec_c(tpd_Data,    tTyp);
    wait;
  end process Test_Lbl;
end TimeSpec_a;
```

Figure 2.3.2.1.3-2 Time Parameters with Two-dimentional Arrays (ch2_dir\tmspec.vhd)

2.3.2.1.4 Anonymous Arrays

Anonymous arrays are arrays that no not belong to any **named** type. Each anonymous array is considered to be one of a kind, whose type is anonymous. **ANONYMOUS ARRAY are ILLEGAL in VHDL.** *(They are legal in Ada).* Figure 2.3.2.1.4 demonstrates an example of an anonymous array.

```
subtype RegSize_Typ is integer range 1 to 4;
type aInt_Typ  is array(RegSize_Typ) of integer;
...
 variable aInt_v    : aInt_Typ; -- belongs to a define type or subtype
 variable Anonymous_v : array(RegSize_Typ) of integer; --   Anonymous Array
```

Figure 2.3.2.1.4 Example of an Illegal Anonymous Array (ch2_dir\anony_ea.vhd)

2.3.2.1.5 Implicit Functions for Array Declarations

When an array type is declared, several implicit functions are automatically declared for operations on this type. These functions are shown in table 2.3.2.1.5.

Table 2.3.2.1.5 Implicit Operations on Arrays[12]

OPERATION	RESTRICTIONS ON LEFT AND RIGHT OPERANDS
Assignments :=, <=	Must be the same size and type.
Relational <, >, <=, >=	Must be one-dimensional discrete arrays of the same type. Per LRM 7.2.2, *[1] for discrete array types, the relation < (less than) is defined such that the left operand is less than the right operand if and only if:* *1. The left operand is a null array and the right operand is a non-null array; otherwise,* *2. Both operands are non-null arrays, and one of the following conditions is satisfied:* *a. The leftmost element of the left operand is less than that of the right; or* *b. The leftmost element of the left operand is equal to that of the right, and the tail of the left operand is less than of the right (the tail consists of the remaining elements to the right of the leftmost element and can be null).* Another way of stating condition 2 is as follows: Each array component is compared, in turn, from left to right, until a difference is found, or until all the components of one or both arrays have been exhausted. If a difference in a corresponding component is detected, then the array possessing the component with the smaller value is considered to be "less than" the other array, regardless of what component may follow. Thus, "DO"<"IF", and (2,5) < (9, 3). Consider the case when all the components of one array have been exhausted and no difference in a corresponding component has been detected. If additional components remain for the second array, then the first array is considered to be "Less than" the second array. Thus, "DO" <"DONE", and (2,5) < (2,5,2,3)

[12] *Rendez Vous with Ada 95,* David Naiditch, John Wiley and Sons Inc. Copyright © 1995 by David J. Naiditch, Reprinted by permission of John Wiley & Sons, Inc.

OPERATION	RESTRICTIONS ON LEFT AND RIGHT OPERANDS
Equality =	Same type, number of components, and all corresponding components are equal. "DONE" = "DONE", (3,1,3) = (3,1,3)
Inequality /=	Each array component is compared, in turn, from left to right, until a difference is found, or until all the components of one or both arrays have been exhausted. If a difference in a corresponding component is detected, then the arrays are not equal. If all the components of one array have been exhausted and no difference in a corresponding component has been detected, and if additional components remain for the second array, then the arrays are unequal. "DONE" /= "DON", (3,1,3) /= (3,1,1)
Concatenation &	Left and right side must be one-dimensional arrays of the same type. A := "VHDL" & " IS" & " GREAT"; -- A is of a string type
Slicing	Array slicing must be one-dimensional array. A(7 **downto** 0) := B(31 **downto** 24);
Logical operator	Predefined operators (and not implicit) for objects of type bit, Boolean, and one dimensional arrays of Bit and Boolean elements include: **not, and, or , nand, nor, xor, xnor**(VHDL'93).
Shift operator	VHDL'93 Bit_Vector shift operators include: (These operators are not implicit operators) **sll, srl, sla, sra, rol, ror**
Array attributes	**A**'left, **A**'right, **A**'high, **A**'low, **A**'range, **A**'reverse_range, **A**'length, (**A**'ascending, -- '93) (where **A** represents the array object)
Type conversion.	An array can be converted to another type of array only when each array has the same shape and size, the same component type, and the same or convertible corresponding index types. Basically, the arrays must be "closely related." **signal** S : Unsigned(7 **downto** 0); **signal** B : Std_Logic_Vector(7 **downto** 0) := "1000_0000"; ... S <= Unsigned(B);

Figure 2.3.2.1.5 demonstrates examples of implicit operations on arrays.

```
library IEEE;
  use IEEE.Std_Logic_1164.all;
entity Array_N is
end Array_N;

architecture Array_a of Array_N is
  subtype A5_Typ      is Std_Logic_Vector(4 downto 0);
  subtype A7_Typ      is Std_Logic_Vector(8 downto 2);
  type    Mem5X5_Typ  is array(4 downto 0) of A5_Typ;

  signal A5_s   : A5_Typ     := (others => '0');
  signal A7_s   : A7_Typ     := (others => '0');
  signal MemA_s : Mem5x5_Typ := (others => (others => '0'));

begin
  Test_Lbl: process
    variable A5_v   : A5_Typ     := (others => '1');
    variable A7_v   : A7_Typ     := (others => '1');
    variable MemA_v : Mem5x5_Typ := (others => (others => '1'));
    variable Int_v  : Integer;
```

```
  begin
    -- Assignment :=, <=
    A5_v := "10101";
    A5_s <= A5_v;

    MemA_v(0) := "10101";
    A7_s(8 downto 4) <= MemA_v(0);

    --equality
    assert not (MemA_v(0) = A5_v)
      report "MemA_v(0) = A5_v"          -- true
      severity note;

    MemA_s <= MemA_v;  -- assignment of whole aggregate
    wait for 10 ns;
    assert not (A5_s = A5_v)  -- comparison
      report "A5_s = A5_v"     -- true
      severity note;
    assert not (MemA_s = MemA_v)
      report "MemA_s = MemA5_v" -- true
      severity note;

    -- Relational Operators  <, <=, >, >=
    A5_s <= "10110";  -- note that A5_v = "10101"
    wait for 10 ns;

    assert not (A5_s > A5_v)   -- "10110" > "10101"
      report "A5_s > A5_v"
      severity note;

    A7_v := "1010100";
    -- A5_v < A7_v because  "10101" < "1010100" -- LRM 7.2.2
    assert not (A5_v < A7_v)
      report "A5_v < A7_v"
      severity note;

    -- Inequality
    assert not (A5_v /= A7_v)
      report "A5_v /= A7_v"
      severity note;

    -- null array < non null array
    assert not (A7_v(2 to 8) < A5_v)
      report "A7_v < A5_v"
      severity note;

    -- Concatenation
    A7_v := A5_v(4) & '1' & A5_v(2 downto 0)  & "00";

    -- Array attributes  (A represents the array object)
    -- A'left, A'right, A'high, A'low, A'range,
    -- A'reverse_range, A'length
    -- A'ascending,  -- '93
    Range_Lbl: for I in A5_v'range loop
      A5_v(A5_v'high - I) := A5_s(I); -- bit reverse
    end loop Range_Lbl;
  end process Test_Lbl;
end Array_a;
```

Null slice in VHDL'87, ERROR in VHDL'93 (See LRM 6.5 and section 2.3)

Figure 2.3.2.1.5 Implicit Operations on Arrays, (arrays\array1.vhd)

2.3.2.1.6 Array Slices and Ranges

Assignment of arrays preserves the left-to-right ordering regardless of the direction. Thus, for the following declaration and signal assignments, the string "1010" is assigned to the elements of *X* from left to right. This means *X(0) = '1', X(1) = '0', X(2) = '1', X(3) = '0'.* Similarly, *Y* is also assigned from left to right, so *Y(3) = '0'* and *Y(0) = '1'*

```
Signal X  : Std_Logic_Vector(0 to 3);      -- CASE 1
signal Y :  Std_Logic_Vector(3 downto 0);  -- CASE 2
X <= "1010";
Y <= "0101";
```

In addition, the assignment *X <= Y* is equivalent to:

```
X(0) <= Y(3); -- the left elements
X(1) <= Y(2);
X(2) <= Y(1);
X(3) <= Y(0); -- the right elements
```

The array range is NOT ignored in determining the index of a loop. For example, for a loop defined as ***for I in 0 to 3 loop*** (or ***for I in X'range loop***) the elements of the arrays would be accessed from left to right. If the same loop were used for *Y*, then the elements of the array would be accessed from right to left instead. The *'reverse_range* attribute would cause the arrays to be accessed in the reverse direction (*3 downto 0* for X, and *0 to 3* for Y).

Note that slices of an array allow for the extraction of certain elements of the array. For example, for the above definition of *X* and *Y*, the following statements are true:

X(1 **to** 2) is "01"
Y(2 **downto** 1) is "10"
X(1 **to** 1) is "0" -- a one element array,
Y(2 **downto** 2) is "1" -- a one element array,

Null slices are slices with no values, but do not necessarily represent an error. There is a difference between VHDL'87 and VHDL'93 in the definition of a *null* slice. Specifically, LRM 6.5 states the following:

VHDL'87

[1](LRM 6.5*) The* slice is a ***null slice*** *if the discrete range is a null range, or if the direction of the discrete range is not the same as that of the object denoted by the prefix of the slice name.*

Constant *DATA : Bit_Vector(31* ***downto*** *0);*

DATA(24 ***to*** *25)* -- a *null* slice --direction
DATA(24 **downto** 25) – a *null* slice,
-- *null* range

VHDL'93

[1] (LRM 6.5*) The* slice is a ***null slice*** *if the discrete range is a null range. It is an error if the direction of the discrete range is not the same as that of the index range of the array denoted by the prefix of the slice name.*

Constant *DATA : Bit_Vector(31* ***downto*** *0);*

DATA(24 ***to*** *25) -- an error*  --direction
DATA(24 **downto** 25) – a *null* slice
-- *null* range

Null slices are NOT SYNTHESIZABLE. Null slices may occur in loops (see barrel shifter example in chapter 3). For synthesis, if null slices could occur, the code must prevent their occurrence by testing the condition that would cause the null slice, and avoid the creation of the null slices.

Other examples of slices include:

```
-- signal X : Std_Logic_Vector(0 to 3);      -- CASE 1
-- signal Y : Std_Logic_Vector(3 downto 0); -- CASE 2
```

X(2 **downto** 1) VHDL'87 null slice because direction of the discrete range is not the same as that of the object denoted by the prefix of the slice name (i.e. Slice direction is downward (**downto**) while slice name (X) is upward (**to**). It is an error in VHDL'93

Y(1 **to** 2) VHDL'87 null slice, VHDL'93 (same as above)

X(1 **downto** 1) VHDL'87 null slice, VHDL'93 (same as above)

Y(2 **to** 2) VHDL'87 null slice, VHDL'93 (same as above)

X(2 **to** 1) Null slice because the discrete range is a null range 2 (is greater than 1)

Y(1 **downto** 2) Null slice because the discrete range is a null range (1 is less than 2)

Null slices are generally NOT synthesizable.

2.3.2.2 Records

[1] A record type is a composite type whose values consist of named elements. These elements can be of the same or different types. A record is referenced by a single identifier. Records are very useful to collapse multiple related elements into a single type so that record objects can be handle as a single objects. They are very useful in Bus Functional Models (BFMs) and testbenches to pass a set of information to another process or another component. Values can be assigned to a record type using aggregates in a manner similar to aggregates for arrays. Figure 2.3.2.2 demonstrates the record declaration, initialization, and assignment onto signals.

> If multiple elements of a record are to be assigned to a SIGNAL of record type, it is more efficient (from a simulation standpoint) to FIRST copy the record onto a local variable, modify the elements of the variable record, and then make a single SIGNAL assignment of the variable to the signal.
>
> ***Rationale:*** *Efficiency is improved when the whole record aggregate is assigned onto a signal instead of assignments of each of the individual components.*

```
record_type_definition ::=
  record
    element_declaration
      { element_declaration }
  end record;

element_declaration ::=
  identifier_list :
    element_subtype_definition ;
```

If an element of a record is an array, then it must be constrained.

Many synthesizers can compile records provided the elements of the record are of type *Bit, Bit_Vector, Boolean, Std_uLogic, Std_uLogic_Vector, Integer*, or subtypes of these types. However, most current synthesizers do not accept record aggregate assignments, and thus require individual assignment of the record elements

```
architecture Record_a of Record_N is
  subtype WordSize_Typ is Integer range 31 downto 0;
  subtype RegSize_Typ  is Integer range 1 to 4;
  subtype Word_Typ is Std_Logic_Vector(WordSize_Typ);
  type TaskMode_Typ is (Normal, DMA, Interrupt);
  type TaskInstr_Typ is (Idle, Read, Write, RTI);
  type Data_Typ is array(RegSize_Typ) of Word_Typ;
  type Task_Typ is record
    Task_Numb   : integer;
    TimeTag     : time;
    Task_Mode   : TaskMode_Typ;
    Instruction : TaskInstr_Typ;
    ADDRESS     : Word_Typ;
    DATA        : Data_Typ;
  end record;
  signal Task_s : task_typ;

begin  --  Record_a
  Record_Lbl : process
    variable Task_v : Task_Typ :=
     (Task_Numb    => 0,
      TimeTag      => now,
      Task_Mode    => Interrupt,
      Instruction  => Idle,
      ADDRESS      => "LLLLLLLLLLLLLLLLLLLLLLLLLLLLLLLL",
      DATA         => (others  => (others => 'Z')));
    constant Task_c : Task_Typ :=
     (Task_Numb    => 0,
      TimeTag      => now,
      Task_Mode    => Normal,
      Instruction  => Idle,
      ADDRESS      => "ZZZZZZZZZZZZZZZZZZZZZZZZZZZZZZZZ",
      DATA         => (others  => (others => '0')));

  begin  --  process Record_Lbl
    Task_v := Task_c;
    Task_v.Address := "00000000000000000000000010101010";
    Task_v.Data := (1 => (others => '0'),
                    2 => "11111111100000000111111110001010",
                    3 => "00000101000011111111100000111111",
                    4 => (31 downto 16 => '0',
                          others       => '1'));

    task_s <= Task_v after 10 ns;  -- ☺ Assignment of the whole aggregate.
    wait;
  end process Record_Lbl;
end Record_a;
```

Figure 2.3.2.2 Record Declaration, Initialization, and Signals Assignment (ch2_dir\recrd.vhd)

2.3.3 Access Type

Access types *(LRM 3.3)* are used to declare values that access dynamically allocated variables. Dynamically allocated variables are referenced, not by name, but by an access value that acts like a pointer to the variable. Access types are used in the TextIO subprograms to read and write text strings. Access types were efficiently used in VHDL in the functional modeling of large memories where fixed memory allocation is not realistic (e.g., a 2**64 x 128 bit RAM). For an example of such a model see *VHDL Answers to Frequently Asked Questions, 2nd Edition*, Ben Cohen, ISBN 0-7923-8115-7, Kluwer Academic Publishers, 1998.

SIGNAL DECLARATION CANNOT BE OF FILE OR ACCESS TYPE.

The variable being pointed to can be one of the following:

1. **Scalar object** (e.g., enumerated type, *integer*). Access types are not very useful as pointers to scalar type since the dynamic allocation of scalar type does not represent any significant savings in memory.
2. **Array objects.** Access types are very useful in array objects when the length of the array is not known beforehand, or the array size can grow during the course of code execution. A good example of such an array is objects of type STRING. Package Std.TextIO defines the type LINE as an access to a string. The procedures READ and WRITE in the same package make use of the access type to store the string data of varying length using objects of LINE types.
3. **Record objects.** Access types are very useful for record objects because they can be used to represent linked lists such as FIFO (First In First Out) or LIFO (Last In First Out) objects.

Figure 2.3.3 represents an example using access types.

```
architecture Access_Beh of Access_N is
  type State_Typ is (Idle, Running, Blocked);
---------------------------------------------------------------------------
--  Access to a scalar object
---------------------------------------------------------------------------
  type AInt_Typ is access integer;
  type AEnum_Typ is access State_Typ;
---------------------------------------------------------------------------
--  Access to an array
---------------------------------------------------------------------------
  type Line_Typ is access string;
---------------------------------------------------------------------------
--  Access to a record
---------------------------------------------------------------------------
  type Rec_Typ;                        -- incomplete type declaration
                                       -- Needed for access declaration
  type ARecPntr_Typ is access Rec_Typ;

  type Rec_Typ is record
    Int_v       : integer;
    IsLabel_v   : boolean;
    NextP_v     : ARecPntr_Typ;  -- pointer to next record
    PreviousP_v : ARecPntr_Typ;  -- pointer to previous record
  end record;
```

```
begin  --  Access_Beh
  ---------------------------------------------------------------------------
  --  Process:   Demo_lbl
  --  Purpose:   Demonstration of Access utilization
  ---------------------------------------------------------------------------
  Demo_lbl : process
    variable AInt1_v : Aint_Typ;    -- initialized to null
    variable Aint2_v : Aint_Typ;
    variable AEnum_v : AEnum_Typ;

    variable Line1_v : Line_Typ;
    variable Line2_v : Line_Typ;

    variable Head_v  : ARecPntr_Typ;     --   initialized to null
    variable Tail_v  : ARecPntr_Typ;
    variable Temp_v  : ARecPntr_Typ;

  begin  --  process Demo_lbl
-----------------------------------------------------------------------------
--  Access type operation on scalars
-----------------------------------------------------------------------------
    Aint1_v := new integer'(5); -- pointer points new integer of value 5
    Aint2_v := new integer;   -- creation of a new storage
                        -- allocation for Aint2_v, initialized to integer'left
    Aint2_v.all := Aint1_v.all;  -- Aint2_v points to integer with value 5
```

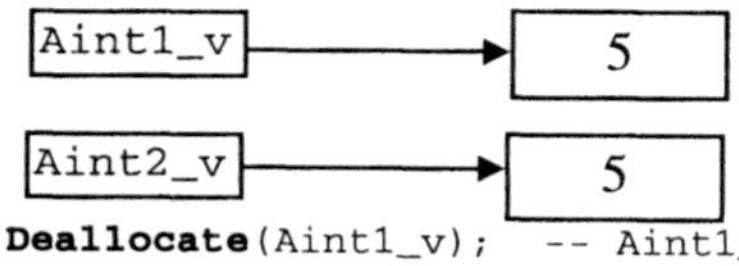

```
    Aint1_v -----> 5
    Aint2_v -----> 5
```

```
    Deallocate(Aint1_v);  -- Aint1_v now points to null
```

```
    Aint1_v -----> null
```

```
    AEnum_v := new State_Typ;  -- pointer points to storage, uninitalized
    AEnum_v.all := Blocked;  -- AEnum_v points to storage with "Blocked" enumeration
-----------------------------------------------------------------------------
--  Access type operation on arrays
-----------------------------------------------------------------------------
    Line1_v := new String'("Hello, This is VHDL");
    Line2_v := Line1_v;  -- Line2_v used to point to null, but now
                         -- points to the same string as Line1_v
    Line2_v := new String(Line1_v'left to Line1_v'right);
    -- Line2_v now points to a new string of the same size as line1_v string

    Line2_v(Line2_v'left to Line2_v'right) :=
      Line1_v(Line1_v'left to Line1_v'right); -- copy of the elements
    Line2_v.all := Line1_v.all;  -- copy all elements -- better approach

-----------------------------------------------------------------------------
--  Access type operations on records
-----------------------------------------------------------------------------
    Head_v := new Rec_Typ;
    Head_v.Int_v          := 7;     -- new record initialized
    Head_v.IsLabel_v      := true;
    Head_v.NextP_v        := null;
    Head_v.PreviousP_v    := null;
```

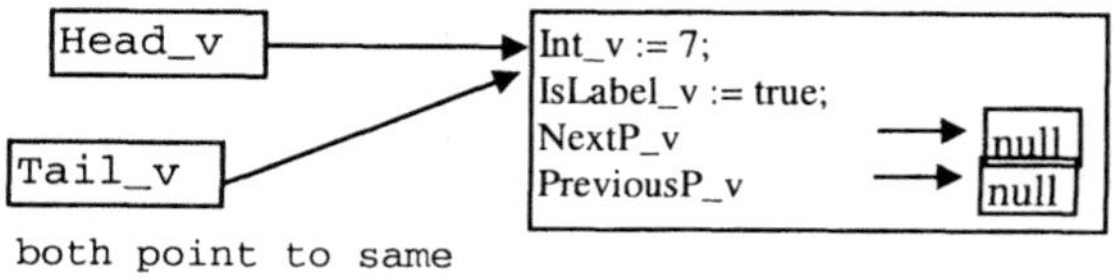

```
    Head_v -----> Int_v := 7;
                  IsLabel_v := true;
    Tail_v ---/   NextP_v      -----> null
                  PreviousP_v  -----> null
```

```
    Tail_v := Head_v;             --   both point to same

    -- Adding a new record to FIFO
    Temp_v := Tail_v;            -- Temp_v points to where Tail_v points to
    Tail_v := new Rec_Typ;
```

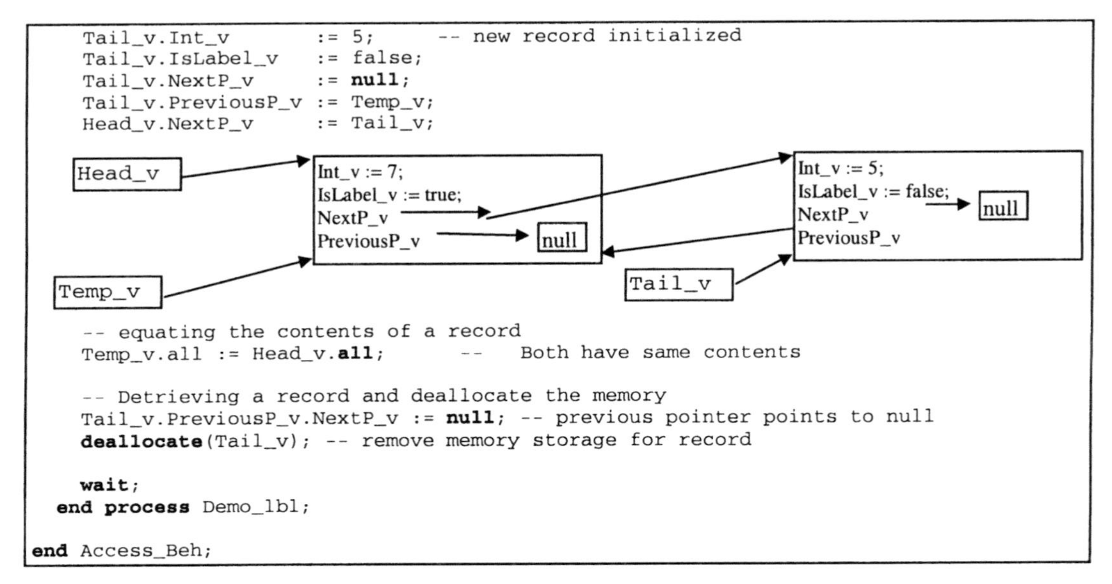


```
    Tail_v.Int_v        := 5;      -- new record initialized
    Tail_v.IsLabel_v    := false;
    Tail_v.NextP_v      := null;
    Tail_v.PreviousP_v  := Temp_v;
    Head_v.NextP_v      := Tail_v;

    -- equating the contents of a record
    Temp_v.all := Head_v.all;       --   Both have same contents

    -- Detrieving a record and deallocate the memory
    Tail_v.PreviousP_v.NextP_v := null; -- previous pointer points to null
    deallocate(Tail_v); -- remove memory storage for record

    wait;
  end process Demo_lbl;

end Access_Beh;
```

Figure 2.3.3 Using Access Types (ch2_dir\acces_ea.vhd)

2.4 FILE 

[1] File types are typically used to access files in the host environment.. File types are used to define objects representing files in the host system environment.. The value of a file object is the sequence of values contained in the host environment. Package TextIO (explained in section 8.2) provides for the definition of file type **TEXT** as "*type TEXT is file of string;*". TextIO package supports human readable IO. It provides procedures to read and write text files. This section elaborates on the use of files without the package TextIO.

Binary files are not however human readable. They require however, an order of magnitude less disk storage. In addition, they simulate significantly faster than text file because no conversion functions are required to either convert between types, or to execute the TEXTIO subprograms. If text files are expected to be read multiple times in the course of simulation runs, it is more efficient to convert those text files into a VHDL binary files using the desired data structures, and then use the binary file in the model simulations. Note that there are no standards in the definition of binary files, thus every vendor's binary format may not be compatible with another vendor's formats.

Examples of binary file type declaration:

```
type StringFile_Typ is file of string;      -- string is the type mark
type IntegerFile_Typ is file of integer;  -- integer is the type mark
type BitVectorFile_Typ is file of  Bit_Vector;
type RecFile_Typ is file of  MyRecord_Typ;
```

The syntax for file declaration and file type definition is as follows:

VHDL'87	VHDL'93
file_declaration ::= file identifier: subtype_indication **is** [mode] file_logical_name; file_logical_name ::= string_expression file_type_definition ::= **file of** type_mark mode ::= **in** | **out**	file_declaration ::= file identifier_list : subtype_indication [file_open_information] ; file_open_information ::= [**open** file_open_kind_expression] **is** file_logical_name file_logical_name ::= string_expression file_type_definition ::= **file of** type_mark

The language implicitly defines the operations for objects of file type. Given the following file type declaration:

***type** FT **is file of** TM;*

the following operations are implicitly declared following the file type declaration:

VHDL'87	VHDL'93
procedure Read(F : **in** FT; Value : **out** TM); -- If FT is an unconstrained array then **procedure** Read(F : **in** FT; Value : **out** TM; Length : **out** natural);	**procedure** Read(F : **in** FT; Value : **out** TM); -- If FT is an unconstrained array then **procedure** Read(F : **in** FT; Value : **out** TM; Length : **out** natural);
procedure Write(F : **out** FT; Value : **in** TM);	**procedure** Write(F : **out** FT; Value : **in** TM);
function EndFile(F : **in** FT) **return** boolean;	**function** EndFile(F : **in** FT) **return** boolean;
	procedure File_Open (**file** F : FT; External_Name : **in** string; Open_Kind : **in** File_Open_Kind := Read_Mode);
	procedure File_Open (Status : **out** File_Open_Status; **file** F : FT; External_Name : **in** string; Open_Kind : **in** File_Open_Kind := Read_Mode);
	procedure File_Close(**file** F : FT);

Some of the language rules about type definition for a file include the following:

1. *[1] The type mark may denote either a constrained or unconstrained type.*
2. *The base type of this subtype must not be a file type or an access type.*
3. *If the base type is a composite type, it must not contain a subelement of an access type.*
4. *If the base type is an array type, it must be one dimensional array type.*

Figure 2.4-2 represents the writing of 10 integer values onto a file using a binary instead of textual representation. This example uses VHDL'93 file format.

```
architecture FileIOInt93_a of FileIO is  -- see finta87.vhd for 87 format
begin  --  FileIO_a
  WriteFile_Lbl  : process
    type IntegerFile_Typ is file of integer;
    file DataOut_f       : IntegerFile_Typ open Write_Mode is "IntOut.txt";
    variable Int_v       : integer := 0;
    variable fStatus_v   : File_Open_Status; -- New buitin type
      -- values: Open_OK, Status_Error, Name_Error

  begin  --  process WriteFile_Lbl
    Write_Lbl: for Count_i in 1 to 10 loop
      Int_v := Int_v + 1;
      Write(F         => DataOut_f,
            Value     => Int_v);
    end loop Write_Lbl;
    wait;
  end process WriteFile_Lbl;
end FileIOInt93_a;
```

Figure 2.4-2 Binary File Write of Integer with NO TextIO (ch2_dir\ finta93.vhd)

Figure 2.4-3 represents the reading of a binary file that holds integers, and the display of the value of those integers onto a signal. This example uses VHDL'93 file format.

```
...
architecture FileWrInt93_a of FileWr is -- see fintb87.vhd for 87 format
  signal Int_s   : integer;
begin  --  FileWr_a
  ------------------------------------------------------------------------
  --  Process:    ReadWriteFile_Lbl
  --  Purpose:    Read a binary file, and display its content on a signal
  ------------------------------------------------------------------------
  ReadWriteFile_Lbl  : process
    type IntegerFile_Typ is file of integer;
    file DataIn_f        : IntegerFile_Typ;
    variable Int_v       : integer;
    variable fStatus_v   : File_Open_Status; -- New buitin type
      -- values: Open_OK, Status_Error, Name_Error
```

```
  begin  --  process ReadWriteFile_Lbl
    File_Open(Status        => fStatus_v,
              F             => DataIn_f,     -- implicit definition
              External_Name => "IntOut.txt",  -- from file declaration
              Open_Kind     => Read_Mode);

    while not Endfile(DataIn_f) loop  -- is implicit built-in:
      -- Read 1 line from the input file
      Read(F        => DataIn_f,
           Value    => Int_v);
      int_s <= Int_v;
      wait for 10 ns;
    end loop;
    wait;
  end process ReadWriteFile_Lbl;
end FileWrInt93_a;
```

Figure 2.4-3 Reading a Binary File (ch2_dir\fintb93.vhd)

Chapter 10 provides additional examples on the application of binary files for testbench designs. Also included in that chapter are examples of writing and reading a binary file of a record type.

2.5 ATTRIBUTES

[1] An attribute is a value, function, type, range, signal, or a constant that may be associated with one or more names within a VHDL description. There are user defined attributes (described later) and predefined attributes *(LRM 14.1)*. There are 5 classes of predefined attributes:

1. Value attributes : Return a constant value
2. Function attributes : Call a function that return a value
3. Signal attributes : Create a new implicit signal
4. Type attributes : Return a type
5. Range attributes : Returns a range.

> **M** 👍 👍 Use attributes of objects instead of fixed values or constant unless the value is known and very unlikely to be modified.
>
> ***Rationale:*** *Attributes creates code that is easier to maintain an reuse.*

Appendix E provides a summary of the predefined attributes with examples. Chapter 5 explains the timing attributes. Figure 2.5 summarizes the predefined non-timing attributes. Table 5.2 in chapter 5 summarizes the predefined timing attributes.

The *'image* is a VHDL'93 attribute that converts scalar objects into strings. This attribute does not exist in VHDL'87, and does work with arrays. Chapter 7 defines the overloaded function *IMAGE* that operates in VHDL'87 and with arrays of types commonly used in synthesis.

Not all predefined attributes are synthesizable (see appendix F).

```
-- file attrprdf.vhd
library IEEE;
  use IEEE.Std_Logic_1164.all;

entity AttributesDemo is
end AttributesDemo;

architecture Attributes_a of AttributesDemo is
  use Std.TextIO.all;

  subtype  BV32_Typ     is Std_Logic_Vector(31 downto 0);
  subtype  BV8_Typ      is Std_Logic_Vector( 7 downto 0);
  subtype  BV0to7_Type  is Std_Logic_Vector(0 to 7);
  subtype  Size_Typ     is Integer range 0 to 15;
  type     Color_Typ    is (Red, Yellow, Blue, Green, Orange, Violet);
  subtype  Primary_Typ  is Color_typ range Red to Blue;
  constant Five_c    : BV32_Typ := X"0000_0005";
                       -- (31 downto 4 => (others => '0'),
                       --   3 downto 0 => "0101");
  constant Delay_c   : time := 100 ns;


  signal   Bus32_s   : BV32_Typ;
  signal   Count_s   : Integer;
  signal   Color_s   : Color_Typ;
```

```
begin
  Attr_Proc: process
    variable Bus32_v      : BV32_Typ;
    variable Temp_v       : Integer;
    variable String32_v   : String(1 to 32);
    variable BV8_v        : Bit_Vector(7 downto 0) := "01001001";
    variable IsValid_v    : Boolean;
    variable Line_v       : Std.TextIO.Line;
    -- ST = Subtype
    -- PT = physical type
  begin
    -----------------+--------+-----------------------------------
    -- Attribute      | Prefix | Result
    -- Name           |        |
    -----------------+--------+-----------------------------------
    -- T'base         | Type / | Base type of T.  Must be prefix to
    --                | subtype| another attribute
    -----------------+--------+-----------------------------------
    -- T'left         | scalar | The left  bound of T, result of type T
    -- T'right        | type/ST| The right bound of T, result of type T
    -----------------+--------+-----------------------------------
    -- T'high         | scalar | The upper bound of T, result of type T
    -- T'low          | type/ST| The lower bound of T, result of type T
    -----------------+--------+-----------------------------------
    -- T'Ascending    | scalar | TRUE if type T is ascending
    --   VHDL'93      | type/ST|

    Temp_v := Size_typ'base'high; -- 2147483647
    Color_s <= Primary_Typ'base'right; -- Violet
    wait for Delay_c;

    Color_s <= Color_Typ'left;  -- Red
    wait for Delay_c;
    if Primary_Typ'ascending then
      report "Primary type is in ascending order"
      severity Note;
    end if;
    -----------------+--------+-----------------------------------
    -- Attribute      | Prefix | Result
    -- Name           |        |
    -----------------+--------+-----------------------------------
    -- T'image(X)     |scalar  | Function that converts
    --   VHDL'93      |type/ST |  scalar object X of type T into string
    -- T'value(X)     |        | Function that converts
    --   VHDL'93      |        |  object X of type string into scalar of type T
    -----------------+--------+-----------------------------------
    -- T'pos(X)       |discrete| Function that returns a universal integer
    --                |/PT/ST  | representing the position number of
    --                |        | parameter X of type T. First position = 0.
    -- T'val(X)       |        | Function that returns of base type T
    --                |        | the value whose position is the universal
    --                |        | integer value corresponding to X.
    -----------------+--------+-----------------------------------
    -- T'succ(X)      |        | Function returning a value of type T
    --                |        | whose value is the position number
    --                |discrete| one greater than the one of the parameter.
    --                |/PT/ST  | It is an error if X = T'high or if
    --                |        | does not belong to the range T'low to T'high
    -- T'pred(X)      |        | Function returning a value of type T
    --                |        | whose value is the position number
    --                |        | one less    than the one of the parameter.
    --                |        | It is an error if X = T'low  or if
    --                |        | does not belong to the range T'low to T'high
    -----------------+--------+-----------------------------------
```

```
-- String32_v := BV32_Typ'image(Bus32_s);
-- illegal, prefix must be scalar, not a composite
STD.TEXTIO.Write(Line_v, time'image(Delay_c)); -- 100 ns
STD.TEXTIO.WriteLine(Output, Line_v);

STD.TEXTIO.Write(Line_v, Color_Typ'image(Color_s));
STD.TEXTIO.WriteLine(Output, Line_v);

STD.TEXTIO.Write(Line_v, integer'image(Count_s));
-- Bus32_v(7 downto 0) := BV8_Typ'value(String32_v);
-- illegal, prefix must be scalar, not a composite

Color_s <= Color_Typ'value("Yellow");  -- Yellow

Temp_v := Color_Typ'pos(Color_s); -- 0 if first position
Temp_v := (Temp_v + 1) mod 6;    -- 0 to 5
Color_s <= Color_Typ'val(Temp_v);

wait for 2 * Delay_c;
-- translate bit vector into an integer

if Color_s = Color_Typ'high then  -- violet
  Color_s <= Color_Typ'pred(Color_s);  -- orange
elsif Color_s = Color_Typ'low then   -- Red
  Color_s <= Color_Typ'succ(Color_s);   -- yellow
else
  Color_s <= Color_Typ'right;  -- Red
end if;
wait for Delay_c;

-----------------+--------+------------------------------------
-- Attribute     | Prefix | Result
-- Name          |        |
-----------------+--------+------------------------------------
-- T'leftof(X)   |        |Function that returns the value that is
--               |discrete|to the left of parameter X of type T.
--               |/PT/ST  |Result type is of type T.
--               |        | Error if X = T'left
-- T'rightof(X)  |        |Function that returns the value that is
--               |        |to the right of parameter X of type T.
--               |        |Result type is of type T.
--               |        | Error is X = T'right
 Color_s <= Color_Typ'leftof(Green); -- blue
 wait for Delay_c;
 Color_s <= Color_Typ'rightof(Green); -- orange
 wait for Delay_c;
```

```
    -- Array* = Any prefix that is appropriate for an array object,
    -- or alias therof, or that denotes a constrained array subtype
    -- (e.g. type, variable, signal)
    -----------------+--------+-----------------------------------
    -- Attribute      | Prefix | Result
    -- Name           |        |
    -----------------+--------+-----------------------------------
    -- A'left(N)      |Array*  |Function that returns the left bound of
    --                |        |of the Nth index range of A.  X is of
    --                |        |type universal integer.  Result type is of type
    --                |        |of the left bound of the left index range of A.
    --                |        | N = 1 if omitted.
    -- A'right(A)     |        |Same as A'left(N), except right bound is returned.
    -----------------------------------------------------------------
    -- A'high(N)      | Array* | Function that returns the upper bound of the
    --                |        | range of A.  Result type is the type of the
    --                |        | Nth index range of A.  N = 1 if omitted.
    -- A'low(N)       |        | Same as A'high(N), e lower bound is returned.
    --------------------------------------------------------------------
    -- A'range(N)     |Array*  | The range of A'left(N) to A'right(N)
    --------------------------------------------------------------------
    -- A'reverse_range(N)
    --                |Array*  | The range of A'right(N) to A'left(N)
    --------------------------------------------------------------------
    -- A'length       |Array*  | returns 0 is array os null.
    --                |        | Else, returns T'pos(A'high(N)) - T'pos(A'low(N) +1
    --                |        | where T is the subtype of the Nth index of A.
    -----------------------------------------------------------------------
    -- A'Ascending    |Array*  | True if Nth index range of A is defined in an
    --                |        | ascending range, else returns false.
    -----------------------------------------------------------------------
    Bus32_s  --  Bit 31 downto 16
     (BV32_Typ'left downto BV32_Typ'right + 16) <= "1111000011110000";
    Bus32_v(Bus32_v'left downto Bus32_v'left - 3) := "0000";

    Temp_v := BV0to7_Type'high; -- 7
    Temp_v := BV0to7_Type'low;  -- 0
    Temp_v := Bus32_s'length; -- 32

    if BV32_Typ'Ascending then
      assert false
        report "BV32_Typ is in ascending order"
        severity Note;
    else
      assert False
        report "BV32_Typ is in Descending order"
        severity Note;
    end if;
    Temp_v := 0;
    ComputeInteger_Lp: for I in BV8_v'range loop -- 7 downto 0
      Temp_v := (Temp_v * 2) + bit'pos(BV8_v(I));
    end loop ComputeInteger_Lp;

    -- Display 8 states in reverse
    Std_Lp: For I in BV8_v'reverse_range loop -- 0 to 7
      Bus32_s(I) <= Std_Logic'val(I);  --('U','X','0','1','Z','W','L','H','-')
                                       --  0   1   2   3   4   5   6   7   8
    end loop Std_Lp;                   -- (HLWZ10XU)

  end process Attr_Proc;
end Attributes_a;
```

Figure 2.5 Using Predefined Non-Timing Attributes (Ch2_dir\attrprdf.vhd)

2.6 ALIASES

An alias *(LRM 4.3.3)* declares an alternate name for all or part of an existing object and enhances readability because it refers to the same object in different ways depending on the context. For example, aliases can refer to fields of an objects where each field has a particular significance (e.g., control, data, Op-code). The syntax for VHDL'87 is as follows:

alias_declaration ::= **alias** identifier : subtype_indication **is** name;

Figure 2.6-1 is an example using alias as defined in VHDL'87. In VHDL'87 aliasing only applies to objects including signals, variables, and constants.

VHDL'93 extends the definition of aliases and adds anything that has been previously declared, except for labels, loop parameters, and generate parameters. Alias'93 provides a very useful feature that is very similar in functionality to the *Ada* "rename" clause. This allows long paths or long names to effectively be renamed.

However, the use of aliases presents two major limitations:

1. Aliases are NOT currently supported in synthesis (per IEEE.1076.6 specification).
2. Simulators cannot trace or view the aliased signals.

The syntax for aliasing'93 is as follows:

alias_declaration ::= **alias** alias_designator [: subtype_indication] **is** name [signature];
alias_designator ::= identifier | character_literal | operator_symbol

Useful rules to non-object aliases for VHDL'93 (see *LRM 4.3.3.2* for complete set of rules):

1. *[1] A subtype indication may not be aliased.*
2. *A signature is required if the name denotes a subprogram (including an operator) or enumeration literal (see example in Figure 2.7-2).*
3. *If the name denotes a type, then implicit declarations for each predefined operator for the type immediately follows the alias declaration for the type and, if present, any implicit alias declarations for literals or units of the type.* Thus, the aliased objects inherit the characteristics of the subtype indicated, including the operators (e.g.,"-", "/=", ">", ..) and any units, if applicable.

Figure 2.6-2 provides an example using aliasing'93.

```
architecture Alias_beh of AliasDemo is
  signal Bus_s : bit_vector(31 downto 0) := X"AB_CD_0234";
  alias  OpCode_s : bit_vector(7 downto 0) is Bus_s(31 downto 24);
  alias  Source_s : bit_vector(3 downto 0) is Bus_s(23 downto 20);
  alias  Dest_s   : bit_vector(3 downto 0) is Bus_s(19 downto 16);
  alias  Data16_s : bit_vector(15 downto 0) is Bus_s(15 downto 0);
begin  --  Alias_beh
  Test_Lbl : process(Bus_s)  -- sensitive to value changes in signal Bus_s
begin  --  process Test_Lbl
    Bus_s  <= x"AB_CD_0955" after 10 ns;
    If OpCode_s = "10101011" then
      If Source_s = "1100" then
        Data16_s <= X"F904";
      end if;
    end if;
  end process Test_Lbl;
end Alias_beh;
```

Figure 2.6-1 Alias with VHDL'87 (ch2_dir\alias_ea.vhd)

```
architecture Alias93_a of Alias93 is
  alias TX_Pkg is Std.TextIO;  -- TextIO used for demonstration
  use TX_Pkg.all;

  alias MLV_Typ is IEEE.Std_Logic_1164.Std_Logic_Vector; -- Multi-Level Vector
  alias ML_Typ  is IEEE.Std_Logic_1164.Std_Logic;        -- Multi-Level
  subtype AddrWidth_Typ is integer range 31 downto 0;
  subtype WordWidth_Typ is integer range 15 downto 0;
  type States_Typ is (InIdle, Running, Waiting, Suspended, Interrupted);
  alias Intrpt is Interrupted [return States_Typ]; -- rename for Interrupted
  signal Addr_s    : MLV_Typ(AddrWidth_Typ);
  signal RwF_s     : ML_Typ;
  signal Data_s    : MLV_Typ(WordWidth_Typ);

begin  --  Alias93_a
  ---------------------------------------
  --  Process:   Test_Lbl
  --  Purpose:   Demonstrates the use of aliases
  -------------------------------------------------------------------------------
  Test_Lbl : process
    variable OutLine_v : TX_Pkg.Line;
    variable States_v  : States_Typ := Intrpt; -- used the alias
  begin  --  process Test_Lbl
    if States_v = Intrpt then
     Addr_s <= X"ABCD_0034" or X"0000_1200";
     if RwF_s = 'U' b
      RwF_s  <= '0';
     end b;
     Data_s <= X"FF02" and X"F001";
     TX_Pkg.Write(OutLine_v, string'("Done test"));
     TX_Pkg.WriteLine(Output, OutLine_v);
    end if;
    wait for 50 ns;
  end process Test_Lbl;
end Alias93_a;
```

Aliasing an enumeration with a signature

Use of Aliased name.

Aliasing enhances documentation by providing path without significantly lengthening the code.

Figure 2.6-2 More Flexible use of Aliasing with VHDL'93 (ch2_dir\alias93.vhd)

EXERCISES

1. Which of the following identifiers are legal for VHDL'87?

```
a.  RdWF        b. RdWF_        d. rdwF1
d.  1Abc        c. Address.10
```

2. Evaluate each of the following expressions. These expressions may be of type integer, real, boolean, character, bit, or bit_vector.

```
a. 16#1#E3   = 1_000        -- True/False?
b. 2#1010#   = 10           -- True/False?
c. 2#1010#   = "1010"       -- True/False?
d. B"1010"   = "1010"       -- True/False?
e. B"10_10"  = "1010"       -- True/False?
f. X"A"      = "1010"       -- True/False?
g. O"64"     = "1100100"    -- True/False?
h. 0.5 < real(1)            -- True/False?
i. 4 = integer(3.6)         -- True/False?
j. 3 = integer(3.7)         -- True/False?
```

3. Evaluate each of the following expression. Indicate if there is an error and why.

```
Test_Lbl: process
  type State_Typ is (Idle, Reading, Writing, Waiting, Stuck);
  variable Int_v     : Integer   := 4;
  variable Flt_v     : Real      := 3.0;
  constant Const_c   : Integer   := 5;
  constant Yes       : Boolean   := true;
  variable IsOn_v    : Boolean   := true;
  variable State_v   : State_Typ;
begin
  Int_v := Int_v + 3;                         --a.
  Int_v := Int_v/3;                           --b.
  IsOn_v := IsOn_v and Yes;                   --c.
  IsOn_v := IsOn_v and not Yes;               --d.
  Int_v  := 3.0 + Const_c;                    --e.
  Flt_v  := 3.0 + 1.0 * Const_c;              --f.
  State_v := State_Typ'right;                 --g.
  if State_v = Stuck then
    State_v := Idle;                          --h.
  end if;
  -- value of State_v ?

  wait;
end process Test_Lbl;
```

4. Write a program (entity/architecture) that computes the area and circumference of a circle. Use the VHDL simulator to examine the results. Assign the results onto 2 signals: Area_s, Circf_s.

5. Given the following type definition, which of the following is illegal?

```
type Cond_Typ is (OK, Broken, Lost, Stolen, Sold);
type ItemCond_Typ is array (1 to 5) of Cond_Typ;
 a.  variable ItemCond_v : ItemCond_Typ := (others => OK);
 b.  variable ItemCond_v : ItemCond_Typ
       := (1       => Lost,
           others => OK);
 c.  variable ItemCond_v : ItemCond_Typ
       := (1 | 2 |4 => OK,
             3 | 5    => Sold);
 d. variable ItemCond_v : ItemCond_Typ
       := (OK, OK, OK, Lost, Sold);
```

6. Define a type declaration that represents four registers of 16 bits each using an array 0 to 3 of words (bit_vector). Define a variable of that type initialized as follows:

```
Register #      Value
    0           0000    in hex
    1           FFFF    in hex
    2           ABCD    in hex
    3           1234    in hex
```

7. Write a program that includes a record type. The components of the record shall include objects of the following types:

```
1. string of 10 characters  (see package Standard)
2. Integer
3. real
4. Enumeration type   (5 colors)
5. Enumeration type   (4 vehicles i.e. car, boat, train, bus)
6. Array vehicles and colors.
```

Includes 2 variables of that record type. Initialize the first variable using an aggregate. Initialize the elements of the other variable using one component at a time. Set the second variable equal to the first variable. (Don't forget the "wait;" statement at the end of the process. Compile and run you program.

3. CONTROL STRUCTURES

This chapter provides the rules and coding guidelines for VHDL control structures that provide alternate paths for the data flow. In synthesis, control structures imply multiplexers, latches, combinational logic and sequential logic (in clocked processes – see chapter 10).

Control structures make use of static and dynamic expressions. These expressions are first defined prior to their use.

3.1 EXPRESSION CLASSIFICATION

There are three kinds of expressions evaluated at different phases of the VHDL compilation/simulation process:

1. **Locally static expressions**: *[1] (LRM 7.4) These are expressions evaluated during analysis of the design unit in which they appear,* such as constants of an architecture.
2. **Globally static expressions:** *[1] These are expressions evaluated as soon as the design hierarchy in which they appear is elaborated,* such as generics and constants that are relations of generics, or generic expressions. Deferred constants are also globally static and are not synthesizable.
3. **Dynamic Expressions:** These are expressions evaluated during initialization or simulation of an architecture and includes variables, or constants of subprograms initialized to either values or attributes of formal parameters of the subprogram.

Figure 3.1 represents some examples of expression classifications.

```
entity Static is
   generic(MaxCount_g : Integer := 5);
 end Static;

architecture Static_a of Static is
   -- Locally static expression
   constant MaxCount_c : Integer := 5; -- MaxCount_c evaluated to 5
  -- Globally static expression
   constant MaxCount2_c : Integer := MaxCount_g / 2;


  function Change(signal Sig_s : Integer) return Integer is
    -- Following constant is "dynamically" defined
    -- and its value depends on the actual parameter
    -- (passed to the formal parameter) during the function call.
    constant Changed_c : Boolean := Sig_s'event;
  begin
     if Changed_c then
       return 100;
     else
       return 0 ;
     end if;
  end Change;    -- end function

  signal  Int_s : integer := 1;

begin
  Test_Lbl: process
    Variable Sum_v    : Integer := 0;
    variable Oprnd1_v : Integer := 0;
    variable Oprnd2_v : Integer := 0;
  begin
    Oprnd1_v := Change(Int_s) + 3;
    Oprnd2_v := 200;
    Sum_v := Oprnd1_v + Oprnd2_v;
    wait;
  end process Test_Lbl;
end Static_a;
```


Figure 3.1 Examples of Expression Classifications (ch3_dir\statc_ea.vhd)

3.2 CONTROL STRUCTURES

In processes, control structures provide alternate paths for the data flow. A control structure might execute one group of statements instead of another group, or execute a group of statements multiple times, or jump over a group of statements. There are three control structures:

1. **if**,
2. **case**,
3. **loop**.

3.2.1 The "if" Statement

*[1] (LRM 8.7) An **if** statement selects for execution one or more of the enclosed sequences of statements, depending on the value of one or more corresponding conditions.* The syntax for the **if** statement is as follows:

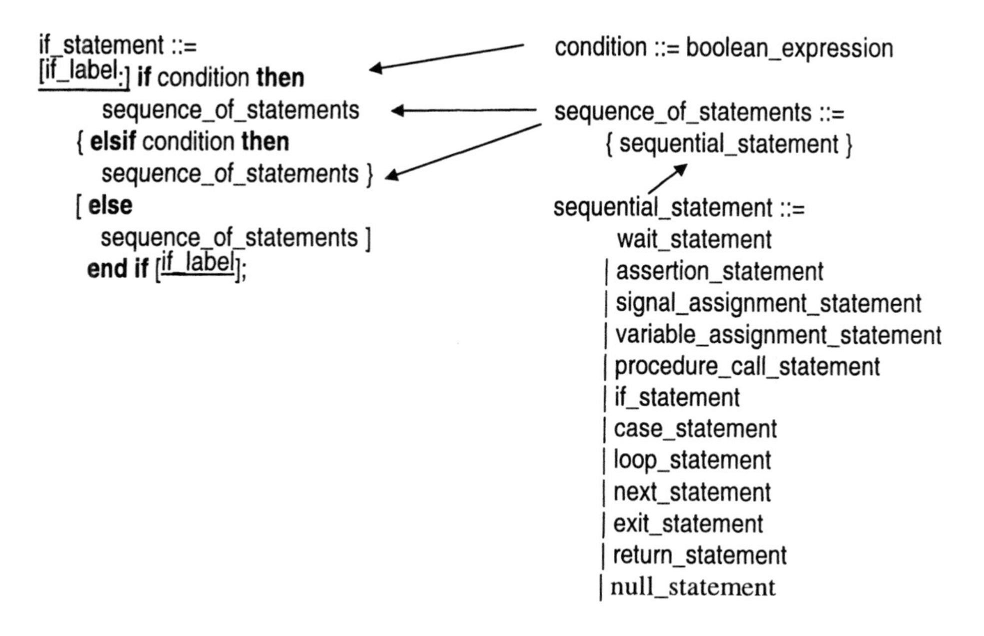

Figure 3.2.1-1 represents the software view of an **if** statement, whereas Figure 3.2.1-2 represents the hardware view.

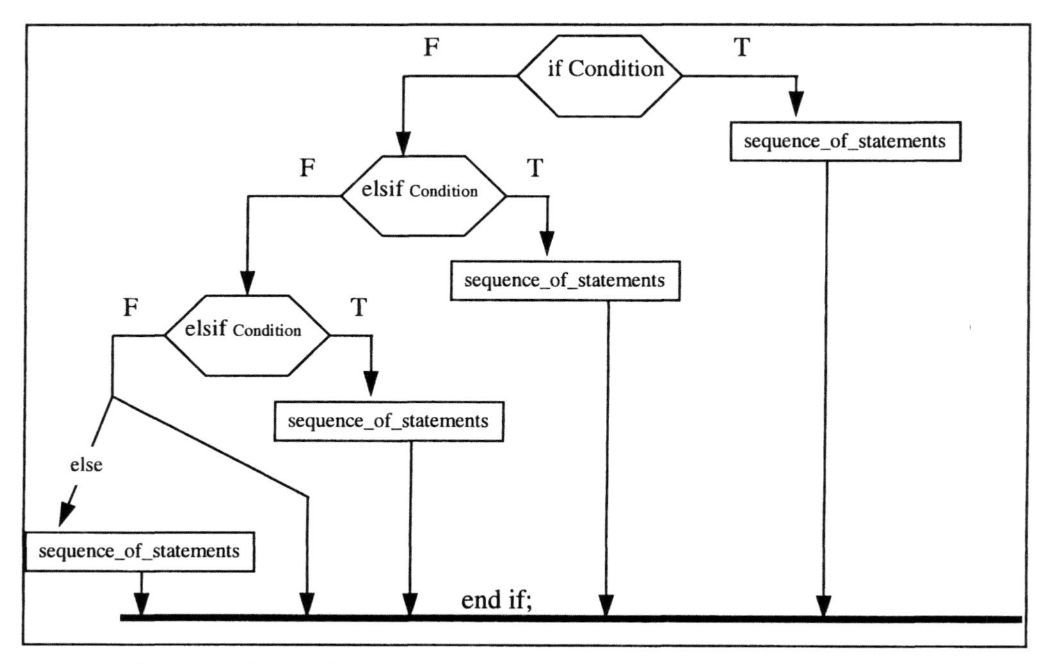

Figure 3.2.1-1 Software Representation of the "if" statement

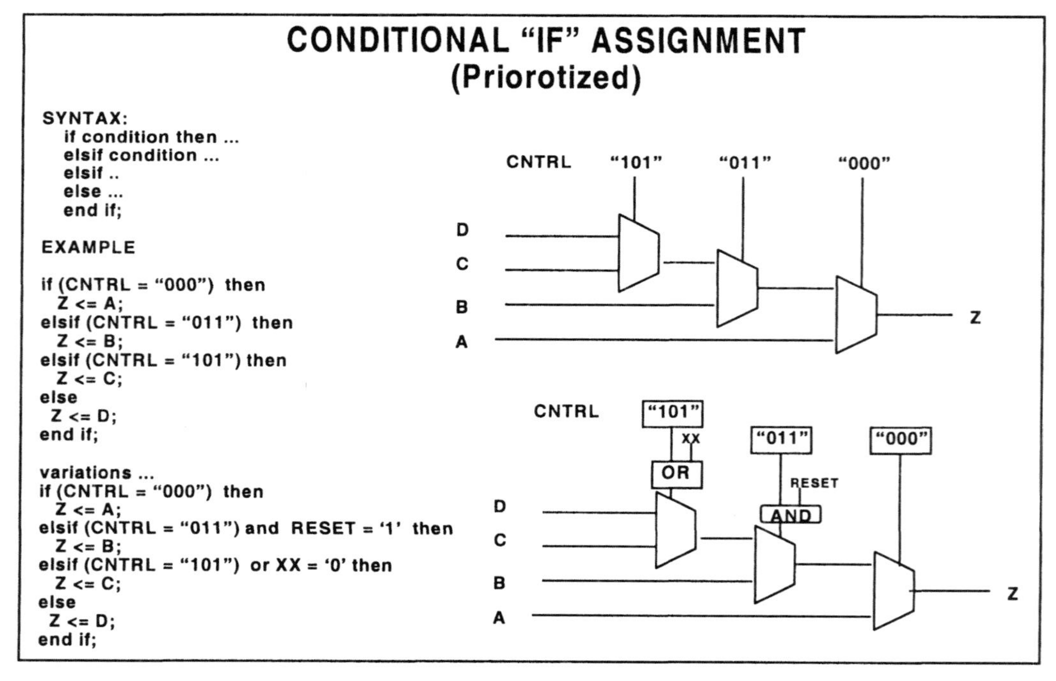

Figure 3.2.1-2 Hardware Representation of the "if" statement

𝔐 Avoid using more than three levels of **if ... else ... end if** statements. If more than three levels are required, encapsulate the inner nested levels with procedure calls. Indent each level of **if** statement. In addition, when using VHDL'93, label each **if** statement.

Rationale: *Limiting the number of levels enhances code readability. Indenting and labeling each level enhances readability. Labeling is not allowed in VHDL'87.*

𝔐 When defining the condition, use parentheses to differentiate levels of operations on the condition. Group the operations in a logical and readable order. For example:

```
if (
    ((Data_s(7) = '1') and (RdWF_s = '1')) or   -- Read condition
    ((Data_s(15 downto 14) = "10")  and (RdWF_s = '0')) or  -- write condition
    (not TriStateON_s'delayed(SysDelay_g))  -- Not tri-state control
   ) then
   Data_s <= X"FFFF";
 end if;
```

Rationale: *This guideline creates more readable code*

Table 3.2.1 demonstrates the use of procedures (see chapter 7) to reduce the number of nested-if levels. VHDL'93 syntax is used in this table.

Table 3.2.1 Use of Procedures to Reduce Number of Nested-if Levels

Three levels of nesting	Translation to 2 levels with procedure call	Sample procedure definition
OneIf_Lbl: **if** Cond1 **then** statements_1T; TwoIf_Lbl: **if** Cond2 **then** statements_2T; ThreeIf_Lbl: **if** Cond3 **then** statements_3T; **else** -- Cond3 statements_3F; **end if** ThreeIf_Lbl; **else** -- Cond2 statements_2F; **end if** TwoIf_Lbl; **else** -- Cond1 statements_1F; **end if** OneIf_Lbl;	OneIf_Lbl: **if** Cond1 **then** statements_1T; TwoIf_Lbl: **if** Cond2 **then** Cond2_Procedure(); **else** -- Cond2 statements_2F; **end if** TwoIf_Lbl; **else** -- Cond1 statements_1F; **end if** OneIf_Lbl;	**procedure** Cond2_Procedure() is **begin** statements_2T; ThreeIf_Lbl: **if** Cond3 **then** statements_3T; **else** -- Cond3 statements_3F; **end if** ThreeIf_Lbl; **end procedure** Cond2_Procedure;

3.2.2 The Case Statement

[1] (LRM 8.8) A case statement selects for execution one of a number of alternative sequences of statements. The syntax for a case statement is as follows:

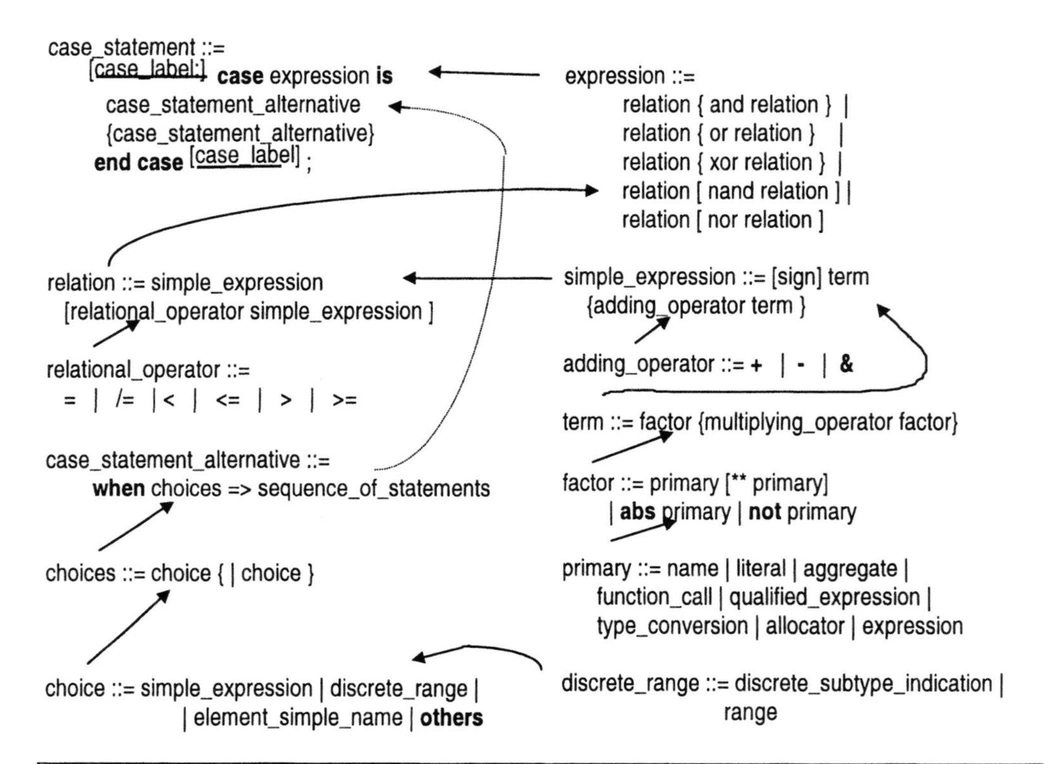

> **M** 👍👍 Choices in a **case** statement should be separated by one blank line and should be indented. The sequence of statements should immediately follow the **case** alternative specification and be indented by at least two spaces.
>
> ***Rationale:*** *Enhances readability by visually the various alternatives to choose from.*

> **M** 👍👍 Keep the expression for the case statement SIMPLE. When the expression of a **case** statement is complex, compute the expression using a variable or a signal, and then use that variable or signal as the expression in the case statement.
>
> ***Rationale:*** *Enhances code readability, and causes the expression to be static (see next subsection).*

Figure 3.2.2-1 represent the software view of a **case** statement, whereas Figure 3.2.2-2 represents the hardware view.

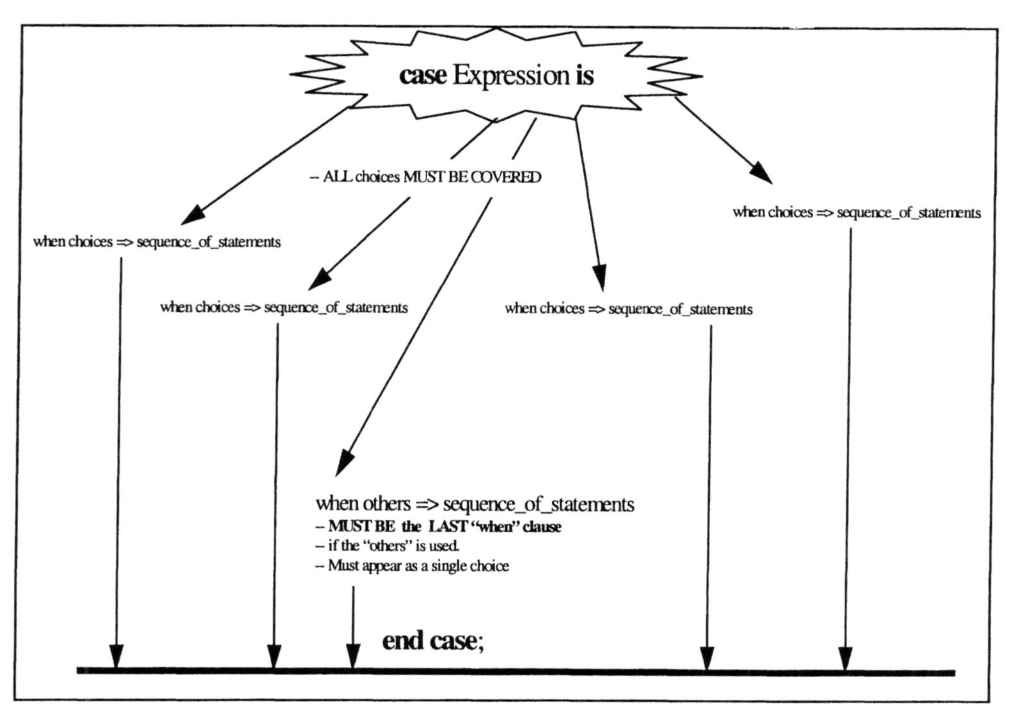

Figure 3.2.2-1 Software Representation of the "case" statement

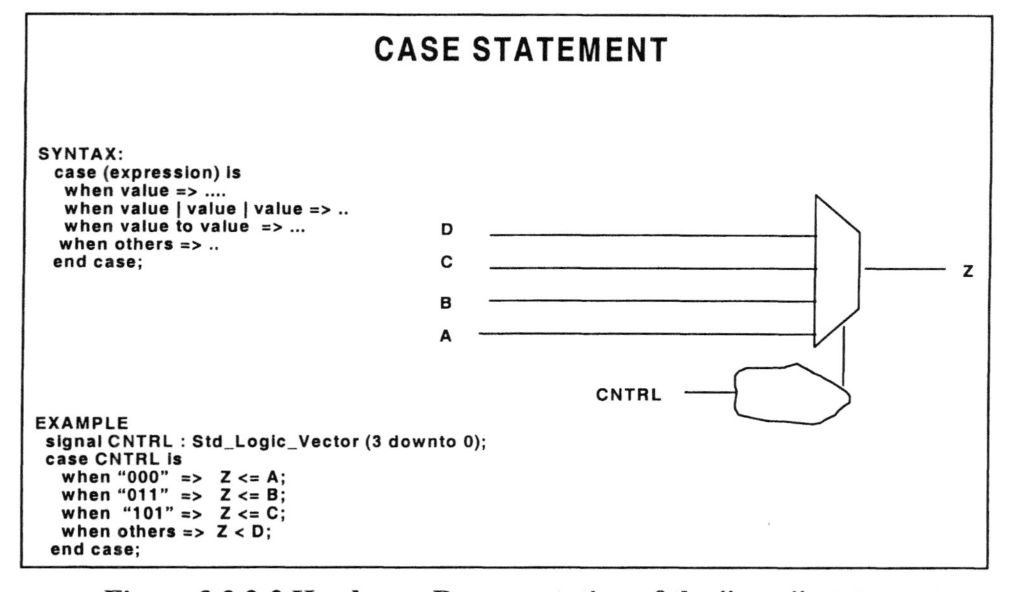

Figure 3.2.2-2 Hardware Representation of the "case" statement

Figure 3.2.2.1 provides an example of the **case** statement that abides by the coding style.

```
architecture CaseDemo_a of CaseDemo is
begin  --  CaseDemo_a
  Demo_Lbl : process
    use Std.TextIO.all;
    use Std.TextIO;                        --  for Text IO operations

    subtype String5_Typ   is string(1 to 5);
    subtype String10_Typ  is string(1 to 10);
    subtype BV5_Typ       is Bit_Vector(1 to 5);

    variable OutLine_v    : TextIO.Line;  -- pointer to string
    variable String5_v    : String5_Typ   := "GREAT";
    variable String10_v   : String10_Typ;
    variable BV5_v        : BV5_Typ       := "10110";
    variable Int_v        : integer       := 5;
  begin  --  process Demo_Lbl
    case BV5_v is                          --  simple primary term
      when  "00000" =>
        TextIO.Write(OutLine_v, string'("BV5_v = 00000"));
        TextIO.WriteLine(Output, OutLine_v);

      when  "11111" =>
        TextIO.Write(OutLine_v, string'("BV5_v = 11111"));
        TextIO.WriteLine(Output, OutLine_v);

      when  others  =>
        TextIO.Write(OutLine_v, string'("BV5_v /= 00000 or 11111"));
        TextIO.WriteLine(Output, OutLine_v);
    end case;                              --  BV5_v


    case +2*Int_v > 10  is
      when true  =>
        TextIO.Write(OutLine_v, string'("Int_v > 10"));
        TextIO.WriteLine(Output, OutLine_v);

      when false =>
        TextIO.Write(OutLine_v, string'("Int_v not > 10"));
        TextIO.WriteLine(Output, OutLine_v);
    end case;                              --  +2*Int_v > 10

    string10_v := string5_v & " HITS";
    case String5_Typ'(string5_v(1 to 3) & String10_v(9 to 10)) is
      when "GREts" =>
        TextIO.Write(OutLine_v, string'("GREts    "));
        TextIO.WriteLine(Output, OutLine_v);

      when "GRETS" =>
        TextIO.Write(OutLine_v, string'("GRETS     "));
        TextIO.WriteLine(Output, OutLine_v);

      when others  =>
        TextIO.Write(OutLine_v, string'("NOT grets "));
        TextIO.WriteLine(Output, OutLine_v);
    end case;  --  String5_Typ'(string5_v(1 to 3) &String10_v(9 to 10))
```

Use type qualifier because compiler doesn't know if the text in parentheses is a **string** or a **bit_vector.**

Use type qualifier because compiler doesn't know size of the concatenation results. Use a variable when expression is complex.

```
    String5_v := string5_v(1 to 3) & String10_v(9 to 10);
    case String5_v is
      when "GREts" =>
        TextIO.Write(OutLine_v, string'("GREts     "));
        TextIO.WriteLine(Output, OutLine_v);

      when "GRETS" =>
        TextIO.Write(OutLine_v, string'("GRETS     "));
        TextIO.WriteLine(Output, OutLine_v);

      when others  =>
        TextIO.Write(OutLine_v, string'("NOT grets "));
        TextIO.WriteLine(Output, OutLine_v);
    end case;                                   --  String5_v

    wait;
  end process Demo_Lbl;

end CaseDemo_a;
```

Use a variable when expression is complex.

Use type qualifier because compiler doesn't know which overloaded TextIO.Write procedure to use: the one for **string** or the one for **Bit_Vector.**

Figure 3.2.2.1 Example of the Case Statement, ch3_dir\case0_ea.vhd

3.2.2.1 Rules for the Case Statement

1. The case expression must be a discrete type[13].
2. Every **POSSIBLE** value of the case expression must be covered in one and only one **when** clause (i.e., cannot duplicate a value in another "when" clause)[14].
3. If the **when others** clause is used, it must appear as a single choice at the end of the case statement[14].
4. Case choice must be a **locally static** expression. Thus, the choices must not be values based on generics or variables or signals.
5. Array case expression must have a **static** subtype. Thus, the type of the case expression must not be based on generics. In addition, the concatenation of bits or characters must be qualified with a subtype because the result of a concatenation is an unconstrained array, and thus not static.

Figure 3.2.2.1-1, 3.2.2.1-2 and 3.2.2.1-3 are examples of the **case** statement with rule violations.

---

[13] *Rendez Vous with Ada*, A Programmer's Introduction, David J. Naiditch, John Wiley & Sons, Inc. Copyright ©1989 by David Naiditch

```
architecture Case_Beh of Case is
  type States_Typ is (Idle, Waiting, Suspended, Running);
begin  --  Case_Beh
  TestCase_Lbl : process
    variable Real_v    : real;
    variable States_v  : States_Typ;
    variable Pos_v     : positive := 10;
  begin  --  process TestCase_Lbl
    Real_v := 3.14;
    States_v := Waiting;
    case States_v is
      when Idle      => null;  -- no statement

      when Waiting | Suspended =>
        States_v := Idle;

      when Running   => null;
    end case;                              --  States_v


    case Pos_v is
      when 1 TO 9     => Pos_v := 1;

      when 10 TO 100 => Pos_v := 100;

      when others     => States_v := Idle;
    end case;                              --  pOS_V

    case States_v is
      when Idle                   => null;

      when Waiting   |
           Suspended |
           Idle                   => Pos_v := 100;--  Idle is already covered

      when Running                => null;
    end case;                              --  States_v

    case Real_v is                         --
      when 1.0 to 10.9 => Real_v := 100.9;

      when others       => States_v := Idle;
    end case;                              --  Real_v


    case States_v is                       --
      when Idle                  => null;  -- missing Running

      when Waiting   |
           Suspended             => Pos_v := 100;--  line 51
    end case;                              --  States_v
    wait;                                  --  suspend process forever
  end process TestCase_Lbl;
end Case_Beh;
```

Rule 2 violation. Every value of case expression must be covered **once.**

Rule 1 violation. Expression must be discrete type

Rule 2 violation. Every value of case expression must be covered

Figure 3.2.2.1-1 Example of the Case Statement with Rule Violations (ch3_dir\case_ea.vhd)

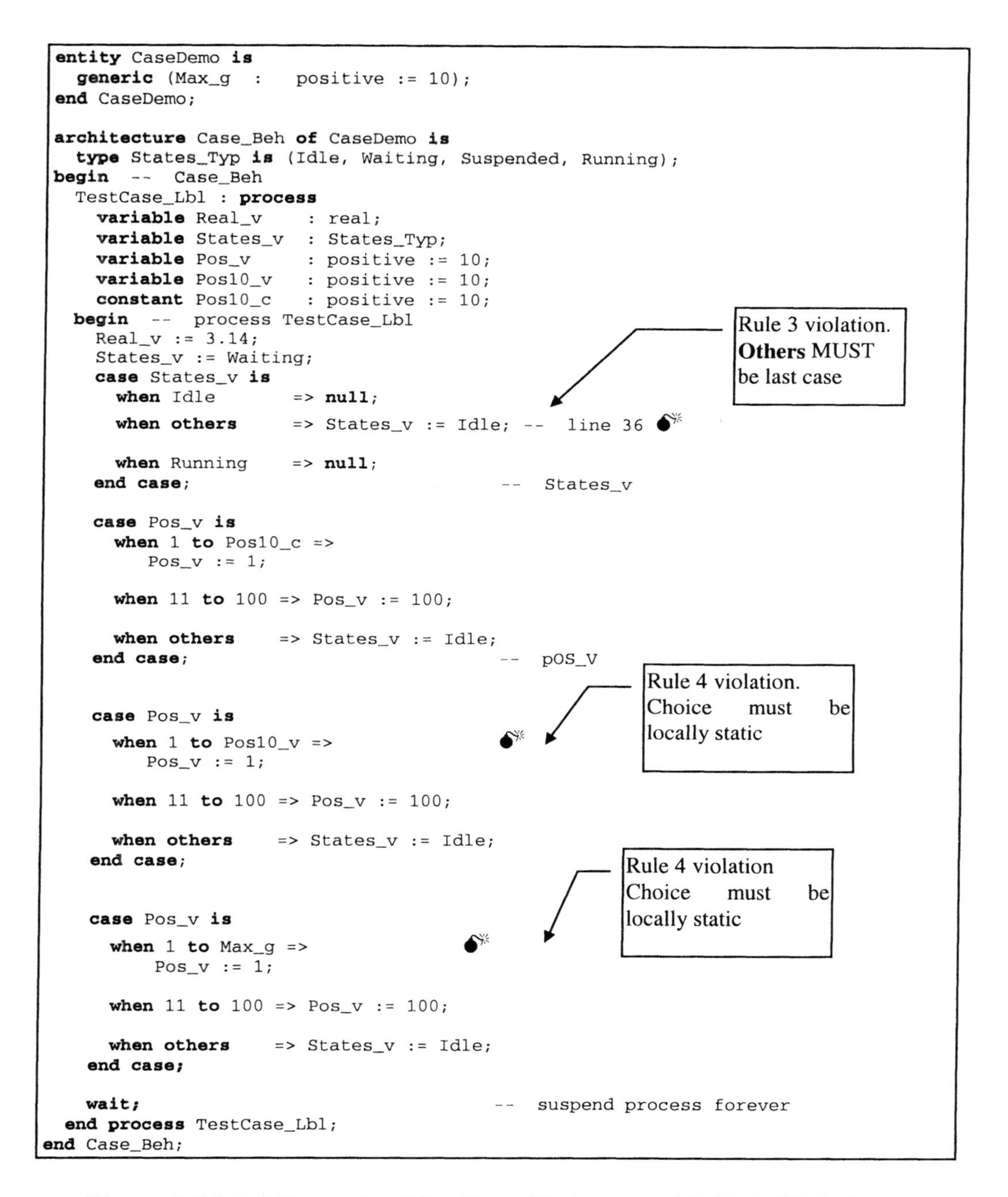

```
entity CaseDemo is
  generic (Max_g  :   positive := 10);
end CaseDemo;

architecture Case_Beh of CaseDemo is
  type States_Typ is (Idle, Waiting, Suspended, Running);
begin  --  Case_Beh
  TestCase_Lbl : process
    variable Real_v     : real;
    variable States_v   : States_Typ;
    variable Pos_v      : positive := 10;
    variable Pos10_v    : positive := 10;
    constant Pos10_c    : positive := 10;
  begin  --  process TestCase_Lbl
    Real_v := 3.14;
    States_v := Waiting;
    case States_v is
      when Idle         => null;
      when others       => States_v := Idle; --  line 36

      when Running      => null;
    end case;                                  --  States_v

    case Pos_v is
      when 1 to Pos10_c =>
         Pos_v := 1;

      when 11 to 100 => Pos_v := 100;

      when others     => States_v := Idle;
    end case;                                  --  pOS_V


    case Pos_v is
      when 1 to Pos10_v =>
         Pos_v := 1;

      when 11 to 100 => Pos_v := 100;

      when others     => States_v := Idle;
    end case;


    case Pos_v is
      when 1 to Max_g =>
          Pos_v := 1;

      when 11 to 100 => Pos_v := 100;

      when others     => States_v := Idle;
    end case;

    wait;                                      --  suspend process forever
  end process TestCase_Lbl;
end Case_Beh;
```


Figure 3.2.2.1-2 Example of the Case Statement with Rule Violations, ch3_dir\case2_ea.vhd

```
architecture Case_Beh of CaseDemo is
  constant        ATEP_c             : string := "ATEP";
  signal          S                  : String(1 to Max_g); -- not static subtype
begin  --  Case_Beh
  Case_Lbl : process
    subtype String2_Typ is string(1 to 2);
  begin  --  process Case_Lbl

    -- case ATEP_c(1) & ATEP_c(3) is    --
    --.

    -- case String'(ATEP_c(1) & ATEP_c(3)) is    --
    --    Array case expression must have a static subtype.

      case String'(ATEP_c(1), ATEP_c(3)) is   --
        when "AE"     =>
          assert false
            report "String constant is AE"
            severity note;
        when others =>
          assert false
            report "String is not AE"
            severity note;
      end case;

    case S is -- Error, S is not of a static subtype --
      when "1234567890" => null;
      when others       => null;
    end case;
    wait;
  end process Case_Lbl;

end Case_Beh;
```

Rule 5 violation. Array case expression must have a static subtype

See note below

Rule 5 violation. Array case expression must have a static subtype

Figure 3.2.2.1-3 Example of the Case Statement with Rule Violations (ch3_dir\caseq_ea.vhd)

When grouping elements in a case expression, the use of the "&" operator is legal, but the use of an aggregate is more efficient when the elements of the aggregate are of the same subtype or type as the elements of the choice in the case alternatives. Thus:

case String2_Typ'(ATEP_c(1) & ATEP_c(3)) **is .. --** is legal
case String'(ATEP_c(1) , ATEP_c(3)) **is** ... -- is more efficient
-- Type or subtype type mark can be used

However, the "&" operator is required when the elements of the grouping are of various lengths. Thus:

subtype BV8_Typ **is** Bit_Vector(7 **downto** 0);
-- ...
case BV8_Typ'(A(7) & B(6 **downto** 1) & C(0)) **is** -- "&" operator is needed
-- with a subtype type mark
--...

3.2.3 Latch Inference

In non-clocked processes, incompletely specified ***if*** and ***case*** statements cause synthesizers to infer latches for the variables and signals being assigned. In other words, in non-clocked processes, if a variable or signal is not assigned prior to ***if*** or ***case*** statement, but it is assigned in some, but not all, of the sequence of statements of that ***if*** or ***case*** statement, then a latch is inferred for that variable or signal. Figure 3.2.3-1 represents an example of a latch inference.

```
-- Data_Out is a Latch
process(SomeSignal)
begin
  if (SomeSignal = "00") then
     Data_Out <= Data_In;
  end if;
end process;
```

```
-- Data_Out is a Latch
process(SomeSignal)
begin
  case SomeSignal is
     when "00" =>
        Data_Out <= Data_In;
     when others => null;
  end case;
end process;
```

Figure 3.2.3-1 Latch Inference in Non-clocked Processes with *if* or *case* Statements

In that example, a latch is inferred for the signal *Data_Out* because is not assigned under all possible conditions. If the tested condition fails then Data_Out must hold the previous value.

In non-clocked processes, two options are possible to avoid a latch in the synthesis of variables or signals.

1. Assign the variables or signals in all the ***if*** or ***case*** sequence of statements.
2. Prior to the entry of the ***if*** or ***case*** statement, assign a default value to the variables or signals. For signals, this method relies on the inertial delay properties where the last signal assignment prior to an implicit or explicit ***wait*** statement is the effective value (see section 5.5) for that signal. Since variables are updated immediately, the last variable assignment prior to an implicit or explicit ***wait*** statement is the effective value for that variables.

Avoid the inference of latches in synchronous designs.

Rationale: *Latches infer feedback. They must be avoided in synchronous designs because they cause difficulties in timing analysis and test insertion applications. Most synthesizers provide warnings when latches are inferred.*

Figure 3.2.3-2 demonstrates the techniques to avoid a latch in non-clocked processes.

<pre>-- Data_Out is a combinational
-- All Conditions covered
process(SomeSignal)
begin
  if (SomeSignal = "00") then
    Data_Out <= Data_In;
  else
    Data_Out <= (others => '0');
  end if;
end process;</pre>
	<pre>-- Data_Out is a combinational
	-- Signal preassigned prior to
	-- case statement
	process(SomeSignal)
	begin
	  Data_Out <= (others => '0');
	  case SomeSignal is
	    when "00" =>
	       Data_Out <= Data_In;
	    when others => null;
	  end case;
	end process;</pre>

Figure 3.2.3-2 Latch Avoidance in Non-clocked Processes with *if* or *case* Statements

3.2.4 Register Inference

In synthesis, as per IEEE 1076.6 Synthesis Interoperability Specification, registers can only be implied in clocked processes. In clocked processes, signal assignments are synthesized as registers whether or not the signals are completely specified in the *if* and *case* statements.

However, in clocked processes, variables may be implied in clocked statements. This is a function of the synthesizer tool. See chapter 10 for further discussions on this topic.

3.2.5 Loop Statement

Loop statements are used to execute a sequence of statements zero or more times. There three forms of loop statements: The simple ***loop***, the ***while*** loop, and the ***for*** loop.
The loop syntax is as follows:

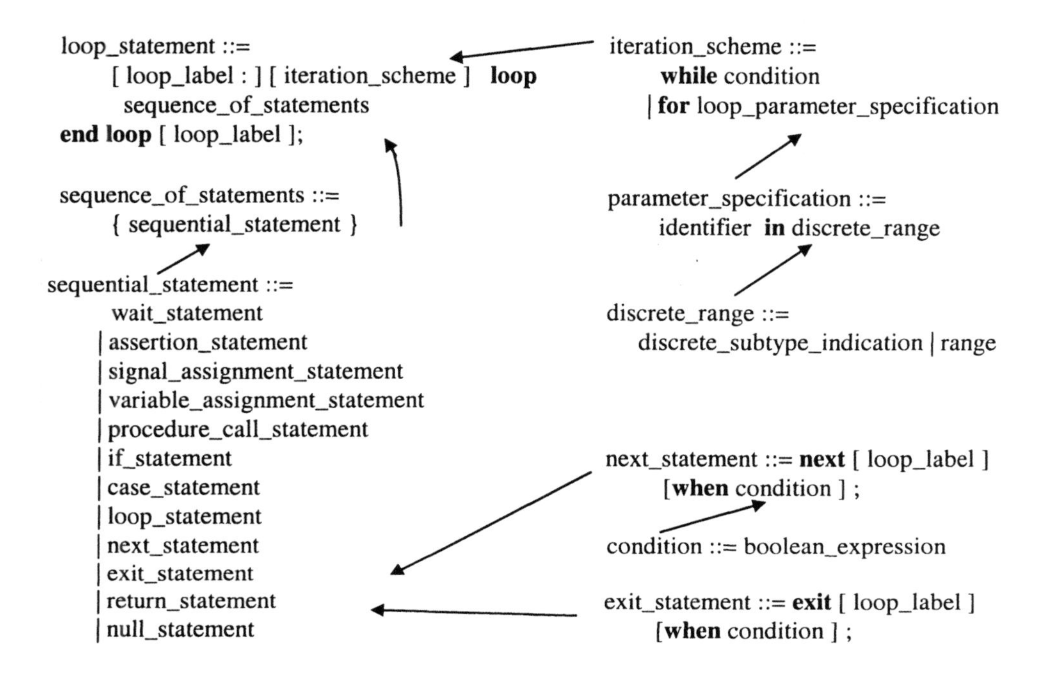

3.2.5.1 The Simple Loop

The simple loop does not have an explicit iteration scheme. The implicit iteration scheme is ***while true***, thus looping forever. The usual way to exit an infinite loop is to use the exit statement. **The simple loop is NOT synthesizable**. Figure 3.2.5.1 demonstrates the use of a simple loop.

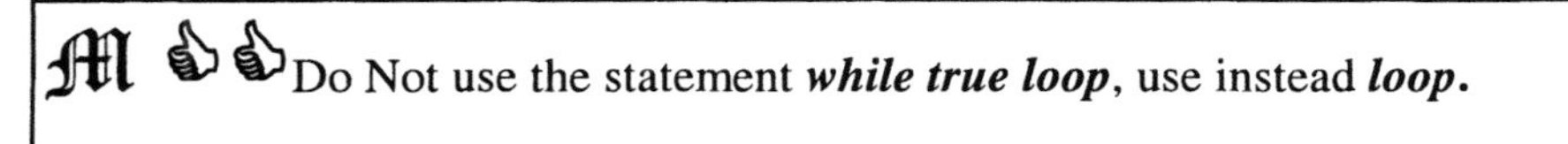

Rationale: While true *statement is redundant. The simple* ***loop*** *is sufficient.*

```
architecture Loop_a of LoopDemo is
begin  --  Loop_a
  Loop_Lbl: process
      variable Max_v   : natural := 10;
      variable Count_v : natural := 0;
    begin
      Max_v := 6;
      SimpleLoop_Lbl: loop
        Count_v := Count_v + 1;
        if Count_v = Max_v then
          exit;  -- loop iterates 6 times, exit current loop
        end if;
      end loop SimpleLoop_Lbl;
      wait;
    end process Loop_Lbl;
end Loop_a;
```

Note use of loop label.

Figure 3.2.5.1 Use of a Simple Loop (ch3_dir\loop_ea.vhd)

3.2.5.2 The while loop

The ***while loop*** iterates as long as the condition expressed in the ***while*** statement is true. This is often used to execute a set of sequential statements while a signal or a variable meets a certain criteria. Note that for a loop with a *[1]* ***while*** *iteration scheme, the condition is evaluated* ***BEFORE*** *each execution of the sequence of statements. If the condition is TRUE, then the sequence is executed, else the iteration scheme is said to be complete and the execution of the loop statement is complete* (i.e., done). **The while loop is NOT synthesizable.** Figure 3.2.5.2 demonstrates the use of a ***while*** loop.

```
...
Loop_Lbl: process
        variable Count_v : natural := 0;
        begin
          -- The process automatically reiterates to the start
          -- after the last statement of the process.
          -- Thus, if the "while loop" fails, the
          -- process gets suspended until the next rising edge of the clock.
          wait until Clk_s = '1'; -- rising edge of Clk_s

          -- If the condition is true in the while loop, this
          -- algorithm counts the number of cycles the condition is true
          SimpleLoop_Lbl: while Receiving_s = '1' loop
            Count_v := Count_v + 1;
            wait until Clk_s = '1'; -- rising edge of Clk_s
          end loop SimpleLoop_Lbl;
      end process Loop_Lbl;
    end While_Beh;
```

Figure 3.2.5.2 Use of a While Loop (ch3_dir\while_ea.vhd)

3.2.5.3 The for loop

The loop parameter type for the ***for*** iteration loop scheme (e.g., the parameter *I* in "***for*** I ***in*** 1 ***to*** 4") is the base type of the discrete range, and is **NOT explicitly defined as a type**. Thus, the type is implicitly defined from the range. Figure 3.2.5.3 is an example for the use of a ***for*** loop statement. This example provides a bit reversal of a bit vector.

```
architecture Demo0_a of For0 is
begin  --  Demo0_a
Demo_Lbl : process
      subtype  DnRange_Typ is natural range 7 downto 0;

      -- below is a more general purpose description by
      -- referring the range to a predefined type.
      -- variable UpWord_v : bit_vector (0 to 7);
      variable DnWord0_v : bit_vector (DnRange_Typ);
      variable DnWord1_v : bit_vector (DnRange_Typ);
  begin  --  process Demo_Lbl
      DnWord0_v := X"84";
      -- This loop is written using attributes so that the algorithm
      -- can be used even if the range type DnRange_Typ is changed.
      -- In the loop below, the loop counter "Index_i" is IMPLICITLY
      --  defined,
      Reverse_Lbl: for Index_i in DnRange_Typ loop   -- 7 downto 0
        DnWord1_v(DnRange_Typ'left - Index_i) := DnWord0_v(Index_i);
      end loop Reverse_Lbl;
      wait;  -- suspend process in this demonstration process.
  end process Demo_Lbl;
end Demo0_a;
```

Figure 3.2.5.3 Bit Reversal with a for loop statement (ch3_dir\for0_ea.vhd)

M 👍👍 When defining the loop parameter specification, either use a type (or subtype) definition, or use predefined object attributes (e.g., PredefinedObject'range, PredefinedObject'length - 1 **downto** 0). Avoid using discrete range (e.g., 1 to 4).

Rationale: *This rule makes the code more reusable and flexible for maintenance.*

3.2.5.3.1 for loop Rules

There are 10 rules for the **for loop**, and they include the following:

1. The loop parameter is **not explicitly** defined, but **implicitly** defined[14].
2. The loop parameter's range is tested at the **beginning of the loop**, not at the end[15].
3. *[1] The loop parameter is an object whose type is the* ***base type of the discrete range.***

Figure 3.2.5.3.1-1 demonstrates the concepts for rules 1, 2 and 3.

---

14 *Rendez Vous with Ada*, A Programmer's Introduction, David J. Naiditch, John Wiley & Sons, Inc. Copyright ©1989 by David Naiditch

```
architecture LoopTest_Beh of LoopTest is
  type States_Typ is (Idle, Running, Stuck);
begin
  Lp_Proc: process
    variable Count_v : natural := 0;
  begin
    -- loop counter States_i is implicitly defined of states_Typ.
    States_Lp: for States_i in States_Typ loop
      case States_i is
        when Idle      =>
          Count_v := Count_v + 1;

        when Running   =>
          Count_v := Count_v + 2;

        when Stuck     =>
          null;
      end case;                              --  States_i
    end loop States_Lp;
    wait;
  end process Lp_Proc;
end LoopTest_Beh;
```

In this example, ALL the cases in the case statement will be covered. The Idle case after the 1st iterationn, the Running case after the 2nd iteration, and the Stuck case after the 3rd iteration.

Figure 3.2.5.3.1-1 The Loop Parameter is an Object whose Type is the Base Type of the Discrete Range, ch3_dir\forrule3.vhd

4. Inside the loop, the **loop parameter is a constant**. Thus, it may be used but **not altered**[15]. (i.e., can be read and compared, but NOT written into). In addition the loop parameter *[1]* ***must not*** *be given as an* ***actual*** *corresponding* ***to a formal of mode out or inout in an association list*** (see chapter 7). Thus, if within the loop a procedure is called, the value of the loop parameter can be passed to that procedure an input, but not as an output, since an output will attempt to modify the loop parameter.

5. The loop **parameter's discrete range may be dynamic**[15]. Example:

```
variable Max_v : integer;
.....
Max_v := 10;
Dynamic_Lbl: for I in 1 to Max_v loop -- Max_v is dynamic (not static)
 ....
end loop Dynamic_Lp;
```

6. The **discrete range** of the loop is **evaluated before the loop is first executed**[15]. *[1] If the discrete range is a null range, then the iteration scheme is said to be complete and the execution of the loop is therefore complete. Otherwise the sequence is executed once for each value of the discrete range (provided the loop is not being left as the result of the execution of a next statement, an exit statement, or a return statement), after which the iteration is said to be complete. Prior to each iteration, the corresponding value of the discrete range is assigned to the loop parameter.*

7. *[1] A **next** statement is used to complete the execution of one of the iterations of an enclosing loop statement.. The syntax for the **next** statement is:*

next_statement ::=
next [loop_label] [**when** condition];

*A **next** statement with a loop label is allowed within the labeled loop and applies only to that loop. A **next** statement without a loop label is only allowed within a loop, applies to the innermost enclosing loop (whether labeled or not). For the execution of a **next** statement, the condition, if present, is first evaluated. The current iteration loop is terminated if the value of the condition is TRUE or if there is no condition.* Figure 3.2.5.3.1-2 is an example using the **next** and **exit** statements.

𝔐 👍 👍 If the condition for the **when** is simple, insert the condition in-line (i.e., in-line with the code). Otherwise compute the condition using a variable or a signal.

***Rationale:** Guideline enhances code readability.*

```
entity ForNextExit is
  generic (Max_g:               Positive := 16);
end ForNextExit;

architecture ForNextExit_a of ForNextExit is
begin  --  ForNextExit_a
  --  Purpose:   Count the number of '1's in bit vector
  --             from left side to the right side, provided the index
  --             is odd (e.g., 31, 29, 27 ) up to Max_g
  DEMO_Lbl : process
    subtype S32range_Typ is natural range 32 downto 1;
    subtype BV32_Typ is bit_vector(S32range_Typ);
    variable BV32_v : BV32_Typ := X"AB86_ABCD";
    variable Count_v  : Natural;      --  initialized to natural'left, = 0
    variable IsEven_v : Boolean;
    variable MaxOut_v : Boolean;
  begin  --  process DEMO_Lbl
    CountOdd_Lbl : for I in BV32_v'range loop
      IsEven_v := I mod 2 = 0;
      MaxOut_v := I < Max_g;
      next CountOdd_Lbl when IsEven_v; --  skip if even
      exit CountOdd_Lbl when MaxOut_v; --  end if limit is reached
      if BV32_v(I) = '1' then
        Count_v := Count_v + 1;
      end if;
    end loop CountOdd_Lbl;
    wait;
  end process DEMO_Lbl;
end ForNextExit_a;
```

If condition is simple, insert the condition in line instead of computing it through a variable

Figure 3.2.5.3.1-2 Example using the next and exit statements (ch3_dir\for1_ea.vhd)

8. *[1] An **exit** statement is used to complete the execution of an enclosing loop statement..* The syntax for the **exit** statement is:

   exit_statement ::=
   exit [loop_label] [**when** condition] ;

   *The completion is conditional if the statement includes a condition. An **exit** statement with a loop label is only allowed within the labeled loop and applies to that loop. An **exit** statement without a loop label is only allowed within a loop, applies to the innermost enclosing loop (whether labeled or not).*
   *For the execution of an **exit** statement, the condition, if present, is first evaluated.* The loop is terminated if the value of the condition is TRUE or if there is no condition.

Figure 3.2.5.3.1-3 demonstrates the use of the **exit** statements and nested loops.

> M Loop statements should be labeled.
>
> ***Rationale:*** *Labels in loop parameters enable better loop control with the **next** and **exit** statements. Labels also enhance readability and maintainability.*

```
architecture NestedFor_a of NestedFor is

begin  --  NestedFor_a
  NestedFor_Proc : process
    variable    Int1_v   : Natural;
    variable    Int2_v   : Natural;
    variable    Int3_v   : Natural;

  begin  --  process NestedFor_Lbl
    Out_Lp : for I in 1 to 10 loop
      next when I mod 2 > 0;          --   skip outer if odd
      Int2_v := Int2_v + 1;
      Inner_Lbl : for X_i in 1 to 3 loop
        next Out_Lp when Int2_v = 2;
        assert false
          report "in inner loop"
          severity Note;
        Int3_v := I + X_i;
        exit Out_Lp when Int2_v = 4;
      end loop Inner_Lp;
    end loop Out_Lp;
    wait;
  end process NestedFor_Proc;
end NestedFor_a;
```

> L Inner loop has visibility on outer loop parameters (e.g. Indx_i).
> Outer loops have NO visibility on inner loop parameters (e.g. X_i)

Figure 3.2.5.3.1-3 Use of the Exit Statements and Nested Loops (ch3_dir\for2_ea.vhd)

9. *[1] The **loop counter only exists within the loop***[15]. Thus at exit of loop, the loop counter has no more significance. Within the loop, the loop counter hides identifiers that exist outside the loop and have the same name as the loop counter. The identifiers that are hidden can be accessible if they are prefixed with the label where they exit followed by a period. Figure 3.2.5.3.1-4 demonstrates this concept.

```
architecture Hide_a of Hide is
begin  --  Hide_a
   Hide_Proc: process
     variable Count_v : Natural := 10;
     variable Other_v : Natural := 0;
   begin
     Count_v := 5; -- Hide_Proc.Count_v := 5
     Lp_Lp: for Count_v in 1 to 10 loop
       Other_v := Other_v + Count_v; -- loop counter is used here
       Other_v := Other_v + Hide_Proc.Count_v; -- process variable is used here
     end loop Lp_Lp;
     wait;
   end process Hide_Proc;
end Hide_a;
```

There are 2 definitions for Count_v:
1. In the process as a variable
2. In the loop as a loop counter

Figure 3.2.3.3.1-4 Loop Counter Only Exits Within the Loop (ch3_dir\hide_ea.vhd)

M 👍👍 Avoid hiding identifiers with loop counter identifiers (i.e., do not name loop counter identifiers as other identifiers visible by the process).

***Rationale:** Hiding identifiers worsens readability.*

10. **FOR** loops may step through the discrete in **reverse range**[15]**.** For example:

```
subtype BV32_Typ is bit_vector(31 downto 0);
   ...
   -- reverse_range operates on arrays
   for I in BV32Typ'reverse_range loop --  0 to 31
```

Note that for synthesis, many synthesizers require that the loop parameter be an integer type (rather than enumerated type). Coding rules are defined by the synthesizer tools. Users should verify the VHDL coding rules for their specific synthesizers.

EXERCISES

1. Which of the following is true about the **if** statement?
   a. The levels of nesting is infinite, and not restricted.
   b. The condition can be of any type.
   c. The **else** clause and **elsif** clause can be intermixed.
   d. All the conditions of the **if** and **elsif** and **else** clauses must be from the same expression of the **if** clause (i.e. like the **case** statement).
   e. Any statement written as a **case** statement can be translated to an **if** statement.
   f. Any statement written as an **if** statement can be rewritten as a **case** statement.

2. Rewrite the following if statements into case statements. Embed this code into an entity/architecture with a single process.

```
if Size_v <= 5 then   -- Size_v is of type natural
  Select_v := Small;   -- Select_v is of an enumeration type
elsif Size_v = 10 then
  Select_v := Medium;
else
  Select_v := Large;
end if;
```

3. Translate and execute the case BV5_v code in file "case0_ea.vhd" using **if** statements.

4. Write and execute a program using **for loop** to compute the integer value of an 8 bit vector array (of subtype bit_vector(7 **downto** 0)). The result is of type natural. Use a constant array to translate the value of the index into a weight

```
(e.g. index of 0 has a weight of 1,
               1              2
               2              4
               3              8 ..).
```

Thus if the value of C is B"101", then its integer value is
1*(LSB)* * **1***(weight of LSB)* + **0** * **2** + **1** *(MSB)* * **4** *(weight of MSB)* = 5
Use attributes instead of hard coded numbers to generalize the code.

5. Write and execute a program using the loop statements which counts the number of ZEROS in the ODD indices (e.g. 1, 3, 5, ..) of a 64 bit array of subtype bit_vector(63 **downto** 0).

6. Which of the statements are TRUE for the "for" loop?

```
a. The loop parameter is implicitly defined.
b. The loop parameter can be modified inside the loop.
c. the loop parameter can be read inside the loop.
d. An inner loop can read an outer loop parameter.
e. an outer loop can read an inner loop loop_parameter.
f. The loop parameter can be of any type
g. The loop parameter discrete range must be static.
h. The loop parameter discrete range may be dynamic.
i. The statement "next outer_loop_label" from an inner loop will
   cause the inner loop to exit, and the outer loop to
   go to the next iteration.
j. The statement "next inner_loop_label" from an inner loop will
   cause the inner loop to exit, and the outer loop to
   go to the next iteration.
k. The statement "next inner_loop_label" from an inner loop will
   cause the inner loop to iterate.
```

7. What is the value of Count_v for architecture ForNextExit_a in file for1_ea.vhd? Run the exercise using the single step simulation control to view the results.

4. DRIVERS

This chapter presents the concepts of drivers and the initialization rules of signals and ports. These concepts are used to prevent the model from yielding unexpected results during simulation. This chapter also explains the resolution function used to resolve bus conflicts when multiple drivers put values onto signals.

4.1 RESOLUTION FUNCTION

Chapter 7 explains how resolution functions (LRM 2.4) are defined. The resolution function and how it relates to a resolved type is explained in this section. This provides a better understanding of the concepts of drivers, and timing in VHDL. A resolution function is a function that examines all the values from each of the drivers on a signal, and resolve the conflict by returning a single value which matches the resolution algorithm. The resolution function is automatically called by the simulator when the types used in the signal or port declarations are of a resolved type. In *std_logic_1164* package, the type *Std_Logic* and *Std_Logic_Vector* are resolved types. A summary of the resolved operations for any two signals of type *Std_Logic* is shown in the Table 4.1.

Table 4.1 Summary of the Resolved Operations for Any 2 Signals Using std_logic_1164

```
TYPE stdlogic_table is array(Std_uLogic, Std_uLogic) of Std_uLogic;
    -------------------------------------------------------------------
    -- resolution function
    -------------------------------------------------------------------
    constant resolution_table : Stdlogic_Table := (
    --      ---------------------------------------------------------
    --      | U    X    0    1    Z    W    L    H    -          |   |
    --      ---------------------------------------------------------
            ( 'U', 'U', 'U', 'U', 'U', 'U', 'U', 'U', 'U' ), -- | U |
            ( 'U', 'X', 'X', 'X', 'X', 'X', 'X', 'X', 'X' ), -- | X |
            ( 'U', 'X', '0', 'X', '0', '0', '0', '0', 'X' ), -- | 0 |
            ( 'U', 'X', 'X', '1', '1', '1', '1', '1', 'X' ), -- | 1 |
            ( 'U', 'X', '0', '1', 'Z', 'W', 'L', 'H', 'X' ), -- | Z |
            ( 'U', 'X', '0', '1', 'W', 'W', 'W', 'W', 'X' ), -- | W |
            ( 'U', 'X', '0', '1', 'L', 'W', 'L', 'W', 'X' ), -- | L |
            ( 'U', 'X', '0', '1', 'H', 'W', 'W', 'H', 'X' ), -- | H |
            ( 'U', 'X', 'X', 'X', 'X', 'X', 'X', 'X', 'X' )  -- | - |
        );
```

This table lookup is used by the resolution function to compute the resolution between two drivers driving values of type *Std_Logic* type. Thus, if one driver asserts a '1' and another drivers asserts a '0' onto the same signal, the result value for that signal is 'X'. Figure 4.1 demonstrates two processes driving different values onto a signal. Because the type of the signal is declared as a resolved type, the resolution function is automatically called by the simulator to compute the resolved value.

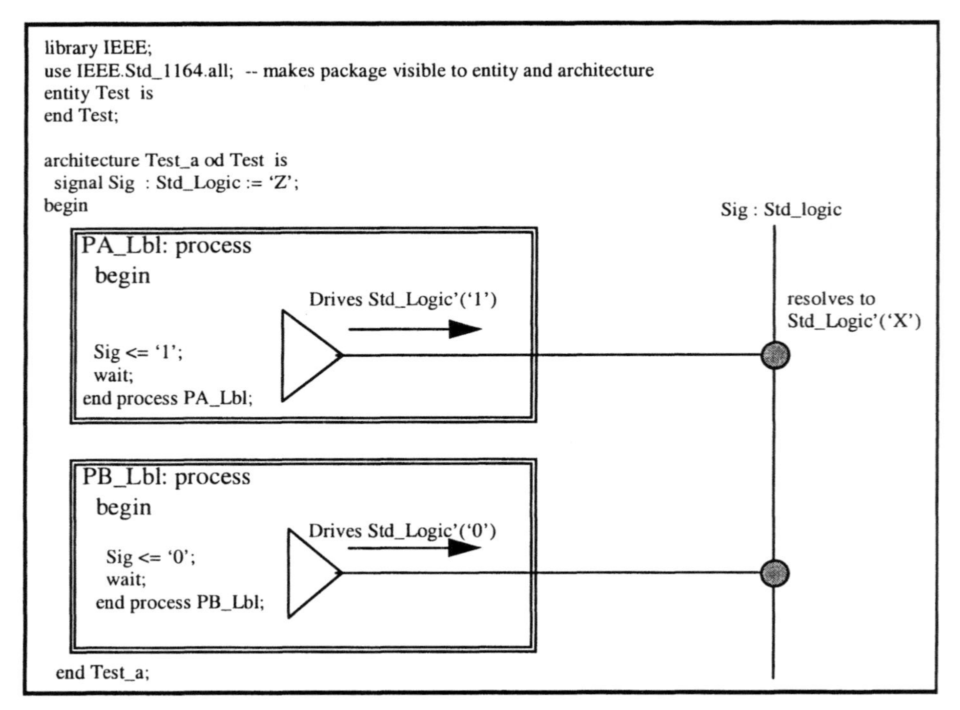

Figure 4.1 Two Processes Driving Different Values onto a Signal

4.2 DRIVERS

4.2.1 Definition and Initialization

[1] (LRM:12.6.1) A driver is a container for a projected output waveform of a signal. Every signal assignment statement in a process (or equivalent process, such as a concurrent signal assignment) defines a set of drivers. A driver can be visualized as a logical hardware device that always sources a value onto a signal. **If a process has at least one signal assignment statement (i.e., "<="), then a single driver (one and only one driver) is automatically created for that signal.** If a process has more than one signal assignment for the same signal, there still is only **ONE driver** created for that signal. Signals defined in an architecture do not have drivers. Each signal assignment is said to be associated with that driver. Drivers are not associated with signal declarations. They are associated with signal assignments.

Figure 4.2.1-1 represents the definition of an entity/architecture with two drivers. This architecture includes a signal called "*A_s*" of type *Std_Logic*, initialized to 'Z'. Also within this architecture are two processes (PA1_Lbl and PA2_Lbl) in which multiple signal assignments (at different times) are made to signal "A_s" of the architecture. Each process thus, create ONE driver to drive signal *"A_s"*.

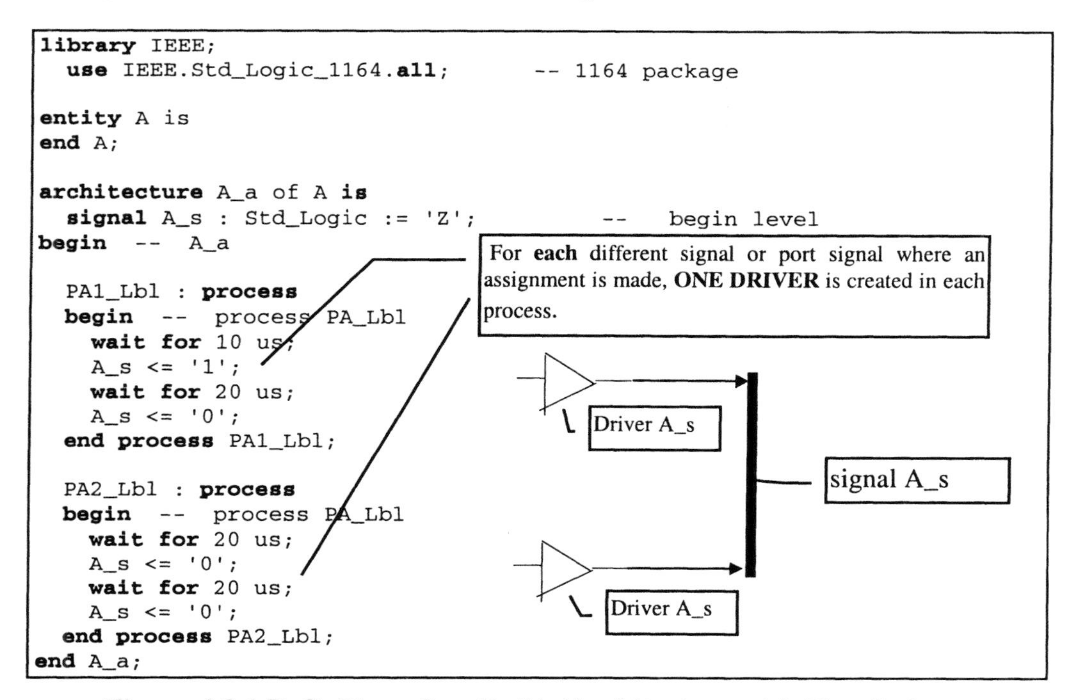

Figure 4.2.1 Definition of an Entity/Architecture with Two Drivers (ch4_dir\drv1_ea.vhd)

[1] (LRM:12.6.1) Each signal assignment is associated with a driver. ***The initial contents of a driver associated with a given signal are defined by the default value associated with the signal.*** *The value of the signal is a function of the current values of its drivers. Each process that assigns to a given signal implicitly contains a driver for that signal. A signal assignment affects only the associated driver(s).*

When using resolved types such as *Std_Logic* or *Std_Logic_Vector*, improperly initialized signals can yield unexpected results during simulation. Table 4.2.1 compares the results of the above model where the signal is initialized to a 'Z' against another model where the signal is initialized to the default value 'U'. Specifically the signal declaration is modified as shown below in file drv1b_ea.vhd:

signal A_s : Std_Logic := 'Z'; -- in file drv1_ea.vhd
signal A_s : Std_Logic; -- defaulted to Std_Logic'left -- 'U' -- in file drv1b_ea.vhd

Table 4.2.1 Comparison of Simulation Results when Drivers are Uninitialized.

Simulation Results, Signal initialized to 'Z' (file drv1_ea.vhd)	Simulation Results, Signal initialized to default value 'U' (file drv1b_ea.vhd)
drivers * -- @ startup # Signal a_s = Z # Driver pa2_lbl = Z # Driver pa1_lbl = Z run 11 us # current time is 11 us	drivers * # Signal a_s = U # Driver pa2_lbl = U # Driver pa1_lbl = U run 11 us # current time is 11 us
drivers * # Signal a_s = 1 # Driver pa2_lbl = Z # Driver pa1_lbl = 1 *The 2 drivers are resolved by resolution function to a '1'* run 10 us # current time is 21 us	drivers * # Signal a_s = U # Driver pa2_lbl = U # Driver pa1_lbl = 1 *The 2 drivers are resolved by resolution function to a 'U'* run 10 us # current time is 21 us
drivers * # Signal a_s = X # Driver pa2_lbl = 0 # Driver pa1_lbl = 1 run 20 us # current time is 41 us write list listfile	drivers * # Signal a_s = X # Driver pa2_lbl = 0 # Driver pa1_lbl = 1 run 20 us # current time is 41 us write list listfile *a 'U' from 0 ns to 20 us. Unexpected results*
ns a_s 0 Z 10000 1 20000 X 30000 0 40000 X	ns a_s 0 U 20000 X 30000 0 40000 X

Ml 👍👍

1. When defining a **synthesizable design DO NOT INITIALIZE PORTS or SIGNALS.**

2. If the initialization is performed through a discrete RESET port, then the model should reflect the initialization reset signal.

3. If the RESET function is not implemented through a discrete RESET port, but rather through initialized scanable registers, then a different approach must be taken for the VHDL model. Specifically, if the model must reflects the functional view of the design (but not the test view through the scan chains) then emulate the initialization through a "phantom" reset port. Some synthesizers provide the capability to direct the compiler to ignore a portion of the VHDL code through pragmas (or directives), thus ignoring for synthesis the phantom RESET code. However, during simulation, all the VHDL code would execute, including the phantom reset code. This emulates the functionality of the synthesized hardware when in non-scanable mode. The model would incorrectly emulate the function of the hardware during scanable mode since the VHDL code would not include the structural view of the scan chain. Figure 4.2.1-2 is an example of the application of the phantom reset function.

Rationale:

1. *Initialization is ignored by synthesizers.*
2. *VHDL model should reflect operation of a discrete RESET signal.*
3. *For tactical (non test) simulation, a discrete RESET control can quickly emulate the reset of the VHDL design. That design can however be synthesizable without modifications. Test simulation requires a structural or more detailed view of the design that models the scan chain.*

Ml 👍👍 In **NON-Synthesizable designs** (e.g., BFMs – see chapter 11) either rely on an **external RESET signal to initialize** the states of the model and its signals (preferred approach), or initialize the ports of an entity and the signals of an architecture to a benign value (other than the default value of 'U' for Std_Logic type signals).

Rationale: *Non-synthesizable designs generally describe a high level model or a BFM used in a testbench environment (see chapter 10).* ***An external RESET signal is the preferred approach*** *to set the model to a known state. However, an alternate approach is to initialize the ports and signals of an architecture.*

```
entity SomeE is
  port
    (
     --   pragma Do_Not_Synthesize
     Reset       : in bit;          <----  The actual pragma is defined by
     --   pragma Synthesize                the synthesizer tool. The
     A           : in  bit;                synthsizer ignores this code .
     Clk         : in bit;
     B           : out bit);
end SomeE;

architecture SomeA_a of SomeE is
  signal C_s     : bit;

begin  --  SomeA_a
  XX_Lbl : process
  begin  --  process XX_Lbl
    wait until Clk'event and CLK = '1';
    --    pragma Do_Not_Synthesize
    if Reset = '1' then                  -- Ignored by Synthesizer.
      B <= '0';                     <--- -- Reset operation
                                         -- is not synthesized.
    else
    --    pragma Synthesize
    B <= not A;                          --   some algorithm
    -- pragma Do_Not_Synthesize
    end if;                         <--- --   Ignored by synthesizer
    -- pragma Synthesize
  end process XX_Lbl;
end SomeA_a;
```

Figure 4.2.1-2 Pragma to Bypass Synthesized VHDL Code (ch4_dir\initlz.vhd)

4.2.2 Creation of Drivers

A driver is created at elaboration time, and exists throughout the lifetime of the model. If a signal is of an array type (e.g., Std_Logic_Vector), and an element of that array selected with the loop index is assigned a value, then drivers are created for every element of the array, regardless of the range of the loop index. The loop index is not elaborated until execution of the loop. At elaboration, drivers are created for every element of the array because the simulator has no knowledge of the loop index range. It must then assume that the index can take any of the allowed values. Figure 4.2.2 demonstrates the issue of multiple drivers in array assignment using the loop index.

```
entity Top is
end Top;

architecture RTL of Top is
   signal A : Bit_Vector(1 downto 0);
   signal X : Bit;
begin
  Bit_Lbl: process(X)
  begin                                     -- One driver on A(0)
    A(0) <= '1';
  end process Bit_Lbl;

  Array_Lbl: process(X)
  begin
    for I in 1 to 1 loop  -- NO PROBLEM WITHOUT THE LOOP AND A(1)
      A(I) <= '0';         --   Drivers for every element of
    end loop;              --   A (e.g,, A(1) and A(0))
  end process Array_Lbl;
end RTL;
```

Figure 4.2.2 Multiple Drivers in Array Assignment using the loop index (drivers\drvbit.vhd)

4.2.3 Drivers and Resolved Signal Types

[1] If more that one process makes a signal assignment to the same signal, then there is a single driver associated with each process. For example, if five processes make signal assignment to signal S, then there are five drivers (one per each process) for signal S. These drivers are created at elaboration and exist during the whole simulation time regardless of when the signal assignment is made.

[1] If more than one driver exists for a signal, then that signal MUST be of a resolved type in order for the simulator to determine the final resolved value on that signal. For example, if one driver asserts a '1' and the other driver asserts a '0', then it is required to use a resolved type to resolve the final value on the signal.

It is an error if more than one driver exists on a signal, and the type of the signal and data types of the values asserted onto the signal is not of a resolved type.

4.2.3.1 Driving Data from multiple Processes onto a Non-Resolved Signal

As mentioned above, it is illegal for multiple processes to assign values onto a common signal of a non-resolved type. This is a common mistake that many VHDL programmers attempt to make. A simple solution to this problem is described below:

1. Allow ONLY ONE process (called the driving process) to drive the desired signal.
2. Use handshaking signals between the other processes (called request processes) and the driving process to request the driving process to send the desired data.

The typical handshaking signals are shown in Figure 4.2.3.1-1.

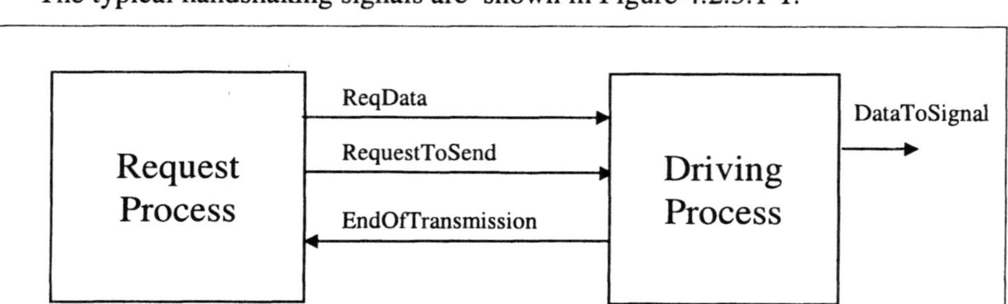

Figure 4.2.3.1-1 Handshaking Signals between a Request Process and a Driving Process

Figure 4.2.3.1-2 represents the handshaking for two processes transferring data onto a non-resolved signal. Figure 4.2.3.1-3 represents the simulation outputs of this model.

```
architecture ReqDrv_a of ReqDrv is
  constant ClkPeriod_c  : time      := 100 ns;
  signal Clk_s          : bit       := '0';
  signal Data_s         : integer;
  signal ReqData_s      : integer;
  signal REQ_s          : boolean;  -- Request To Send
  signal EOT_s          : boolean;  -- End Of Transfer

begin  -- ReqDrv_a
  ---------------------------------------------------------------------
  -- Concurrent statement for clock emulation
  ---------------------------------------------------------------------
  Clk_s  <= not Clk_s after ClkPeriod_c;

  ---------------------------------------------------------------------
  -- Process:   Request_Lbl
  -- Purpose:   Request from the driving process to send an integer
  --        :   value onto a non-resolved signal (data_s)
  ---------------------------------------------------------------------
  Request_Lbl : process
    variable Int_v : integer := 100;
  begin  -- process Request_Lbl
    wait until Clk_s'event and Clk_s = '1'; -- rising edge of clock
    Int_v := Int_v + 1;
    ReqData_s  <= Int_v;
    Req_s  <= True;
    wait on EOT_s until EOT_s;
  end process Request_Lbl;
  ---------------------------------------------------------------------
  -- Process:   Driving_Lbl
  -- Purpose:   Sends data onto Data_s and also passes data from the
  --        :   Request process onto Data_s.
  ---------------------------------------------------------------------
  Driving_Lbl : process
    variable Int2_v : integer := 1000;
    variable ThenTime_v : time;
  begin  -- process Driving_Lbl
    ThenTime_v := now;            -- remember the start time of this process
    wait until Clk_s'event and Clk_s = '1'; -- rising edge of clock
    Int2_v := Int2_v +1;
    Data_s  <=  Int2_v;
    Work_Lbl : for Worl_i in 1 to 6 loop  -- emulate work done by this process
```

```
      wait until Clk_s'event and Clk_s = '1'; -- rising edge of clock
    end loop Work_Lbl;
    -- must now detect if the request process wants
    -- this process to send data
    -- Next statement states:
    --    If the time since the last activity (or signal assignment) of signal
    --    Req_s is less than the current time minus
    --    the last time that was checked then  => send requested data
    --     (see chapter 5 for signal attributes)
    if Req_s'last_active < (now - ThenTime_v) then
      Data_s  <= ReqData_s;
      EOT_s   <= true, false after 1 ns; -- pulse for easier debugging
    end if;
  end process Driving_Lbl;
end ReqDrv_a;
```

Figure 4.2.3.1-2 Handshaking for 2 Processes Transferring Data onto a Non-Resolved Signal (ch4_dir\reqdrv_ea.vhd)

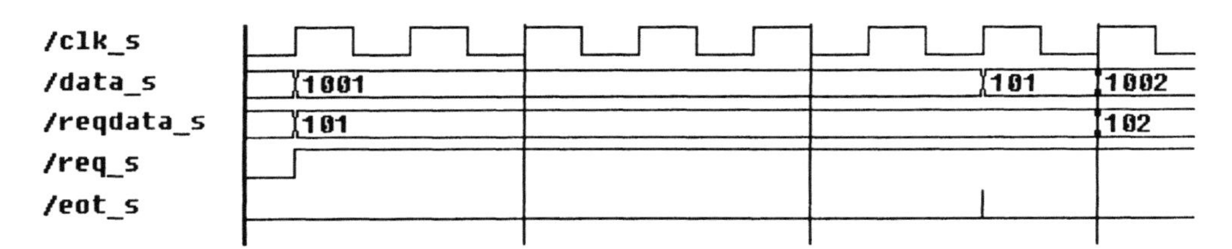

Figure 4.2.1.1-3 Simulation Outputs of Handshaking Model

4.3 PORTS

The previous **initialization recommendations** apply not only to **signals**, but also to **PORTS**, but also for different reasons. This section explains the initialization of ports and signals in hierarchical designs that incorporates components.

[1] (LRM 4.3.2) *In an interface signal declaration appearing in a port list, the default expression defines the default value(s) associated with the interface signal or its subelements. The **value**, whether implicitly or explicitly provided, is used to determine the **initial contents of drivers**, if any, **of the interface signal**.* In other words, the initial value of the driver is the initialized value of its associated port.

[1] (LRM:12.6.2) The kernel process determine two values for certain signals during any given simulation cycle. The kernel process is the process performing the actual signal assignment.

1. **The driving value**: value as sourced
2. **The effective value**: Resolved value.

These two statements are very significant because improper initializations of ports, particularly by using the default values for interfaces of type *Std_Logic*, can cause unexpected simulation results. Figure 4.3-11 demonstrates the concept of port and signal initialization, as per LRM'93. Figure 4.3-2 represents the code for this model.

At initialization, the initial value of a driver from within a kernel process is the default value of its associated signal or port for that driver. Thus, in the example shown in Figure 4.3, the initial value for the driver of signal *A_s* in *PA1_Proc* process is the default value of the signal *A_s*, or 'Z' in this case. Similarly, the initial value for the driver for port *pA2* in *PA2_Proc* process is the default value of the port pA2, or a 'U'.

The default value for the sources of port *pA1* and *pB* in the default value for those ports ('U').

Once the signals and ports are initialized, every process is executed until suspended. Events and signals (and ports) are scheduled for updated, and time is reset to zero. Simulation is then ready to start. During simulation the scheduled updates propagate as required. In this example, port pA1 is updated in one delta time as a result of the concurrent signal assignment. Signal *B_s* is associated with pA1, and is thus updated at the same time as pA1. Figure 4.3-2 provides a simulation listing for the initialization process.

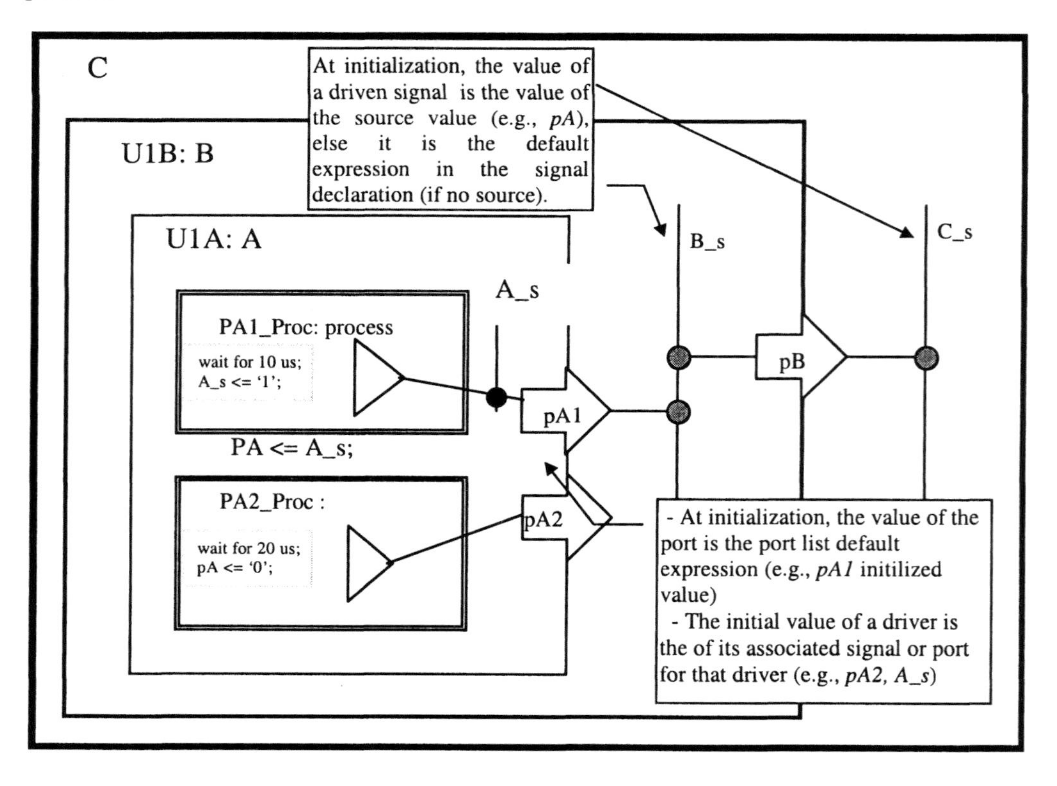

```
library IEEE;
  use IEEE.Std_Logic_1164.all;      -- 1164 package
entity A is
  port
    (pA1        : out Std_Logic;
     pA2        : out Std_Logic);
end A;

architecture A_a of A is
  signal A_s : Std_Logic := 'Z';
begin  --  A_a
  PA1_Proc : process
  begin  --  process PA1_Proc
    wait for 10 us;
    A_s <= '1';
  end process PA1_Proc;

  pA1 <= A_s;  -- Concurrent signal assignment

  PA2_Proc : process
  begin  --  process PA2_Proc
    wait for 20 us;
    pA2 <= '0';                          --   conflict on bus
  end process PA2_Proc;
end A_a;

-------------------------------------------------------------------------------
--  Definition of level B of hierachy, which included
--  component A
-------------------------------------------------------------------------------
library IEEE;
use IEEE.Std_Logic_1164.all;
entity B is
  Port
    (pB          : out Std_logic);
end B;

architecture B_a of B is
  signal B_s    : Std_Logic := 'Z';

  component A
    port
    (pA1       : out Std_Logic;
     pA2       : out Std_Logic);
  end component;
begin  --  B_a
  U1: A
    port map
      (pA1       => B_s);

  pB <= B_s;                            --  Signal to port
end B_a;
```

```
-------------------------------------------------------------------------------
--   Definition of level C of hierachy, which included
--   component B
-------------------------------------------------------------------------------
library IEEE;
use IEEE.Std_Logic_1164.all;          -- 1164 package

entity C is
end C;

architecture C_a of C is
  signal C_s      : Std_Logic := '1';         --  Initalized

  component B
    port
     (pB          : out Std_Logic);
  end component;

begin  --  C_a
  U1: B
    port map
      (pB       => C_s);                    --  Formal to Actual mapping
end C_a;
```

Figure 4.3-2 Source of Drivers During Initialization(Ch4_dir\DrvTest.vhd)

ns	delta	pa1	pa2	a_s	pb	b_s
0	+0	U	U	Z	U	U
0	+1	Z	U	Z	U	Z
0	+2	Z	U	Z	Z	Z

Figure 4.3-3 Initialization Listing during Simulation

Note that a port or signal of type Std_Logic initialized to a 'U' can lock the value of that signal to a 'U' when there are more than one driver driving that signal. This is the result of the resolution function that always yields a 'U' when any of the signal involved in the resolution function is a 'U'.

An OUT or INOUT or BUFFER port is a "SOURCE" of data. If such a port is declared, it provides a source of data to signals at the next level of hierarchy whether or not there exist a driver inside its architecture.

EXERCISES

1. Given the following entity architecture, draw a picture of the various drivers (similar to Figure 4-1). Give a name to each driver.

```
library IEEE;
  use IEEE.Std_Logic_1164.all;        -- 1164 package

entity A is
  port
    (A_p     : out Std_Logic);
end A;

architecture A_a of A is
  signal A_s : Std_Logic := 'Z';          --    begin level
begin  -- A_a

  PA1_Proc : process
  begin  -- process PA1_Proc
    wait for 10 us;
    A_s <= '1';
  end process PA1_Proc;

  PA2_Proc : process
  begin  -- process PA2_Proc
    wait for 20 us;
    A_s <= '0';
  end process PA2_Proc;

  PA3_Proc : process
  begin  -- process PA3_Proc
    A_p <= A_s;
    wait on A_s;                          -- sensitive to change on A_s
  end process PA3_Proc;
end A_a;
```

2. For the above architecture, what are the values of the drivers at time 0 ns, prior to simulation? Provide name of driver and value. What is the value of Signal A_s and port A_P.

ELEMENT	VALUE @ 0 ns	VALUE @ 1 delta
A_p		
A_s		
Driver ___		
Driver ___		
Driver ___		
Driver ___		
Driver ___		
Driver ___		
Driver ___		

3. For the above architecture, what is the final value of signal *A_s* and port *A_p* at 10 us, and 20 us?

ELEMENT	VALUE
A_p	@ 10 us
A_s	@ 10 us
A_p	@ 20 us
A_s	@ 20 us
A_p	@ 30 us
A_s	@ 30 us

4. Simulate the design, and verify your results.

5. VHDL TIMING

This chapter explains how a VHDL simulator interprets timing descriptions. Timing is one of the concepts that makes VHDL different from other programming languages like *C, Pascal*, or *Ada*. This chapter introduces **signal attributes** that provide a wealth of timing and value information about signals and ports. The **wait** statement is explained because of its importance in providing synchronization in processes. The simulation engine from initialization through simulation demonstrates the effects of the **wait** statement, **drivers** on signals, and the concepts of **delta times**. Delta times are often an area of errors and confusion. Guidelines are provided in the design of architectures with delta times. **Inertial** and **transport** delays are presented because of their impact in the timing projection of values onto signals. The significance of inertial delay in synthesis is also explained.

5.1 SIGNAL ATTRIBUTES

Signal attributes are often used as synchronization mechanisms for concurrent statements and as modeling parameters for the creation of models. Three VHDL terms are defined because of their relationship to attributes and timing.

1. **EVENT:** *[1] An event is a change in the current value of a signal, which occurs when the signal is updated with its effective value.* Thus if signal S changes from '1' to a '0', or from an 'H' to a '1', it is considered an event.

2. **ACTIVE:** *[1] A signal is said to be active when it acquires an updated value during a simulation cycle regardless of whether the updated value is the same or different from the previous value.* Thus, if signal S is updated form a '1' to a '1', the signal is considered

"active" at the update time. Every time a signal becomes active, it is considered a new "transaction".

Remember! Every time you get a paycheck it is considered an **ACTIVITY** or a **TRANSACTION**. However, if your check is **DIFFERENT** than the previous paycheck, you may think of it as an **EVENT** (particularly if it is 10 X your normal paycheck).

3. **IMPLICIT SIGNAL:** An implicit signal is a signal **not declared** in the VHDL design, but **implicitly** declared as the result of the following attributes: **'stable, 'quiet, 'delayed, 'transaction**. If any of those attributes are used in a design, then an implicit signal is automatically generated by the simulator.

   It is an error if implicit signals are read from within a subprogram (see section 7.2.4). This is because all signals (implicit or explicit) are created at elaboration time, whereas subprograms local objects (e.g., variables) are created at runtime. Since a subprogram may be called by various processes with different actual signal parameters, it is not possible to create the implicit signals read from within subprograms during elaboration.

Table 5.1 provides a summary of the VHDL signal attributes, with an explanation about their operation and typical usage.

Table 5.1 Summary of the VHDL Signal Attributes

S'event	**Function** returning a **Boolean** that identifies if signal S has a new value assigned onto this signal (i.e., value is different that last value). **if** Clk'event **then** ... -- if Clk just changed in value then ... **wait until** Clk'event **and** Clk = '1'; -- rising edge of clock
S'active	**Function** returning a **Boolean** that identifies if signal S had a new assignment made onto it (whether the value of the assignment is the SAME or DIFFERENT. **if** Data'active **then** ... -- New assignment of Data -- ☺ **wait on** Data'active; -- 💣 **wait until** Data'active; -- ☺
S'transaction	**Implicit signal** of type **bit** created for signal S when *S'transaction* is used in the code. This implicit signal is NOT declared since it is implicitly defined. *[1] This signal toggles in value (between '0' and '1') when signal S had a new assignment made onto it* (whether the value of the assignment is the SAME or DIFFERENT. The user should NOT rely on its VALUE. -- Process resumes when ReceiveData gets a new signal -- assignment of same or different value. **wait on** ReceivedData'transaction;

S'delayed(T)	**Implicit signal** of the same **base type as S**. It represents the value of signal S delayed by a time Tn. Thus, *[1] the value of S'**delayed**(T) at time Tn is always equal to the value of S at time Tn - T.* For example, the value of S'delayed(5 ns) at time 1000 ns is the value of S at time 995 ns. If time is omitted, it defaults to 0 ns. **Subtype** BV2_Typ is Bit_Vector(1 to 2); [Data at last delta time & Data at current time] .... **wait on** Data'transaction; **case** BV2_Typ'(Data'Delayed & Data) **is** **when** "X0" => **... --** from X to 0 transition **when** "10:" => ... -- from 1 to 0 transition **when others => ... --** **end case;**
S'stable(T)	**Implicit signal** of **Boolean** type. *[1] This implicit signal has the value TRUE when an event (change in value) has NOT occurred on signal S for T time units, and the value FALSE otherwise.* If time is omitted, it defaults to 0 ns. **if** Data'stable(40 ns) **then** -- met set up time
S'quiet(T)	**Implicit signal** of **Boolean** type. *[1] This implicit signal has the value TRUE when the signal has been quiet* (i.e., no activity or signal assignment*) for T time units, and the value FALSE otherwise.* If time is omitted, it defaults to 0 ns. **if** Data'quiet(40 ns) **then** -- Really quiet, not even an assignment of -- the same value during the last T time units
S'last_event	**Function** *returning the [1] amount of **time** that has elapsed since the last event* (change in value*) occurred on signal S. If there was no previous event, it returns Time'high* (The maximum value for time). **variable** : TsinceLastEvent : time; -- .. TsinceLastEvent := Data'last_event;
S'last_active	**Function** returning the *[1] amount of **time** that has elapsed since the last activity* (assignment*) occurred on signal S. If there was no previous event, it returns Time'high.* **variable** : TsinceLastEvent : time; -- .. TsinceLastEvent := Data'last_active;

<table>
<tr><td>S'last_value</td><td>Function [1] returning the previous value of S, immediately before the last change of S. The return type is the base type of S.

```
Subtype BV2_Typ is bit_Vector(1 to 2);
....
wait on Data'transaction;
  case BV2_Typ'(Data'last_value & Data) is  -- Data @ last value
    when "X0" => ... -- from X to 0 transition
    when "10" => ... -- from 1 to 0 transition
    when others => ... --
 end case;
```

Data at last change &
Data at current time</td></tr>
</table>

In synthesis, the only allowed signal attribute is *'event* used in the definition of clocked processes. All the other attributes are not allowed. However, signal attributes are used in the definition of testbenches.

Attributes that represent a signal can be used as signals, whereas attributes that return a function can only use the returned value. The following examples demonstrates this concept:

```
if Clk'event then                    -- 'event is a function          ☺
                                     -- of type Boolean

wait until Clk'event and Clk = '1';  -- 'event is a function          ☺
                                     -- of type Boolean

wait on Clk'event;                   -- 'event is NOT a signal        💣

wait on Data'transaction             -- 'transaction is a signal      ☺

if Data'transaction then             -- 'transaction is not Boolean   💣

if Data'transaction'event then       -- 'event is a function          ☺
                                     -- of type Boolean
                                     -- (equivalent to Data'active)
```

M 👍👍 If possible, avoid the use of implicit signals defined through the use of the attributes **'transaction, 'delayed, 'stable, 'quiet.** Attempt to use timing attributes with function calls instead.

Rationale: *Signals are expensive in simulation.*

If a composite (array or record) signal is resolved at the composite level, then a change to one element causes an **activity** on the whole composite. However, if a composite signal is non-resolved, then a change to one element causes an **activity** on the element of the composite, but not on the whole composite. This issue is related to a concept called atomicity, which states that any **signal associated with a resolution function is atomic**. In addition, any scalar signal, with or without a resolution function, is also atomic. A composite signal is a collection of atomic signals. Note that *Std_Logic_Vector* type is defined as an unconstrained array of *Std_Logic*, and thus each subelement is resolved independently, so they act as separate signals.

Figure 5.1-1 represents a model that makes use of timing attributes to verify positive setup and hold time of a data signal with respect to a clock signal. Figure 5.1-2 represents the simulation outputs using Model Technology ModelSim tools.

```
architecture SetHold_a of SetHold is
  signal        D_s              : Std_Logic := '0';
  signal        Clk_s            : Std_Logic := '0';

begin -- SetHold_a
-------------------------------------------------------------------------------
-- Process:   ClockGen_Lbl
-- Purpose:   Generates clock pulses
-------------------------------------------------------------------------------
  ClockGen_Lbl : process
  begin -- process ClockGen_Lbl
    Clk_s <= not Clk_s;
    wait for 50 ns;
  end process ClockGen_Lbl;

-------------------------------------------------------------------------------
-- Process:   MakeD_Change_Lbl
-- Purpose:   Cause D_s to change in value to test setup and Hold
-------------------------------------------------------------------------------
  MakeD_Change_Lbl : process

  begin -- process MakeD_Change_Lbl
    D_s  <= '0' after   1 ns,  -- assignment of different values at
            '1' after  80 ns,  -- different times.
            '0' after 120 ns,
            '1' after 195 ns,
            '0' after 202 ns;
    wait for 300 ns;
  end process MakeD_Change_Lbl;

-------------------------------------------------------------------------------
-- Process:   SetHoldCheck_Lbl
-- Purpose:   Provides the setup and hold checks
-- Inputs:    D_s and Clk_s
--            Process is sensitive to a change on D_s or Clk_s.
--            Thus, process "fires" on an event on either D_s or Clk_s.
--            Once fired, process executes until the end, then
--            it wait for another event on either of those two signals.
-- Outputs:   Annunciation of error on transcript window.
-------------------------------------------------------------------------------
  SetHoldCheck_Lbl : process (D_s, Clk_s)
    use Std.TextIO.all;                     --   localized to this process only
    use Std.TextIO;      -- allows use of package name without library name
    constant SetUpTime_c   :  time := 10 ns;
    constant HoldTime_c    :  time := 5 ns;
    variable Violation_v   :  boolean;
    variable ErrorSet_v    :  string(1 to 7);
    variable ErrorHld_v    :  string(1 to 7);
    variable IsSetError_v  :  boolean := false;
    variable IsHoldError_v :  boolean := false;
    variable OutLine       :  TextIO.Line;
```

```
  begin
    --                            ->|   |<- hold
    --         --+    <-- setup ---->+-----------------------
    --   CLK     |___________________|
    --
    --         -----+                  +--------------
    --   D_S        |__________________|
    --
    --
    --   Check rising edge of clock
    --   This model demonstrates the use of attributes, and checks
    --   that the clock went from a '0' to '1'.  If the clock is
    --   guarantee to always transition between '1' and '0', then
    --   the check of "Clk_s'last_value ='0'" is not necessary.
    if Clk_s'event and Clk_s ='1' and Clk_s'last_value = '0' then
      --   Check setup time
      Violation_v := D_s'last_event < SetUpTime_c;
      if Violation_v then
        ErrorSet_v := "  Setup";
        IsSetError_v := true;
      end if;
    end if;


    -- Check Hold violation
    -- When data transitions, check if clock to a '1'
    if D_s'event and Clk_s'last_value = '0' then
      Violation_v := Clk_s'last_event < holdTime_c;
        if violation_v then
          ErrorHld_v := "   Hold";
          IsHoldError_v := true;
        end if;
    end if;

    --   Error reporting, Setup
    if IsSetError_v then
      IsSetError_v := false;                --   reset flag
      TextIO.Write(Outline, now);  -- now is a predefined function returning time
      TextIO.Write(OutLine, ErrorSet_v);            --   setup error
      TextIO.Write(OutLine, string'(" time violation on "));
      TextIO.Write(OutLine, string'(" signal D_s by "));
      TextIO.Write(OutLine, SetUpTime_c - D_s'Last_event);
      TextIO.Writeline(Output, Outline);
    end if;

    --   Error reporting, Hold
    if IsHoldError_v then
      IsHoldError_v := false;               --   reset flag
      TextIO.Write(Outline, now);
      TextIO.Write(OutLine, ErrorHld_v);            --   hold error
      TextIO.Write(OutLine, string'(" time violation on "));
      TextIO.Write(OutLine, string'(" signal D_s by "));
      TextIO.Write(OutLine, HoldTime_c - Clk_s'Last_event);
      TextIO.Writeline(Output, Outline);
    end if;
  end process SetHoldCheck_Lbl;
end SetHold_a;
```

'last_value needed if signal Clk is of type Std_Logic and Clk could transition from 'H' to '1' or from 'Z' to '1'

-- Could have also used
Violation_v := not D_s'**stable**(SetUpTime_c);

No check done here if D_s changes from 'H' to '1' or from 'L' to a '0'. Could add :
if (D_s = '1' or D_s = '0') **and** -- current value
(D_s'last_value = '1' or -- previous value
D_s'last_value = '0') **then** -- do the check

Figure 5.1-1 Setup and Hold Model (ch5_dir\sethold.vhd)

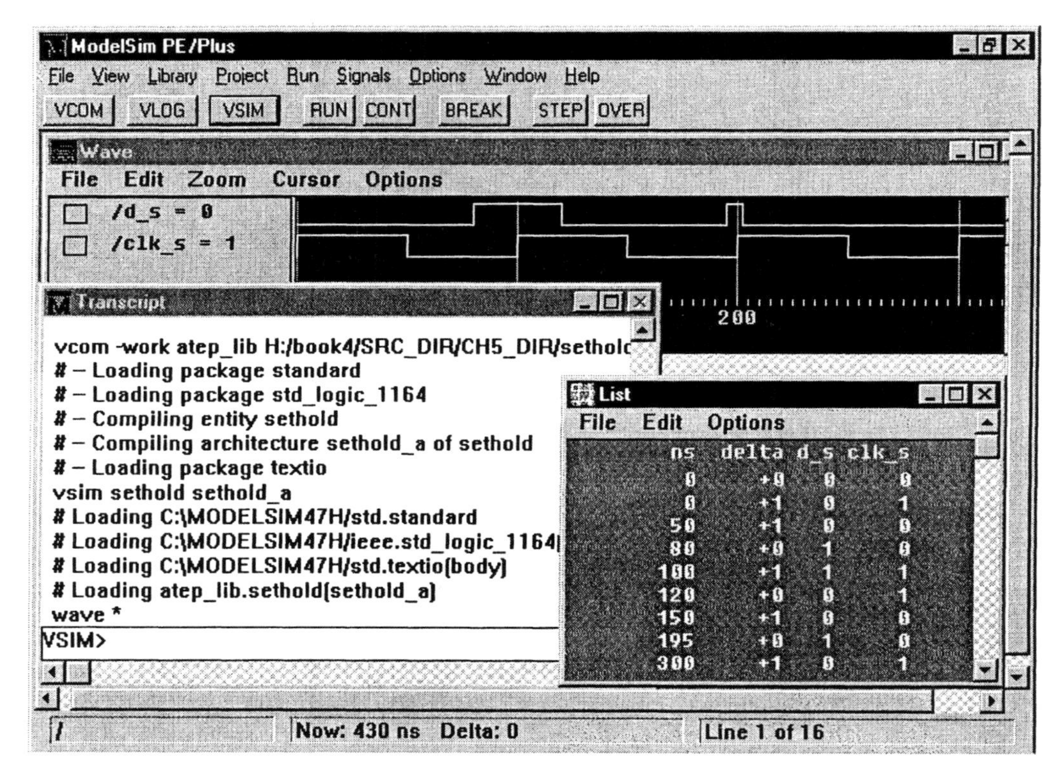

Figure 5.1-2 Setup and Hold Model Simulation Outputs

5.2 THE "WAIT" STATEMENT

The **wait** statement is a sequential statement in a process or a procedure (sequential or concurrent (see chapter 6)) and *[1] (LRM 8.1) causes the suspension of a process statement or a procedure.* The syntax is as follows:

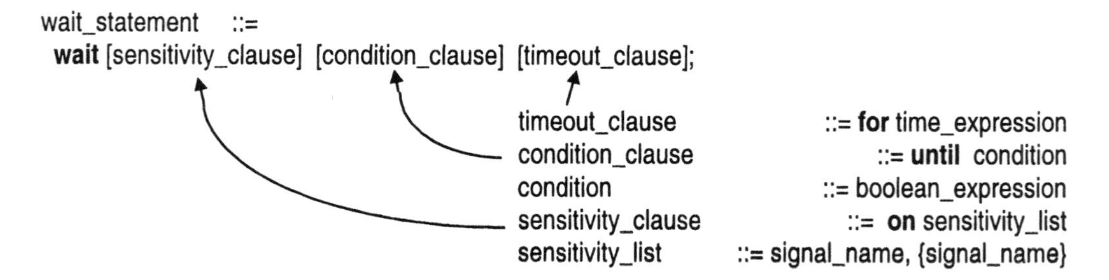

The **wait** statement is the synchronization scheme for all concurrent statements. When a process encounters a wait statement, whether it is an explicit or implicit wait (e.g., process with a sensitivity clause), the process gets suspended until the condition to wake up the process occurs. Each of the variations of the wait statement, and its implications is described in the sections below.

5.2.1 Delta Time

One of the requirements of simulation is **order independence** in the execution of each of the concurrent statements (e.g., processes, concurrent signal assignments). Another requirement is that within any simulation time the **integrity on any signal must be maintained**; thus, a signal cannot change within any simulation time. For example, if a swap of values occurs between two signals (in same or separate processes) the value of the assignment cannot occur within the same simulation time. Otherwise, incorrect results may occur.

To achieve these goals, the simulator tags the signals assigned with zero delays and schedules them for an update when ALL the current processes have completed execution within the current time period. This time is called a **delta time**. A **delta time is an infinitesimal amount of time greater than zero, but yet represents an effective zero time.** The effective value of a delta time is zero when added to a discrete amount of time. Thus if a signal assignment is made at time "105 ns + 3 delta time" with a discrete delay of 1 ns (e.g., SomeSignal <= Value after 1 ns;), the new value of the signal is posted at 105 ns + 1 ns + 0 delta time. This is because 3 * 1 delta time = 0 ns. When the simulator has no more work in the current time, simulation time advances to the next time a task must be performed (e.g., signal update, process scheduling).

The concept of delta time is necessary model to emulate the concurrency in real hardware.
It is re-examined in section 5.4, the simulation engine. Table 5.2.1 demonstrates the need for this requirement by simulating a swap of two values that occur in zero time.

Table 5.2.1 Signal Integrity Within a Simulation Time

SWAP WITH NO DELTA TIME	**SWAP WITH DELTA TIME**
A <= B; **B <= A;** Signal assignments with zero delay. Assume ***A*** has a value of '0' Assume ***B*** has a value of '1' If assignment is immediate then ***A*** would get a '1' ***B*** would get the new value of ***A***, or a '1' **INCORRECT SWAP !!!**	**A <= B;** **B <= A;** Signal assignments with zero delay. Assume ***A*** has a value of '0' Assume ***B*** has a value of '1' ***A*** retains value of '0', but is scheduled to have current value of ***B*** ('1') in one delta time from current time. ***B*** retains value of '1', but is scheduled to have current value of ***A*** ('0') in one delta time from current time. When time advances to now + 1 delta, ***A*** signal is updated to the scheduled value ('1') . Also, ***B*** signal is updated to the scheduled value ('0'). **CORRECT SWAP !!!**

Figure 5.2.1-1 demonstrates the code for the swap of values between two signals updated in clocked processes . Figure 5.2.1-2 represents the synthesis RTL view of this swap

model. In synthesis, any signal update within a clocked process represents a register. Figure 5.2.1-3 represents a simulation of the swap model. The *LIST* window in the simulation shows that at time 10 ns + 0 delta, *qA_r* has a value of '0', and *qB_r* has a value of '1'. At that time, the swap is scheduled to occur, but the values are not updated until 10 ns + 1 delta time. The output ports (A, B) are updated one delta time later at 10 ns + 2 delta times as a result of the concurrent signal assignments.

```
entity Swap is
  port(A   : out   Bit;
       B   : out   Bit;
       Clk : in    Bit);
end Swap;

architecture Swap_a of Swap is
  signal QA_r  : Bit := '0';
  signal QB_r  : Bit := '1';
begin
  A <= QA_r;
  B <= QB_r;

  QA_Proc : process
  begin
    wait until Clk = '1';
    QA_r <= QB_r;
  end process QA_Proc;

  QB_Proc : process
  begin
    wait until Clk = '1';
    QB_r <= QA_r;
  end process QB_Proc;
end Swap_a;
```

Internal feedback signals because *out* ports cannot be read. Initialization used for simulation, and is ignored in synthesis.

Concurrent signal assignments to connect feedback signals to outputs

Execution of concurrent statements requires order independence. Otherwise, concurrency cannot be properly simulated.

Figure 5.2.1-1 Swap of Values Between Two Signals (ch5_dir\swap.vhd)

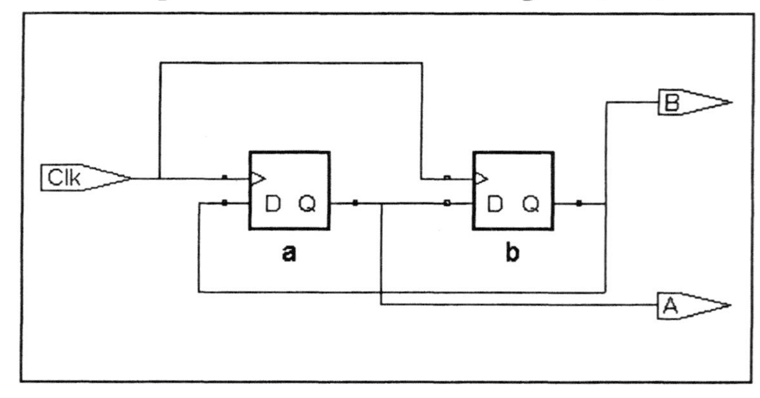

Figure 5.2.1-2 Synthesis RTL View of Swap Model *(RTL View Generated with Synplify from Synplicity)*

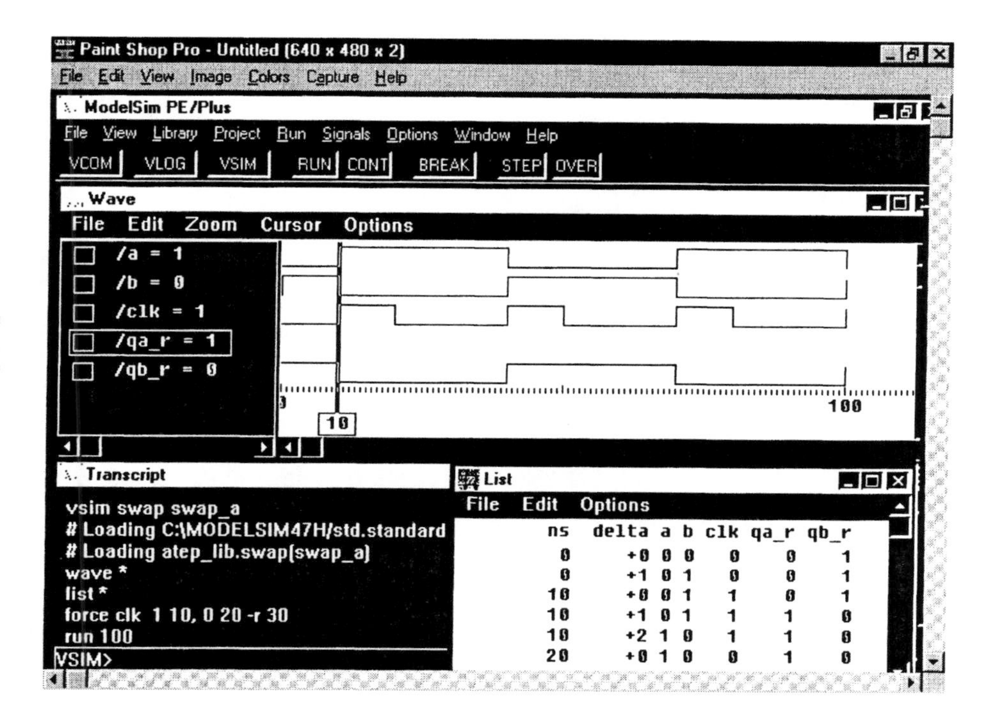

Figure 5.2.1-3 Simulation of Swap Model

5.2.2 wait on sensitivity_list

The ***wait on*** *sensitivity_list* suspends a process until the occurrence of another event on one of the signals in the sensitivity list. Thus, the process with a *wait on S1, S2;* will be suspended until an event (i.e., a change in value) occurs on either *S1* or *S2*. If it is desired to wait until a new transaction (or update) of a signal, one could use the **wait on *S'transaction*** statement,

5.2.3 wait until condition

The ***wait until*** *condition* suspends the process until the condition following the *until* is satisfied. If there is NO sensitivity list in the ***wait until*** clause, then there is an **IMPLICIT** sensitivity list on **ALL** the **SIGNALS** expressed in the condition clause. Thus, the statement:

wait until Clk = '1';

causes the process to have an implicit sensitivity to CLK. The above statement is equivalent to:

wait on Clk **until** Clk = '1';

The process will then be suspended until an event occurs on Clk. If an event occurs, the

process checks that *CLK = '1'*, and will resume if the condition matches, else it will be suspended again.
The next two statements are equivalent because both statement have an implied ***wait on*** *CLK,* which represents a *Clk'event.*

```
wait until Clk'event and Clk = '1';  -- statement #1
wait until Clk = '1';                -- statement #2
```

Thus, the *Clk'event* in statement #1 is redundant. It also slows down the simulator to a minor extent because when an event occurs on signal CLK because the simulator must perform the ***and*** function of *Clk'event* with *Clk = '1'.* This ***and*** function is not necessary in statement #2. However, statement #1 is more explicit than statement #2 and represents the "classical" way of defining a rising edge of a clock for synthesis. Some synthesis tools require the clock form with the *'event.* This book uses the classical definition of a rising edge of a clock. The reader may choose statement #2 style without any ill effects, particularly if the synthesizer tool can accept this type of clock template.

The following examples demonstrates the use of the ***wait until*** statement with signal and variables in the Boolean expression:

```
...
 signal Data_s   : Bit_Vector(1 to 4);
 signal Enable_s : Boolean;
Xyz_Lbl: process
 variable MyVar_v  : Integer;
begin
 wait until Data_s = "1011" and Enable_s and MyVar_v < 30;
```

This process **wakes up when an event occurs on Data_s or Enable_s**. After it wakes up, the process checks the condition (*Data_s = "1011"* ***and*** *Enable_s* ***and*** *My_Var < 30*). If this condition is true, the process resumes, else it gets suspended again.

Note that the following statement is always suspended because there is not an implicit signal to wake it up.

```
wait until MyVar_v  < 30;   -- MyVar_v  is a variable
```

Figure 5.2.3 demonstrates the incorrect use of a ***wait until*** clause that suspends a process indefinitely.

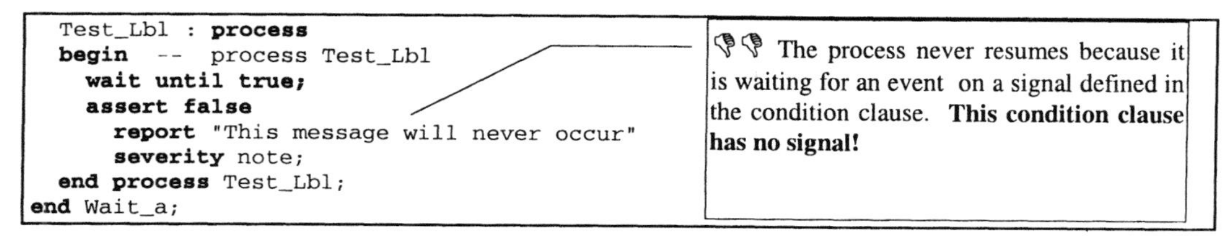

```
  Test_Lbl : process
  begin  --  process Test_Lbl
    wait until true;
    assert false
      report "This message will never occur"
      severity note;
  end process Test_Lbl;
end Wait_a;
```

The process never resumes because it is waiting for an event on a signal defined in the condition clause. **This condition clause has no signal!**

Figure 5.2.3 Incorrect Use of a "wait until" which Suspends a Process Indefinitely (ch5_dir\wait1.vhd)

Consider the following variation to the ***wait until*** statement:

wait on Enable_s **until** Data_s = "1011" **and** Enable_s **and** MyVar_v < 30;

This statement with the sensitivity clause will be suspended until an event occurs on Enable_s ONLY (thus an event on Data_s is ignored). Once the process awakes, it will then check that the condition after the ***until*** clause is true. If it is, the process will resume, else, it will be suspended. This is VERY DIFFERENT than the previous statement (***wait until*** *Data_s = "1011"* ***and*** *Enable_s* ***and*** *My_Var < 30;*)

Once the sensitivity clause is explicitly defined, the implicit sensitivity signals in the ***wait until*** clause are lost.

5.2.4 wait for time_expression

The ***wait for*** *time_expression* statement suspends a process until the elapsed time in the time expression. Thus, the statement ***wait for*** *50 ns;* suspends a process for 50 ns. The process resumes unconditionally afterward. The statement

wait until Data_s = "1011" **and** Enable_s **and** My_Var < 30 **for** 100 ns;

suspends the process until either of the following condition occurs:

1. Event occurs on either Data_s or Enable_s signals and the condition *Data_s = "1011" and Enable_s and My_Var < 30* is true

2. 100 ns has elapsed.

Thus the process resumes if either condition 1 or condition 2 discussed above has occurred.

The ***wait for*** ***0 ns;*** statement causes the process to be suspended for one delta time, thus allowing signals assignments made in the current process to propagate. See Section 5.5 for guidelines on the use of ***wait for*** *0 ns* clause.

Table 5.2.4 provides a summary of sample ***wait*** expressions with their implied sensitivity list, conditions tested upon activation, and timeout time once activated.

Table 5.2.4 Summary of Sample Wait Expressions

STATEMENT	SENSITIVITY LIST	CONDITION	TIMEOUT
wait;	none**	true	time'**high**
wait on S1, S2;	S1, S2	true	time'**high**
wait until Clk'event;	Clk	Clk'event	time'**high**
wait until Clk'event **and** Clk = '1';	Clk	Clk'event **and** Clk = '1'	time'**high**
wait on S1 **until** CLK = '1';	S1	Clk = '1'	time'**high**
wait on S2 **for** 100 ns;	S2	true	100 ns
wait on S1, S2 **until** Clk = '1' **for** 10 us;	S1, S2	Clk = '1'	10 us
wait on CLK **for** VAR * 1 ns;	Clk	true	Var * 1 ns
wait on S1 **until** VAR = 10;	S1	VAR = 10	time'**high**
wait until VAR = 10;	none**	VAR = 10	time'**high**
wait on S1'transaction;	S1'transaction	true	time'**high**
wait for 100 ns;	none	true	100 ns
wait on S1'event; -- 💣	**ERROR** (*S1'**event** is not a signal)*	true	time'**high**
wait on S1'active; -- 💣	**ERROR** (*S1'**active** is not a signal)*	true	time'**high**
wait on S1'delayed(100 ns)	S1'delayed (100 ns)	true	time'**high**
wait on S1'delayed(100 ns)'transaction	S1'delayed (100 ns) 'transaction	true	time'**high**

** Process never wakes up because there is no signal in sensitivity list

M 👍👍 In a process, avoid loops or branches when the loop or the branch is not broken with a **wait** statement. For example (also stored in file "ch5_dir\badloop.vhd):

```
Bad_Lbl: process
  if Reset_s = '1' then
    A_s <= '0';
  else
    BadLoop_Lbl: while (Enable_s = '1') loop
      wait until clk = '1';
    end loop BadLoop_Lbl;
  end if;
-- wait until clk = '1';
end process Bad_Lbl;
```

If Reset = '1', A_s <= '0', Process gets restarted immediately without suspension. Time never adances.

If Reset = '0' and Enable_s = '0' then process gets restarted immediately without suspension. Time never adances.

The **FIX**. A wait statement is needed here (Shown here commented out to emphasize problem)

Rationale: *The code tends to compile correctly, but will hang the simulator if the path without the **wait** statement is taken.*

5.3 SIMULATION ENGINE

The simulation engine from initialization through simulation is first examined through the example shown in Figure 5.3-1. This demonstrates the effects of drivers on signals, and the concepts of delta time.

```
...
architecture Timing_a of Timing is
  signal S1   : Std_Logic := '0';
  signal S2   : Std_Logic := '0';
begin
  P1 : process
  begin
    wait for 10000 ns;
    S2 <= '1';
    S1 <= '1';
    wait;
  end process P1;

  P2: process
  begin
    S1 <= not S1 after 1 ns;
  wait on S2;
  end process P2;
end Timing_a;
```

Two drivers created:
1. P1_S1 for signal S1
2. P1_S2 for signal S2

One driver created:
P2_S1 for signal S1

Figure 5.3-1 Effects of Drivers on Signals (ch5_dir\timing.vhd

Since there are two separate signal assignments in process *P1*, two drivers are created: One driver for signal assignment *S2* named *P1_S2*, and one driver for signal assignment *S1* named *P1_S1*. Similarly, since process *P2* has one signal assignment, driver *P2_S1* is created. **At elaboration, those drivers take the value of the initial value of the signals they are connected to**, or '0' in this case since *S1* and *S2* are initialized to '0'.

At initialization, every process is executed until it is suspended. Specifically,

1. Process *P1* is suspended IMMEDIATELY upon entry because of the ***wait for 10000 ns*** statement. This process is thus scheduled to be resumed at time 10 us. This scheduling is demonstrated in Figure 5.3-3 in the line labeled "NEXT EVENT TIME QUEUE", or the **event queue**.

2. Process *P2* is also executed at initialization, and schedules an assignment of the "not of the current value of *S1*" onto *S1* after 1 ns. This assignment is demonstrated in the Figure under "signal driver *P2_S1*" or **signal driver queue** that shows the driving values of the signal at various time units.

After initialization, **simulation is started** and time is **reset to 0 ns**. At the current simulation time, the values of the drivers for each of the resolved signals are resolved (i.e., evaluated) to determine the signal value. At time 0 ns, *P1_S1* drives a '0', and *P2_S1* drives a '0', thus the signal value is resolved to a '0'. Since no process is scheduled to be executed at time 0, time progress to the NEXT time an activity is scheduled. This time is 1 ns.

At time 1 ns, driver *P2_S1* takes on the assigned value stored in the signal driver queue (*P2_S1* becomes a '1'). The drivers for signal *S1* resolve to an 'X' because *P1_S1* drives a '0' while *P2_S1* drives a '1'. Driver *P1_S2* is the sole driver with a value of '0'.

Time then progresses to the **next activity in the event queue**, which is 10 us. Process P1 resumes because of the expiration of the *wait for 10000 ns*, and assigns a '1' to driver *P2_S2* after 0 ns + 1 delta time, and assigns a '1' to *P2_S1* after 0 ns + 1 delta time. The one delta time is necessary because the simulator needs to retain the current value of the drivers (to maintain process order independence). If the values of those drivers were to change instantly, then the process that was executed first by the simulation engine will modify the value of its drivers and signals. Other processes that depend on the values of the modified signals will see a different value than if the processing order were different. This creates a great hazard since the simulation engine must maintain order independence. The delta time concept solves this problem because delta time is another time, beyond the current time.

At time 10 us + 1 delta time, *S2* changes value, thus creating an event on *S2*. Since process *P2* is sensitive to *S2*, process *P2* executes and assigns on the *P2_S1* driver a value of '0' at time 10 us + 1 ns. Note that even though current time is 10 us + 1 delta, an assignment of a signal of at discrete time delay wipes out all the delta times from the equation since delta time have a net value of 0. Thus:

10000 ns + 1 delta + 1 ns = 10,001 ns (with NO delta time).

Figure 5.3-2 represents the simulation "list" file for this model.

```
      ns   delta s1 s2
       0      +0  0  0
       1      +0  X  0
   10000      +1  1  1
   10001      +0  X  1
```

Figure 5.3-2 Simulation "List" File for Timing Model

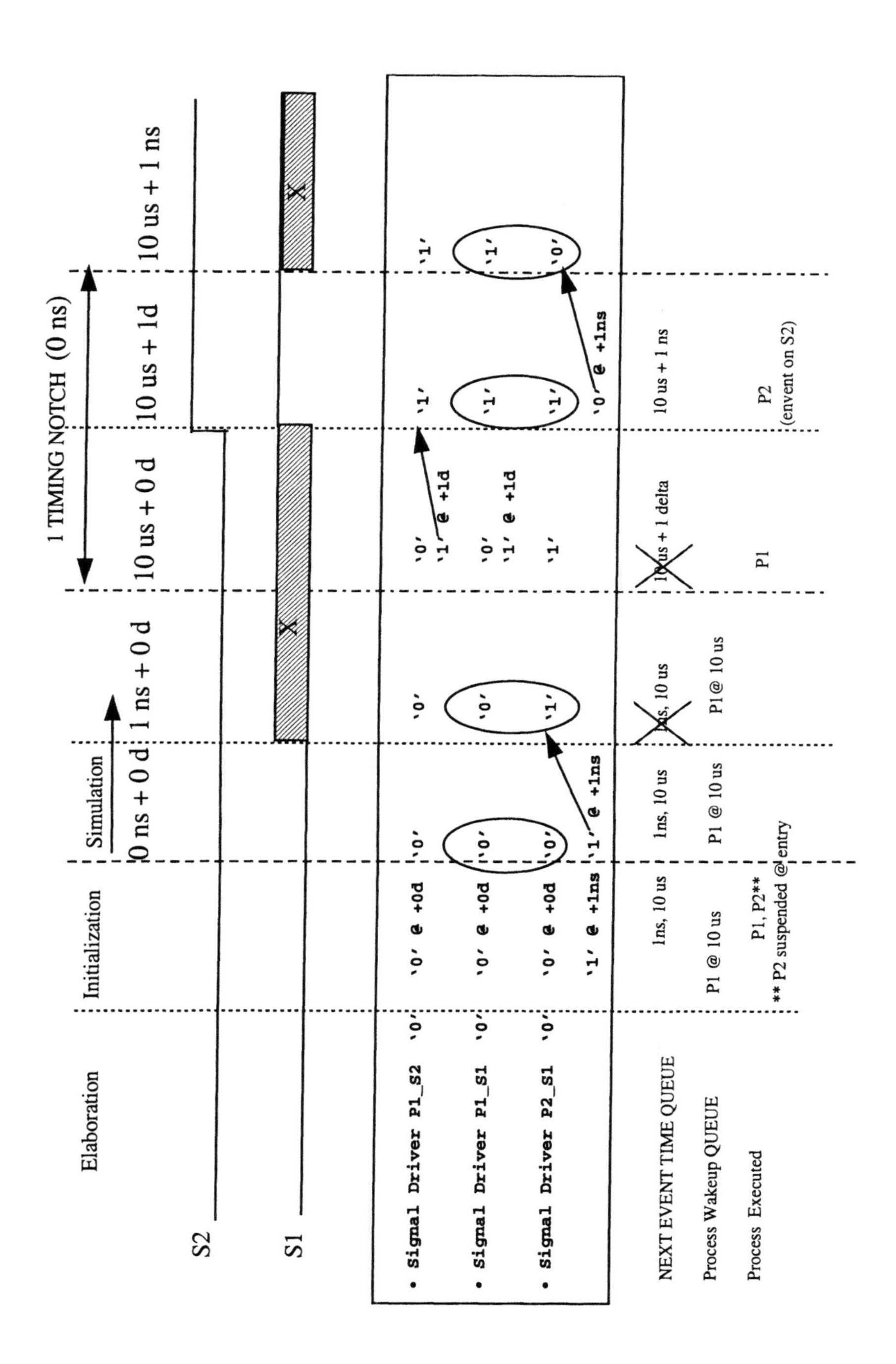

Figure 5.3-3 Timing for Architecture "Timing_a" of Entity "Timing"

5.4 MODELING WITH DELTA TIME DELAYS

Consider the modeling of circuit shown in Figure 5.4-1, assuming an ideal flip-flop with 0 ns delay. Assume that in this circuit *Din2* is synchronous, and a change in signal *Din2* can occurs at the same time as *Q*.

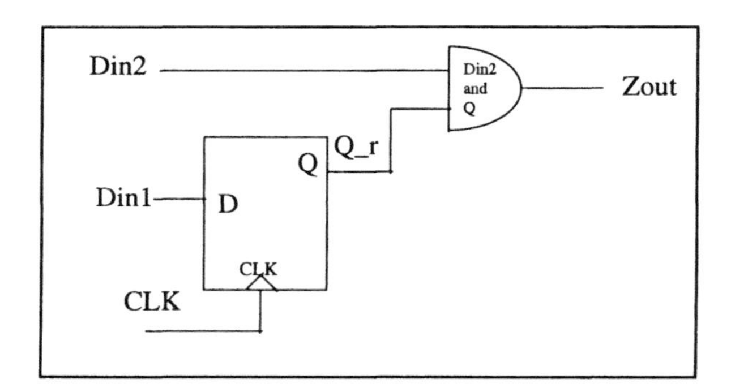

Figure 5.4-1 Modeling of Flip-Flop Followed by and "And" Gate Function

Four modeling approaches are explored in this chapter because they reflect different modeling styles.

5.4.1 Wait for 0 ns Method

This style uses a ***wait for** 0 ns;* statement to allow a signal to propagate or be updated onto the timing queue. This approach is demonstrated in Figure 5.4.1.

> M 👎 👎 Avoid the use of ***wait for** 0 ns;* statement to solve signal propagation through delta times
>
> ***Rationale:*** *1)"**Wait for** 0 ns;" is not synthesizable. 2) Many signals in an architecture change in value after several delta times, thus the number of "**wait for** 0 ns;" statements must be tuned to the number of delta times. This practice is difficult to control and maintain.*

```
architecture Wait0_a of Wait0a is
  signal      Din1_s   : bit := '0'; -- Input 1
  signal      Din2_s   : bit := '0'; -- Input 2
  signal      Zout1_s  : bit := '0'; -- Output for option 1
  signal      Clk_s    : bit := '0'; -- Clock
  signal      Q_r     : bit := '0'; -- Q1 Flip-flop, option 1
```

```
begin  -- Wait0_a
  ------------------------------------------------------------------------
  --  Process:   Circuit_Lbl
  --  Purpose:   Demonstration of use of wait for 0 ns
  --             Circuit defined in a single process.
  --             ** AVOID THIS APPROACH **
  ------------------------------------------------------------------------
  Circuit1_Lbl : process
  begin  -- process Circuit1_Lbl
    wait until Clk_s'event and Clk_s = '1';  --  rising edge of Clk_s
    Q_r <= Din1_s;                             --  default time = 0 ns
    wait for 0 ns;                             --  Wait for Q to propagate
    Zout1_s <= Q_r and Din2_s;
  end process Circuit1_Lbl;
...
end Wait0_a;
```

Figure 5.4.1 Wait for 0 ns Model (ch5_dir\wait0a.vhd)

5.4.2 Concurrent Statements Method

This method uses multiple concurrent statements sensitive to signal events, thus avoiding the need for *wait for 0 ns;* statements. Figure 5.4.2 demonstrates this concept for the design shown in Figure 5.4-1. This model is synthesizable.

```
architecture Wait0b_a of Wait0b is
  signal        Din1_s   : bit := '0'; -- Input 1
  signal        Din2_s   : bit := '0'; -- Input 2
  signal        Zout1_s  : bit := '0'; -- Output for option 1
  signal        Clk_s    : bit := '0'; -- Clock
  signal        Q_r     : bit := '0'; -- Q1 Flip-flop, option 1

begin  -- Wait0b_a
--------------------------------------------------------------------------
--  2 separate processes to accomplish same thing.
--  ** GOOD APPROACH FOR SYNTHESIS **
--------------------------------------------------------------------------
  Circuit2a_Proc: process
  begin  -- process Circuit3a_Lbl
    wait until Clk_s'event and Clk_s = '1';  --  rising edge of Clk_s
    Q_r <= Din1_s;                             --  default time = 0 ns
  end process Circuit2a_Proc;

  -- Process is sensitive to  change on Q_r and Din2_s
  Circuit1b_Proc: process (Q_r, Din2_s)
  begin  -- process Circuit3b_Lbl
    Zout1_s <= Q_r and Din2_s;
  end process Circuit1b_Proc;
```

Figure 5.4.2 Concurrent Statements Method (ch5_dir\wait0b.vhd)

5.4.3 Use of Variables Method

Variables must be used with caution because they may imply latches and registers (see chapter 13). However, variables are very useful in behavioral models because they require significantly less storage than signals, and are very efficient in the modeling of storage elements (e.g., registers, memories). Figure 5.4.3 demonstrates the application of variables for the design shown in Figure 5.4-1.

```
architecture Wait0c_a of Wait0c is
  signal        Din1_s    : bit := '0'; -- Input 1
  signal        Din2_s    : bit := '0'; -- Input 2
  signal        Zout1_s   : bit := '0'; -- Output for option 1
  signal        Clk_s     : bit := '0'; -- Clock
begin  --  Wait0c_a
  ------------------------------------------------------------------------
  --  Process:     Wait0_Lbl
  --  Purpose:     Demonstrate use of a variable to avoid wait for 0 ns.
  --               Reset of FF is not used here, but should be included
  --               when designing synthesizable hardware.
  -------------------------------------------------------------
  Wait0_Lbl : process(Clk_s, Din1_s, Din2_s)
    variable Q_v  : bit;  -- FF holding the Q output.
  begin  --  process Wait0_Lbl
    if Clk_s'event and Clk_s = '1' then
      -- implied register in variable because variable is in a
      -- clocked process and variable is read before it is modified
      -- (i.e. old registered value is used).
      Zout1_s <= Q_v and Din2_s;
      Q_v := Din1_s;  -- Q_V FF is updated
    end if ;

    -- NOTE: The next statement is good VHDL, but generates
    -- code which is NOT NECESSARILY SYNTHESIZABLE because
    -- some synthesizers do allow the reading of a variable if that
    -- variable is assigned a value inside a clocked process.
    Zout1_s  <= Q_v and Din2_s;
  end process Wait0_Lbl;
...
end Wait0c_a;
```

For behavior (thumbs up) / For synthesis (thumbs down)

Figure 5.4.3 Variables in Process (ch5_dir\wait0c.vhd)

5.4.4 VITAL Tables

VHDL Initiative Toward ASIC Libraries (VITAL) provides a set of packages and a methodology to easily define state machines. Code developed in VITAL is currently not synthesizable, but can be very concise because of the use of predefined tables. VITAL is briefly explained in chapter 13.

5.5 INERTIAL / TRANSPORT DELAY

[1] (LRM 8.4) ***Inertial delay*** *is a delay model used for switching circuits. Thus, a pulse whose duration is shorter than the switching time of the circuit will not be transmitted. Inertial delay is the default mode for signal assignment statements.* Inertial delay is also the default delay assumed in synthesis. Section 3.1.3 specifies, in synthesis, the application of inertial delay to avoid the creation of latches in non-clocked processes.

[1] ***Transport delay*** *is an optional delay for signal assignment. Transport delay is characteristic of hardware devices (such as transmission lines) that exhibit infinite frequency response: any pulse is transmitted, no matter how short its duration.*

In the circuit shown in Figure 5.5, a pulse with a width of less than the RC time constant will not be seen at the output. This delay is called **inertial delay** with a **reject** of pulses less than or equal to the RC time constant.

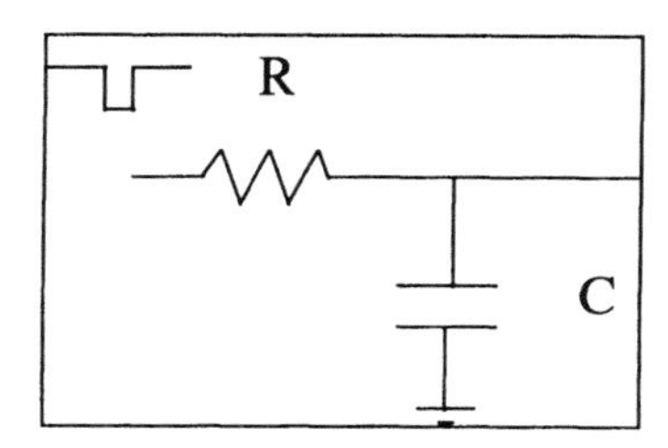

Figure 5.5 Circuit Exhibiting Inertial Delay with Pulse Rejection

5.5.1 Simulation Engine Handling of Inertial Delay

5.5.1.1 Simple View

Each signal driver queue maintains a list of the driver values at specific time units. When an inertial signal assignment is made, all driver values in the queue past the current time are deleted. All old transactions are deleted.

Helpful Reminder! **Inertial delay is like the army, LAST COMMAND from commander GOES, ALL previous commands from commander are forgotten.**

5.5.1.2 Updating Projected Waveforms per LRM 8.4.1

The syntax for a target assignment is:

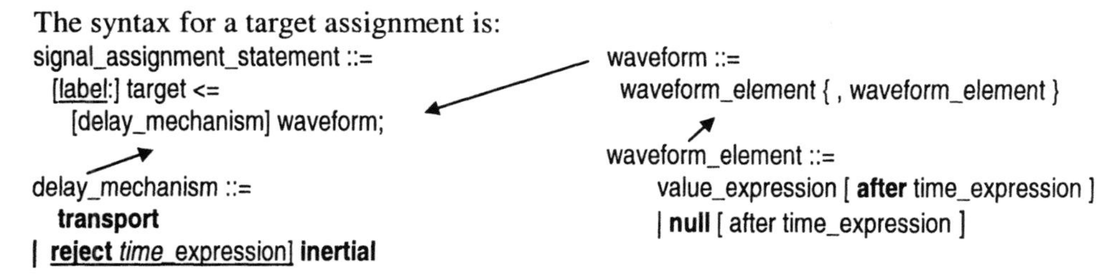

MA 👍 👍 Avoid using the **null** in the waveform element. Use the value 'Z' from package *IEEE.Std_Logic_1164* for type *Std_Logic* or *Std_Logic_Vector.*

Rationale: *The* ***null*** *form of the waveform element is used to specify that the driver of the signal is to be turned off so that it stops contributing to the value of the target. It is non-synthesizable. It can only be used for guarded signals. The null waveform cannot be used in concurrent signal assignments. Ina addition, it is illegal to assign null waveform to non-guarded signals. Instead, use the assignment of 'Z' defined the package IEEE.Std_Logic_1164 in type Std_Logic and Std_Logic_Vector to emulate a tri-state, or weak contribution.*

Figure 5.5.1.2-1 is a VHDL code demonstrating the three ways that waves can be projected in time.

```
architecture Wave_a of Wave is
  signal        S1                 : Natural;
  signal        S2                 : Natural;
  signal        S3                 : Natural;
begin  --  Wave_a

  Demo_Proc : process
  begin  --  process Demo_Lbl
    S1 <= 10   after 4    ns,
          1    after 8    ns,
          2    after 15   ns,
          100  after 100  ns;

    S1 <= 2    after 18   ns,
          25   after 25   ns,
          65   after 65   ns;

    S2 <= transport 10   after 4    ns,
                    1    after 8    ns,
                    2    after 15   ns,
                    100  after 100  ns;

    S2 <= transport 2    after 18   ns,
                    25   after 25   ns,
                    65   after 65   ns;

    S3 <= reject 12 ns inertial 10   after 4    ns,
                                1    after 8    ns,
                                2    after 15   ns,
                                100  after 100  ns;

    S3 <= reject  5 ns inertial 3    after 18   ns,
                                25   after 25   ns,
                                65   after 65   ns;

    wait;
  end process Demo_Proc;
end Wave_a;
```

First waveform projection onto signal S1 in Demo_Proc process

Second waveform projection onto signal S1 in Demo_Proc process

Figure 5.5.1.2-1 VHDL Code Projected Output Example and List Simulation Results (ch5_dir\wave_ea.vhd)

The following discussion presents a timeline of the projected waveforms for the code shown in Figure 5.5.1.2-1, and explains how VHDL rules are applied to predict the projected waveforms. The syntax used for the diagrams are as follows:

"●" represents a value in the first waveform projection.
"⊗" represents a canceled value in the waveform projection.
"□" represents a new value in the waveform projection (the second projection)

The first value projections for signals S1, S2, and S3 are as follows:

time -->

10 1 2 100 (Value)

4 8 15 100 (time in ns)

The update for a projected output waveform is as follows:

1. For any of the delay mechanism, *[1] all old transactions after the time of the first new transaction are deleted.* The second projection has a first new assignment at time 18 ns from the current time. Thus, **all previous projections after time 18 ns from the current time are deleted** (*shown with cross and a "* ***D1*** *" (for Delete #1).* Thus, *S1*, *S2*, and *S3* will have the following timeline (Updated values are not done yet)

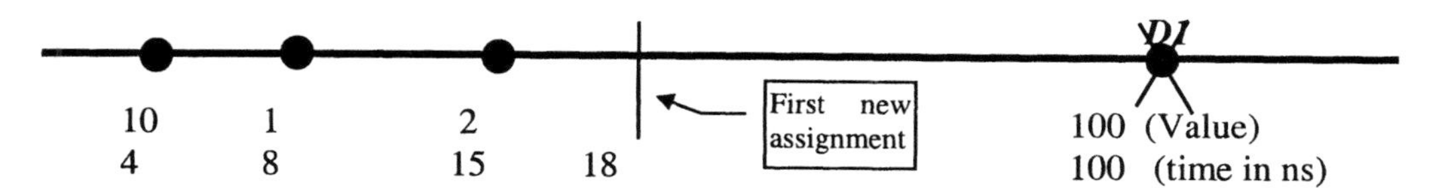

2. *[1] The new transactions are appended and marked.* Thus, S1, S2, and S3 will have the following timeline.

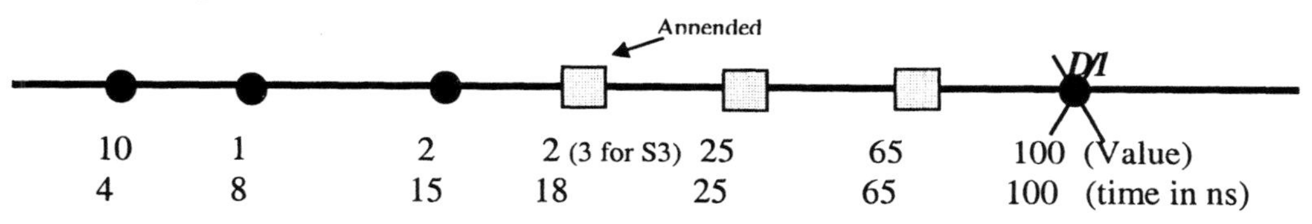

3. *[1] For inertial delay, any old transaction that precedes a marked transaction and is of the same value as the first new transaction is kept. This step is recursive until either there are no more old transactions, or the value of the old transaction is different that the first marked transaction. All other old transactions are deleted* (shown with cross and a *"D2")*. For **transport** delay, this step is **not performed.** Thus, signal *S1* will be marked as shown since it is inertial. Signal *S2* would be the same as shown in the previous step.

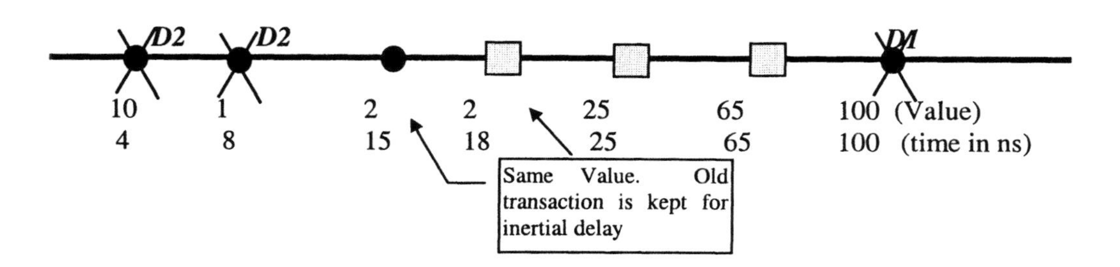

4. *For **reject - inertial** delay, any old transaction that precedes a marked transaction and is of value as the first new transaction is kept. This step is recursive until either there are no more old transactions, or the value of the old transaction is different that the first marked transaction. In addition, any **old transaction that falls within the reject region is deleted*** (shown with cross and a " **D3"**). *The reject region is the time zone from between the current time, and current time less the reject time expression (negative time zones are ignored).* Thus, signal S3 will be marked as shown since it reject - inertial.

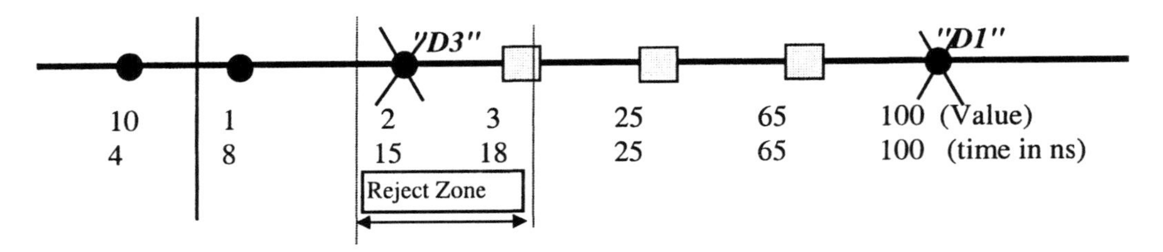

[1] The deleted transactions are removed from the projected waveforms. The final projections for the example shown in Figure 5.5.1.2-1 are shown in 5.5.1.2-2 through 5.5.1.2-4.

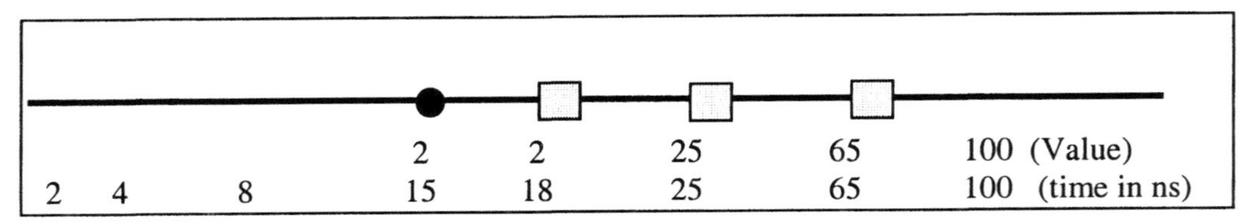

Figure 5.5.1.2-2 Inertial delay Time Projections

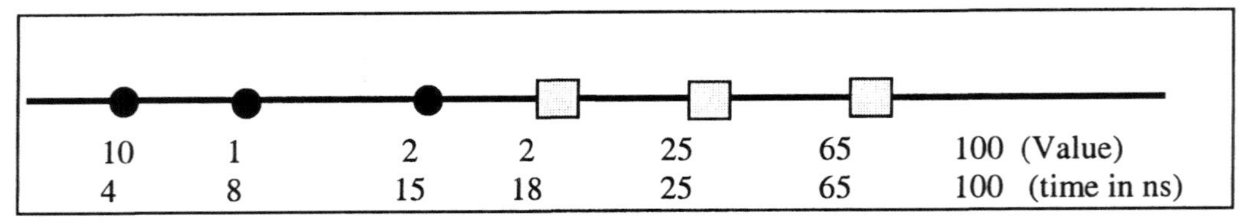

Figure 5.5.1.2-3 Transport Delay Time Projections

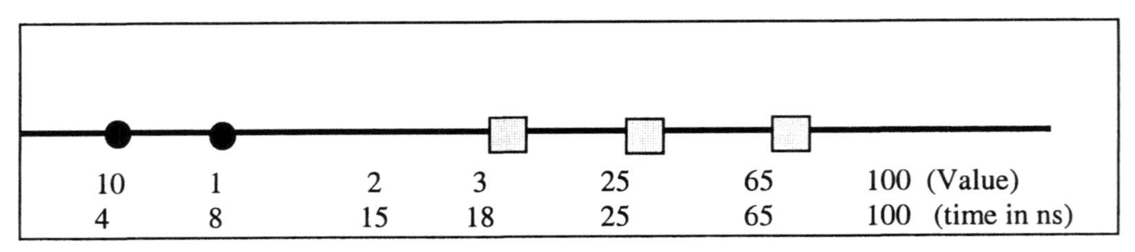

Figure 5.5.1.2-4 Reject 12 ns Inertial Time Projections

Figure 5.5.1.2-5 and Figure 5.5.1.2-6 represent the simulation timing waveform and list views of the model.

/s1 = 0 | 0 | 2 | 25 | 65
/s2 = 0 | 0 | 10 | 1 | 2 | 25 | 65
/s3 = 0 | 0 | 10 | 1 | 3 | 25 | 65

Figure 5.5.1.2-5 Simulation Timing Waveform of the Wave Model

ns	delta	s1	s2	s3
0	+0	0	0	0
4	+0	0	10	10
8	+0	0	1	1
15	+0	2	2	1
18	+0	2	2	3
25	+0	25	25	25
65	+0	65	65	65

5.5.1.2-6 List Views of the Wave Model.

EXERCISES

1. Fill-in the table below given the following wait statements. Identify which signals the wait is sensitive to (sensitivity list), the condition tested, and the timeout period. If a field does not apply, put a "-". If an entry is illegal, put ILLEGAL. If an entry causes the simulator to stop, put STOP. S , Clk are signals of type bit, VAR is a natural variable.

STATEMENT	SENSI-TIVITY LIST	CONDITION	TIMEOUT
wait until Clk'event;			
wait on Clk'event **until** CLK = '0';			
wait for Clk'event;			
wait on S **until** CLK = '0';			
wait on S **until** CLK = '0' **for** 100 ns;			
wait on S **for** 1000 ns;			
wait on S1, S2 **until** Clk = '0' **for** 10 us;			
wait;			
wait for CLK until VAR = 1;			
wait on S until VAR = 10;			
wait until VAR = 10;			
wait on S 'quiet(10 ns);			
wait on S 'stable(10 ns);			
wait on S 'delayed(10 ns);			
wait on S'event;			
wait on S'delayed(10 ns)'transaction;			
wait until S'delayed(10 ns) = '1';			

2. Write and simulate an empty entity with an architecture where
   1. ALL signals are of type *Std_Logic*. (library IEEE; use IEEE.Std_Logic.1164.all;)
   2. Include the following signals: *Q, CLK*
   3. Write a Flip-flop process so that
      - When *CLK* transitions from '0' to a '1' the value of *Q* is inverted
      - When *CLK* transitions from 'H' to '1', nothing happens
      - When *CLK* transitions from 'X' to anything, or from anything to 'X' assert a message *(assert False report "Error in clock" severity note;)*

Write a process that creates a clock waveform so that *CLK* has the following values:

'0' at 0 ns, '1' at 100 ns, 'Z' at 200 ns, 'H' at 300 ns, '1' at 400 ns,	'0' at 500 ns, '1' at 600 ns, 'X' at 700 ns, '0' at 800 ns

3. Explain the simulation output waveforms for the following entity/architecture.

```
---------------------------------------------------------------------------
--
--    Project       :  ATEP
--    File name     :  hmwrkch5.vhd
--    Title         :  Homework for chapter 5
--    Description   :  Projected waveform
--
---------------------------------------------------------------------------
entity HmWrkCh5 is
  port (Out1            : out     natural);
end Hmwrkch5;

architecture HmWrkCh5_a of HmWrkCh5 is

begin  --  Hmwrkch5_a
  -------------------------------------------------------------------------
  --  Process:   Test_Lbl
  -------------------------------------------------------------------------
  Test_Proc  : process

  begin  --  process Test_Proc
    Out1 <= 50 after 100 ns;
    wait for 50 ns;
    Out1 <= 100;
    wait for 500 ns;
    Out1 <= 500 after 50 ns;
    wait for 100 ns;
    Out1 <= 1000 after 100 ns;
    Out1 <= 1050 after 200 ns;
    wait for 10 ns;
    Out1 <= 2000;
    wait;
  end process Test_Proc;
end Hmwrkch5_a;
```

Simulation results:

ns	delta	out1
0	+0	0
50	+1	100
600	+0	500
660	+1	2000

6. ELEMENTS OF ENTITY/ARCHITECTURE

This chapter provides a complete definition of a component. All the concurrent statements of architecture are examined.

6.1 VHDL ENTITY

[1] (LRM 1.1) An entity declaration represents the definition interface between a given design and the environment in which it is used. It may also specify declarations and statements that are part of the design. A given entity declaration may be shared by many design architectures. Thus, an entity declaration can potentially represent a class of design architectures, each with the same interface. The syntax for an entity is as follows:

```
entity_declaration ::=
    entity identifier is
     entity_header
     entity_declarative_part
    [begin
     entity_statement_part]
    end [ entity_simple_name ];

entity_header ::=
    [ formal_generic_clause ]
    [ formal_port_clause ]

entity_declarative_part ::=
    { entity_declarative_item }

entity_declarative_item ::=
      subprogram_declaration
    | subprogram_body
    | type_declaration
    | subtype_declaration
    | constant_declaration
    | signal_declaration
    | file_declaration
    | alias_declaration
    | attribute_declaration
    | attribute_specification
    | disconnection_specification
    | use_clause
```

```
entity_statement_part ::=
    { entity_statement }

generic_clause ::=
    generic ( generic_list ) ;

generic_list ::= generic_interface_list

generic_map_aspect ::=
    generic map ( generic_association_list )

entity_statement ::=
      concurrent_assertion_statement
    | passive_concurrent_procedure_call
    | passive_process_statement

port_clause ::=
    port ( port_list ) ;

port_list ::= port_interface_list
port_interface_list ::=
    identifier_list : mode subtype_indication
        := static_expression ]
```

[1] A passive process is a process statement where neither the process itself, nor any procedure of which the process is a parent, contains a signal assignment statement. Figure 6.1-1 represents an example of legal declarations in an entity. Figure 6.1-2 demonstrates the scope of visibility rules.

M 👍 👎

1. If the subtype of a port interface is an **array** (e.g., *Std.Bit_Vector* or *IEEE.Std_Logic_Vector*), and if it is desired to create a component that is "**universal**" (i.e., independent of vector width) then AVOID constraining the size of the array in the port interface list.
2. If the subtype of a port interface is an **array**, and if either the design is a **synthesizable** design or the size of the array is **known** with little likelihood of change, then use a fixed size for the array size definition. Use a predefined subtype to define the array range. Generics may be used to allow flexibility in the width of the array.

Rationale: *1. Unconstrained arrays are more "universal" in defining the port interface type for arrays. In addition, they create models that require much less effort on verifying the validity of all possible array lengths. However* ***caution*** *must be used when handling those ports. The architecture must use attributes to determine the size of the arrays. The architecture must typically use internal signals to represent the port interfaces. The direction and bound for those signals are defined in the declaration of those internal signals. Concurrent signal assignments are necessary to connect the internal signals to the ports. The size of the ports is automatically determined when the component is instantiated and actual parameters are associated with formal parameters.*

2. Using predefined subtypes facilitates the use of attributes and the definition of ranges in loop constructs.

Figure 6.1-3 demonstrates the declaration and architectural coding style of a component with unconstrained arrays.

```
library IEEE;
  use IEEE.Std_Logic_1164.all;

entity NtyDemo is
  generic (MaxCount_g:           natural := 10;
           MaxDelay_g:           time := 100 ns);
  port (ClK           : in     Std_Logic;
        Data          : in     Std_Logic_Vector(31 downto 0);
        ResetF        : in     Std_Logic;
        Count         : inout  Std_Logic;
        TC            : out    Std_Logic);

  type      States_Typ is (Idle, Ready, Running, Blocked);

  constant  Limit_c     : natural := 16;
  -- 
  alias OpCode_p    : Std_Logic_vector( 7 downto  0) is Data(31 downto 24);
  alias Source_p    : Std_Logic_vector( 3 downto  0) is Data(23 downto 20);
  alias Dest_p      : Std_Logic_vector( 3 downto  0) is Data(19 downto 16);
  alias Data16_p    : Std_Logic_vector(15 downto  0) is Data(15 downto  0);

  use std.TextIO.all;
  use std.TextIO;
  file  Error_f     : TextIO.Text is out "error.rpt";
  file  Instr_f     : TextIO.Text is in  "instr.txt";

  signal TermCount_s : Std_Logic;
  signal Flag_s      : Std_Logic;

begin
  assert not OpCode_p = "11111111"
    report "Illegal Operation Code"
    severity warning;

  assert false
    report "Model for a XYZ Counter"
    severity note;
end NtyDemo;
```

These statements allow the access to all of the declarations defined in Std_Logic_1164 package located in library IEEE. This package becomes visible by ALL architectures of this entity.

Organize the port list by mode (in, out) and then by function or sorted alphabetically.

This entity_declarative_part is visible by ALL architectures of this entity. In general, it is best to limit the scope of visibility to where objects are used

Even though legal, local signal declarations here are not recommended

This type of error monitoring is usally done by an external monitor entityarchitecture with processes because it tends to be complex.

This message is displayed ONLY at the beginning of a simulation for every architecture using this entity. Recommended only for designs where the message is desired prior to simulation.

Figure 6.1-1 Example of Legal Entity Declarations (ch6_dir\entity_e.vhd)

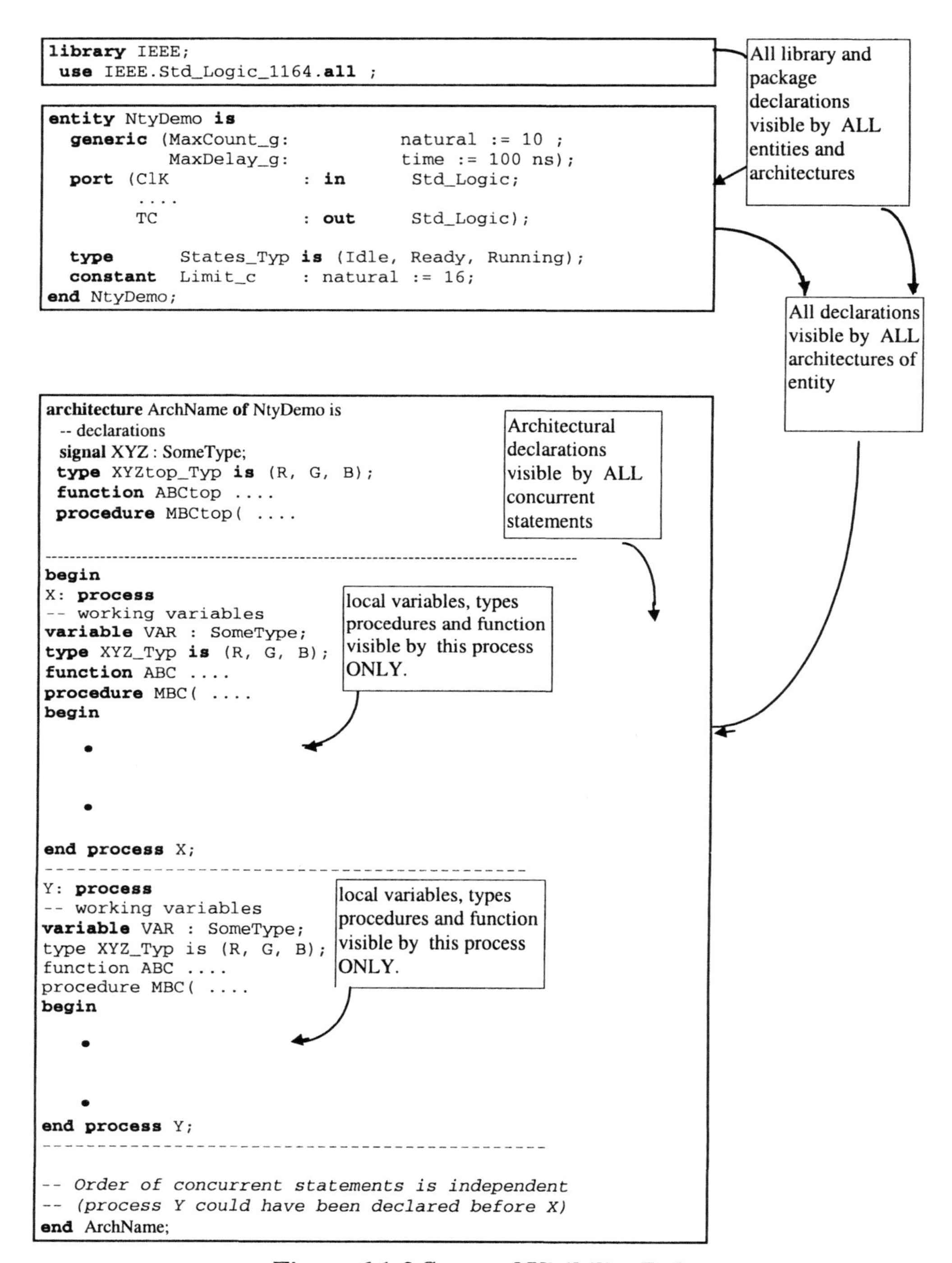

Figure 6.1-2 Scope of Visibility Rules.

```
entity Invert is
  generic (Width_g:              natural := 16);  -- not used in this example
  port (In_p   :  in     Bit_Vector;
        Out_p  :  out    Bit_Vector);
end Invert;

architecture Invert_a of Invert is
  signal Out_s   : Bit_Vector(Out_p'length - 1 downto 0);
  signal In_s    : Bit_Vector(In_p'length - 1 downto 0);
begin  --  Invert_a
  ------------------------------------------------------------
  --  Process:    Special_Lbl
  --  Purpose:    If MSB of In_s = '1' then invert all remaining bits else
  --                Set MSB, and reset remaining bits to '0'
  --------------------------------------------------------------------------
  Special_Lbl : process (In_s)
  begin  --  process Special_Lbl
    if (In_s(In_s'left) = '1') then
      Out_s <= In_s(In_s'left) & not In_s(In_s'left - 1 downto 0);
    else
      Out_s(In_s'left) <= not In_s(In_s'left) ;
      Bits_Lbl  : for Bit_i in In_s'left - 1 downto 0 loop
         Out_s(Bit_i)   <= '0';
      end loop Bits_Lbl;
    --  Out_s(In_s'left - 1 downto 0) <= (others => '0');
    end if;
  end process Special_Lbl ;

  Out_p <= Out_s;    --  Concurrent statements

  In_s <= In_p; --  Concurrent statements
end Invert_a;
-------------------------------------------------------------------------
-- TEST BENCH Demonstrating component instantiation, and
-- implicit definition of length of bit vector.
entity TB is
end TB;

architecture TB_a of TB is
  component Invert
    port(
          In_p  : in    Bit_Vector;
          Out_p : out   Bit_Vector
    );
  end component;

  signal In_p  : Bit_Vector(7 downto 0);
  signal Out_p : Bit_Vector(7 downto 0);
begin  --  TB_a
  U1_Invert: Invert
    port map (
    In_p    => In_p,
    Out_p   => Out_p
    );

  In_p <= X"01",
          X"85" after 100 ns,
          X"05" after 200 ns,
          X"F1" after 300 ns;
end TB_a;
```

Unconstrained arrays

Size and bounds of array redefined using attributes of ports

Use of attributes to define selected bits or slice.

Alternate method for setting other bits. See file Invert2.vhd

Width of component ports determined by type of the ACTUAL parameters

Figure 6.1-3 Declaration and Use of a Component with Unconstrained Array (ch6_dir\invert.vhd)

M 👍 👎 Organize the port list consistently either by mode (in, out) and then by function or sorted alphabetically.

Rationale: *Enhances maintainability and readability.*

M 👍 👎 Avoid local signal declarations in an entity.

Rationale: *Local signals belong in architectural declarations and not in an entity because the signals may not be used by all architectures of that entity. If the entity signals are meant as a means of communication between multiple instantiations of the same component (entity/architecture pair), it is preferred to either use external ports or global signals defined in package (see chapter 8). Global signals provide channels of communication between components, but are more readable and maintainable because the path can be provided when global signals are used. However, global signals are not synthesizable.*

M 👍 👎 For complex non-synthesizable designs, assert a message in the entity statement part. This message should provide information about the design, such as design name, disclaimers, etc. Note that an abuse of these messages would clutter the messages displayed at the beginning of a simulation. Do not use the assert statement to perform design checks.

Rationale: *The message is displayed ONLY at the beginning of a simulation for every architecture using this entity, and thus clarifies points that need to be made to users of the model. Testbenches are responsible for performing design checks.*

M 👍 👍 In the declarations of packages, entities, architectures, procedures, and concurrent statements, it is best to **limit the scope of visibility to where objects are used**.

Rationale: *Good software practice .*

6.2 VHDL ARCHITECTURE

[1] (LRM 1.2) The architectural body is a body associated with an entity declaration to describe the internal organization or operation of a design entity. An architectural body is used to describe the behavior, data flow, or structure of a design entity. The syntax for the architectural body is as follows:

```
architecture_body ::=
    architecture identifier of entity_name is
                architecture_declarative_part
    begin
                architecture_statement_part
    end [ architecture ] [ architecture_simple_name ] ;

architecture_declarative_part ::= ----------------------------------
    { block_declarative_item }

block_declarative_item ::=
    subprogram_declaration              | component_declaration
    | subprogram_body                   | attribute_declaration
    | type_declaration                  | attribute_specification
    | subtype_declaration               | configuration_specification
    | constant_declaration              | disconnection_specification
    | signal_declaration                | use_clause
    | shared_variable_declaration       | group_template_declaration
    | file_declaration                  | group_declaration
    | alias_declaration

architecture_statement_part ::= --------------------------------------------------------------------
    { concurrent_statement }
```

The architecture declarative part is defined in other sections of this book. This section focuses on the concurrent statements.

[1] Concurrent statements are used to define interconnected blocks and processes that jointly describe the overall behavior or structure of a design. Concurrent statements execute asynchronously with respect to each other, and thus, the **order of definitions is irrelevant**. Synchronization between concurrent statements can be achieved with signals through the sensitivity clauses or wait statements on those signals. A concurrent statement can be any of the following:

```
concurrent_statement ::=
    process_statement
    | concurrent_signal_assignment_statement
    | component_instantiation_statement
    | concurrent_procedure_call
    | generate_statement
    | concurrent_assertion_statement
    | block_statement
```

6.2.1 Process Statement

[1] (LRM 9.2) A process statement defines an independent sequential process representing the behavior of some portion of the design. A process is the most often used concurrent statement because it provides a wide range of flexibility as a result of its sequential nature. The process allows for **synchronization** with other concurrent statements through signals as a result of either its **explicit sensitivity clause** or the ***wait* statement** with its implicit or explicit sensitivity clause. The syntax for a process is given below.

```
process_statement ::=
[process_label :] [postponed] process
[(sensitivity_list)]
        process_declarative_part
        begin
         process_statement_part
        end [postponed] process
                [process_label];

process_statement_part ::= -------
     { sequential_statement}

sequential_statement ::=
        wait_statement
      | assertion_statement
      | signal_assignment_statement
      | variable_assignment_statement
      | procedure_call_statement
      | if_statement
      | case_statement
      | loop_statement
      | next_statement
      | exit_statement
      | return_statement
      | null_statement

process_declarative_part ::=
     { process_declarative_item }

process_declarative_item ::=
     subprogram_declaration
     | subprogram_body
     | type_declaration
     | subtype_declaration
     | constant_declaration
     | variable_declaration
     | file_declaration
     | alias_declaration
     | attribute_declaration
     | attribute_specification
     | use_clause
```

Rules about processes:

1. ***[1] If a sensitivity list appears following the reserved word process, then the process statement is assumed to contain an implicit wait statement as the last statement of the process part.*** Thus a process that has an explicit sensitivity list always has exactly one (implicit) *wait* statement in it at the end of the process. Figure 6.2.1-1 demonstrates two identical process statements, one with a sensitivity list, and its equivalent process without a sensitivity list.

Process with explicit sensitivity list	Equivalent process with a wait statement

`XYZ_Proc : process (S1, S2)`
`begin`
`  S1 <= '1';`
`  S2 <= '0' after 10 ns;`
 	
`end process XYZ_Proc;`
	`XYZ_Proc : process`
	`begin`
	`  S1 <= '1';`
	`  S2 <= '0' after 10 ns;`
	`  wait on S1, S2;`
	`end process XYZ_Proc;`

Figure 6.2.1-1 Process with and Without a Sensitivity List

2. **A process with a sensitivity clause MUST NOT contain an EXPLICIT *WAIT* statement**. Figure 6.2.1-2 demonstrates that a process with a sensitivity clause may not have a wait statement.

```
XYZ_Proc: process (S1, S2) -- Sensitivity clause
  begin
   S1 <= '1';
   S2 <= '0' after 10 ns;
   wait for 100 ns;          -- NO wait state in a process with a sensitivity clause
  end process XYZ_Proc;
```

Figure 6.2.1-2 Process with Sensitivity Clause may not have Wait Statements

3. *[1]* ***Only static signal names for which reading is permitted may appear in the sensitivity list of a process statement.*** *A static signal name is a slice name whose prefix is a static name and whose discrete range is a static discrete range.* Figure 6.2.1-3 demonstrates examples of static signal names.

4. *[1]* ***The execution of a process statement consists of the repetitive execution of its sequence of statements.*** *After the last statement in the sequence of statements of a process is executed, execution will immediately continue with the first statement in the sequence of statements.* Figure 6.2.1-5 demonstrates this principle through the use of a Linear Feedback Shift Register (*LFSR*) that acts as a 32-bit pseudo-random number generator of type *Std_Logic_Vector*. Figure 6.2.1-4 demonstrates the algorithm graphically. The pseudo-random number generator uses the polynomial:

   $X := X^{31}$ xor X^{30} xor X^{31} xor X^{10} xor X^{0} ; -- non-VHDL code
   -- X is a 32 bit register, bit 31 = MSB, bit 0 = LSB

   Figure 6.2.1-6 demonstrates some of the simulation results.

```
entity ProcRules is
  port (In1    :  in      Bit_Vector(31 downto 0);
        Out1   :  out     Bit_Vector(31 downto 0);
        Out2   :  out     Bit_Vector(31 downto 0);
        Out3   :  out     Bit_Vector(31 downto 0));

end ProcRules;

architecture ProcRules_a of ProcRules is
  signal        Test_s          : natural := 5;

begin  -- ProcRules_a
  ------------------------------------------------------------------------------
  --  Process:   Demo_Lbl
  --  Purpose:   OK process with a sensitivity clause
  ------------------------------------------------------------------------------
  Demo1_Lbl : process (In1)
  begin  -- process Demo_Lbl
    Out1 <= not In1;
  end process Demo1_Lbl;

  ------------------------------------------------------------------------------
  --  Process:   Error_Lbl
  --  Purpose:   Demonstrate that a process with a sensitivity clause
  --             MUST NOT contain an EXPLICIT WAIT statement
  --             Demonstrate that only static signal names for which
  --             reading is permitted may appear in the sensitivity list
  --             of a process statement.
  ------------------------------------------------------------------------------
  Error2_Lbl : process (Out1)
  begin  -- process Error_Lbl
    Out2 <= not In1;
    wait for 100 ns;
  end process Error2_Lbl;

  Error3_Lbl : process (In1(31 downto Test_s))
  begin  -- process Error_Lbl
    Out3 <= not In1;
   end process Error3_Lbl;
end ProcRules_a;
```

Error2_Lbl : process (Out1)
Cannot read output: out1.
wait for 100 ns;
Wait statements not allowed in process with sensitivity list.

Error3_Lbl : process (In1(31 downto Test_s))
Signal name in sensitivity list is not static.

Figure 6.2.1-3 Process Rules (ch6_dir\procrule.vhd)

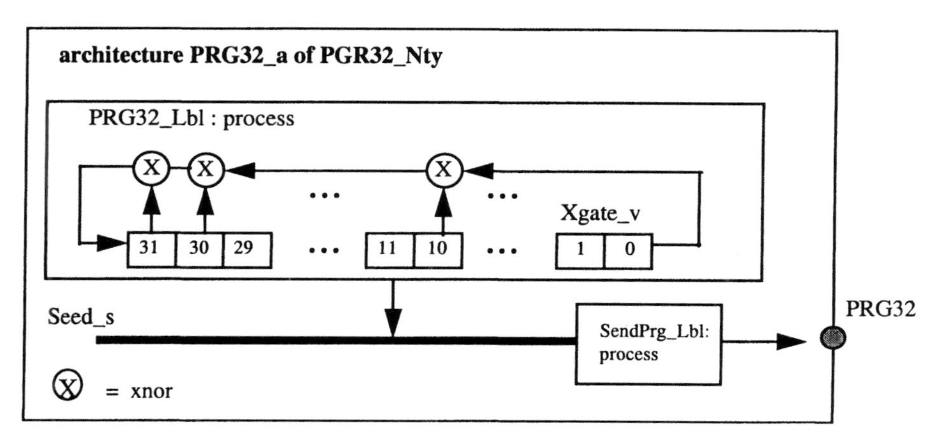

Figure 6.2.1-4 Graphical Representation of Pseudo-Random Number Generator Algorithm

```
library IEEE;
  use IEEE.Std_Logic_1164.all;

entity PRG32 is
  generic (Seed_g      :  Std_Logic_Vector(31 downto 0)
      := "10000000000011111000000000000110");
  port (PRG32          :  out   Std_logic_vector(31 downto 0));
end PRG32;

architecture PRG32_a of PRG32 is
  signal          Seed_s             : Std_Logic_Vector(31 downto 0) := Seed_g;
begin   --   PRG32_a
  --  Purpose:   Demonstrate generation of a pseudorandom number generator
  --             with the seed stored in a signal.
  ---------------------------------------------------------------------------
  PRG32_Lbl  : process
    variable Xgate_v   : STD_LOGIC;         -- Output of xor gate begin
  begin   --   process PRG32_Lbl
    Xgate_v := Seed_s (31) xor Seed_s (30) xor
               Seed_s (10)  xor Seed_s (0);

    Seed_s <=  not Xgate_v & Seed_s (31 downto 1);

    wait for 100 ns;

  end process PRG32_Lbl;


  --  Purpose:   Transfers number to output
  SendPrg_Lbl  : process (Seed_s)
  begin   --   process SendPrg_Lbl

    PRG32  <= Seed_s;

  end process SendPrg_Lbl;
end PRG32_a;
```

REPEAT

Repeat on sensitivity of signal (implied wait on Seed_s)

Figure 6.2.1-5 32-Bit Pseudo-Random Number Generator (ch6_dir\prg32_ea.vhd)

ns	delta	prg32	seed_s	ns	delta	prg32	seed_s
0	+0	XXXXXXXX	800F8006	500	+1	F4007C00	7A003E00
0	+1	800F8006	4007C003	500	+2	7A003E00	7A003E00
0	+2	4007C003	4007C003	600	+1	7A003E00	BD001F00
100	+1	4007C003	A003E001	600	+2	BD001F00	BD001F00
100	+2	A003E001	A003E001	700	+1	BD001F00	DE800F80
200	+1	A003E001	D001F000	700	+2	DE800F80	DE800F80
200	+2	D001F000	D001F000	800	+1	DE800F80	6F4007C0
300	+1	D001F000	E800F800	800	+2	6F4007C0	6F4007C0
300	+2	E800F800	E800F800	900	+1	6F4007C0	B7A003E0
400	+1	E800F800	F4007C00	900	+2	B7A003E0	B7A003E0
400	+2	F4007C00	F4007C00	1000	+1	B7A003E0	5BD001F0

Figure 6.2.1-6 32-Bit Pseudo-Random Number Generator Simulation Results.

6.2.2 Concurrent Signal Assignment Statements

[1](LRM 9.5) ***A concurrent signal assignment statement represents an equivalent process statement that assigns values to signals.*** The concurrent signal assignment is a useful concurrent statement that can be used to define data flow in behavioral modeling. It is also very useful in describing the behavior of combinational logic (or logical clouds) as an input to a logic synthesizer. Logic synthesizers minimize the equations defined in the concurrent signal assignments and produce a circuit with all the constraints imposed on the design. In testbenches concurrent statements can serve to generate test waveforms. **A concurrent signal assignment is implicitly sensitive to ALL the signals used in the waveforms**. The syntax for concurrent signal assignment is as follows:

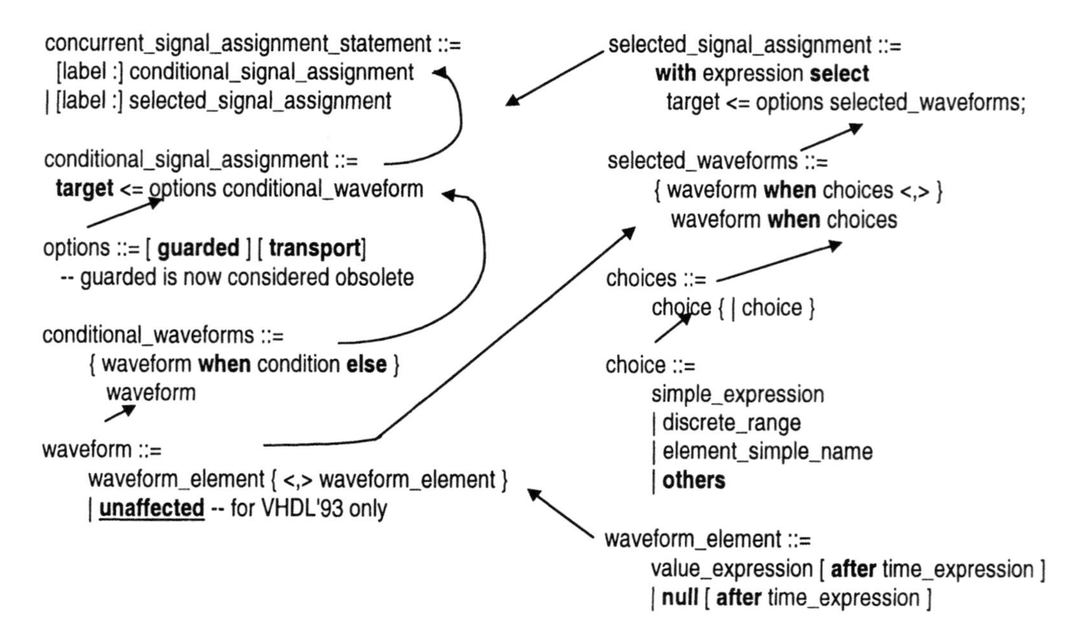

6.2.2.1 Conditional Signal Assignment

The conditional signal assignment (LRM 9.5.1) *[1] is one of the forms of the concurrent signal assignment. The conditional signal assignment represents a process statement in which the signal transform is an **if** statement. For a given conditional signal assignment, there is an equivalent process statement corresponding to it. If the conditional signal assignment statement is of the form:*

```
target <= options waveform1 when condition 1 else
                  waveform2 when condition 2 else


                 waveformN-1 when conditionN-1 else
                 waveformN when condition N;
```

then the signal transform in the corresponding process statement is of the form:

```
if condition1 then
  wave_transform1
elsif condition2 then
  wave_transform2


elsif conditionN-1 then
  wave_transformN-1
else
  wave_transformN
end if;
```

wave_transform is of the corresponding process statement is of the form
target <= [delay_mechanism] waveform_element1, waveform_element2, ... waveform_elementN;

*If the waveform is of the form **unaffected*** (VHDL'93 only) *then the wave in the corresponding process statement is of the form **null**.* The **unaffected** offers the advantage that no activity occurs on the signal if the condition is true. For example:

```
S <= unaffected when CS_F = '1' else
     Mem(Address);
```

6.2.2.2 Selected Signal Assignment

*[1](LRM9.5.2) The selected signal assignment represents a process statement in which the signal transform is a **case** statement. If the selected signal assignment is of the form:*

```
with expression select
   target <= options  waveform1 when choice_list1,
                      waveform2 when choice_list2,
                      ...
                      waveformN when choice_listN;
```

then the signal transform in the corresponding process statement is of the form:

```
case expression is
  when choice_list1 => wave_transform1
  when choice_list2 => wave_transform2
  ...
  when choice_listN => wave_transformN
end case;
```

Figure 6.2.2-1 provides examples of concurrent signal assignment statements.

```
...
architecture ConcSig_a of ConcSig is

begin  --  ConcSign_a
  ----------------------------------------------
  --  conditional_signal_assignment
  ----------------------------------------------
  A1Out <= not In1   when OpCode_p = "00000000"              else
           In1       when OpCode_p(7) = '1' and In2 = '0'   else
           In3(0);


  A2Out <= In2     after 10 ns  when Source_p = "1111"          else
           In1     after 200 ns when Source_p = "0000"          else
           not In2 after 15  ns,
               '0' after 20  ns,
               '1' after 100 ns,
               In2 after 200 ns;


   A4out <= (In1 and In2) or In3(31);   -- no conditions attached

  ------------------------------
  --  selected_signal_assignment
  ------------------------------
  with OpCode_p select
    A3Out(31) <=
        In1 or   In2 after 100 ns when "00000001",
        In1 and  In2 after 90  ns when "00000010",
        In1 nand In2 after 95  ns when "00000011",
        In1 nor  In2 after 96  ns when "00000100",
        In1 xor  In2 after 98  ns when "00000101",
        In1                       when others;
end ConcSig_a;
```

A1Out is sensitive to signals:
OpCode_p, In1, In2, In3(0)

A2Out is sensitive to signals:
Source_p, In1, In2

A3Out is sensitive to signals:
OpCode_p, In1, In2

Figure 6.2.2-1 Concurrent Signal Assignment Statements Examples (ch6_dir\concsig.vhd)

Figure 6.2.2-2 represents another conditional signal assignment example that is useful for generating test vectors into signals. This format can be used efficiently when the data is available in textual format, and needs to be converted by a program to a VHDL stimulus format. The simulation results are shown in Figure 6.2.2-3.

```
architecture Aggr_a of Aggr is
  subtype  Reg6_Typ is Bit_Vector(5 downto 0);
  signal A_s    : Bit;
  signal B_s    : Bit_Vector(3 downto 0);
  signal C_s    : Bit;
begin  --  Aggr_a
   -- Assign an array of bit to an array of bits
   -- Thus, left hand side must be bits,
   -- Right hand side is an array of bits
   -- Type qualifier is needed since compiler does not know if
   -- right hand side is bits or string
  (A_s, B_s(3), B_s(2), B_s(1), B_s(0), C_s)
                    <= Bit_Vector'("110001") after 100 ns,
                       Bit_Vector'("100010") after 200 ns,
                       Bit_Vector'("000101") after 300 ns,
                       Bit_Vector'("101000") after 400 ns,
                       Bit_Vector'("010001") after 500 ns,
                       Bit_Vector'("111110") after 600 ns;

end Aggr_a;
```

Could also use a subtype for type mark, thus:
Reg6_Typ'("110001") after 100 ns, ...
is legal

Figure 6.2.2-2 Aggregates in concurrent statements (ch6_dir\concaggr.vhd)

ns	a_s	b_s	c_s
0	0	0000	0
100	1	1000	1
200	1	0001	0
300	0	0010	1
400	1	0100	0
500	0	1000	1
600	1	1111	0

Figure 6.2.2-2 Simulation Results for model shown in Figure 6.2.2.-2

6.2.3 Component Instantiation Statement

A component represents an entity/architecture pair. Instantiations of components in architectures is a method to define hierarchy because architectures of components can have within them other components. *[1] (LRM 5.2.1.2) A component instantiation statement defines a subcomponent of the design entity in which it appears, associates signals or values with the ports of that subcomponent, and associates values with generics of that subcomponent.* Thus, a component instantiation is equivalent to plugging a hardware component into a board, and making the electrical connections between the pins of the component and the signals of the circuit board. The component association syntax is:

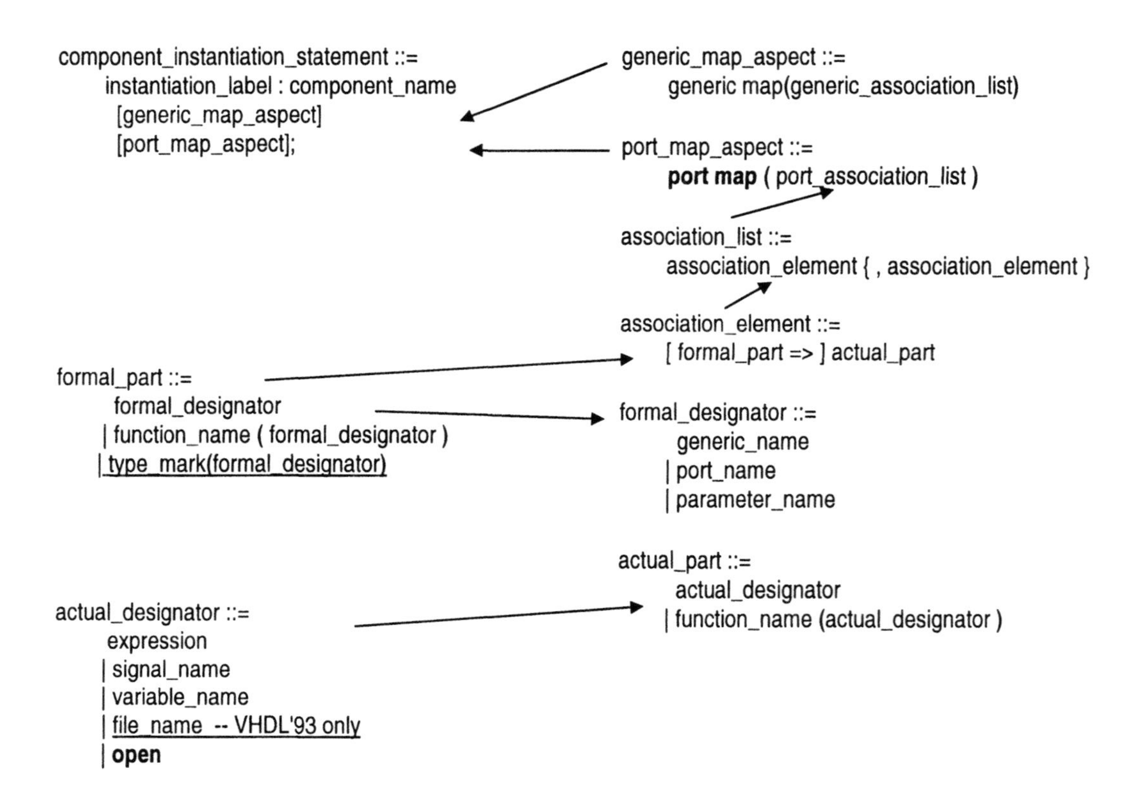

Prior to instantiating a component, the component must be declared in either the current architecture or in a package. In a component declaration, if the component identifier represents the entity name, and if there is only one architecture for this entity, then no configuration is needed to identify which architecture is attached to this entity. This is because the simulator defaults to the only architecture that is defined. Remember, a component represents an entity/architecture combination pair, in which a configuration binds an architecture to a component. Configurations are covered in chapter 9. The syntax for a component declaration is as follows:

```
component_declaration ::=
   component identifier
     [local_generic_clause ]  -- Copy of the entity generic declaration
     [local_port_clause ]     -- Copy of the entity port declaration
   end component ;
```

Figure 6.2.3-1 represents a down counter. Figure 6.2.3-2 represents a simple testbench to test this counter, and utilizes a component instantiation for the counter and two concurrent signal assignments to emulate the clock and the Reset signal.

```
entity Counter is
  generic (Modulus_g    : positive := 4;
           CountDly_g   : time     := 10 ns);
  port (Reset      : in     bit;
        Clk        : in     bit;
        Count      : out    integer);
end Counter;

architecture DownCounter_Beh of Counter is
  signal Count_s : integer := 0;
begin
   -- Process (conconcurrent statement)
   DownCounter_Lbl: process
   begin
     -- wait for rising edge of internal signal clock
     wait until Clk = '1';  -- more on "wait" later
     if (Reset = '1') then
       Count_s <= 0;
     else
       Count_s <= (Count_s - 1) mod Modulus_g;
     end if;
   end process DownCounter_Lbl;

   -- concurrent signal assignment
   Count   <= Count_s after CountDly_g;
end DownCounter_Beh;
```

Figure 6.2.3-1 Down Counter Component (ch6_dir\contr_ea.vhd)

```
entity bnch1 is
end Bnch1;

architecture Bench_a of Bnch1 is                  Component Declaration.
  constant ClkPeriod_c : time := 100 ns;          (Could have been done in
                                                   a package)
  -- Component declaration
  component Counter          -- Note suffix ""
    generic          -- Like constants, can be modified by configuration
      (Modulus_g    : integer := 4;
       CountDly_g   : time    := 5 ns); -- default value, if not modified
                                        -- generics are optional
    port
      (Reset        : in     bit;
       Clk          : in     bit;      -- port_name   : direction type;
       Count        : out    integer); -- ports are optional
  end component;

  -- Declarations go here  (signals, constants, types, subprograms)
  -- All declarations defined here are visible by the architecture
  signal Reset  : bit := '0';
  signal Clk    : bit := '0'; -- Type is defined in package Standard
  signal Count  : integer;

begin                                              Component instantiation
  U1_Counter: Counter                              (a concurrent statement)
    generic map
      (Modulus_g    => 4,      -- Named notation. Also "," as delimiters
       CountDly_g   => 6 ns)   -- default value, if not modified
    port map
      (Reset        => Reset,
       Clk          => Clk,
       Count        => Count);
```

```
  -- Concurrent signal assignments
  Clk  <= not Clk  after ClkPeriod_c / 2;

  Reset  <= '1' after ClkPeriod_c,
            '0' after 2 * ClkPeriod_c;

end Bench_a;
```

Figure 6.2.3-2 Testbench for Counter Component (ch6_dir\bnch1_ea.vhd)

M Use named notation in component association list.

Rationale: *Enhanced readability*

6.2.3.1 Port Association Rules

6.2.3.1.1 Connection

Per LRM 1.1.1.2, [1] *if a formal port is associated with an actual port, signal, or expression (VHDL'93 only), then the formal port is said to be connected. If a formal port is instead associated with the reserved word* ***open****, then the formal is said to be unconnected.* This is equivalent to expressing that in a printed circuit board, if a pin of a device *(the formal of the component)* is wired (or associated) to a trace of the board (*the actual*), then that pin is connected to the printed circuit board. However, if that pin is associated to **open,** then that pin is unconnected.

[1] A port of mode ***in*** *may be unconnected or unassociated only if its declaration includes a default expression. A port of any mode other than* ***in*** *may be unconnected or unassociated, as long as its type is not an unconstrained array type. It is an error if some of the subelements of a composite formal ports are unconnected and others are either unconnected or unassociated.* Figure 6.2.3.1.1 demonstrates the port association concepts. In synthesis, initialization of ports, signals, and variables is not allowed.

```
entity OpenComp is
  port (InPlain      :  in      integer;
        InInit       :  in      integer := 0;
        InOutPlain   :  inout   integer;
        InOutInit    :  inout   integer := 10;
        BufferPlain  :  buffer  integer;
        OutPlain     :  out     integer;
        InBVU        :  in      Bit_Vector;  -- unconstrained array
        InOutBVU     :  inout   Bit_Vector;
        InBVInit     :  in      Bit_Vector(7 downto 0)   -- -- constrained array
                                       := X"AB"
        );

end OpenComp;
```

```
entity OpenTB is  --  Testbench
begin
end OpenTB;

architecture OpenTB_A of OpenTB is
  signal TestBV_s : Bit_Vector(7 downto 0);
  signal Int_s   : integer := 100;
  component OpenComp -- Component declaration       Component Declaration.
    port(
        InPlain                    : in     integer;
        InInit                     : in     integer := 0;
        InOutPlain                 : inout  integer;
        InOutInit                  : inout  integer := 10;
        BufferPlain                : buffer integer;
        OutPlain                   : out    integer;
        InBVU                      : in     Bit_Vector;
        InOutBVU                   : inout  Bit_Vector;
        InBVInit                   : in     Bit_Vector(7 downto 0)
                                              := X"AB"
    );
  end component;

begin
  U1_Open: OpenComp             -- Component instantiations
    port map (

        InPlain                 => open,  -- uninitialized IN port
        InInit                  => open,  -- OK per LRM 1.1.1.2
        InOutPlain              => open,  -- OK per LRM 1.1.1.2
        InOutInit               => open,  -- OK per LRM 1.1.1.2
        BufferPlain             => open,  -- OK per LRM 1.1.1.2
        OutPlain                => open,  -- OK per LRM 1.1.1.2

        InBVU                   => open,  -- Error per LRM 1.1.1.2
                                          -- Unconstrained array

        InOutBVU                => open,  -- Error per LRM 1.1.1.2
                                          -- Unconstrained array

        InBVInit                => open
    );

  U2_Open: OpenComp
    port map (

        InPlain                 => 20,   -- error for VHDL'87 only      '87
                                         -- ☺'93 (20 is an expression)
        InInit                  => Int_s,
        InOutPlain              => Int_s,
        InOutInit               => open,  -- OK per LRM 1.1.1.2
        BufferPlain             => open,  -- OK per LRM 1.1.1.2
        OutPlain                => open,  -- OK per LRM 1.1.1.2
        InBVU                   => TestBV_s,
        InOutBVU                => TestBV_s(3 downto 0),

        -- Formal InBVInit must not be associated with OPEN
        -- when subelements are associated individually.

        InBVInit(7)             => open,  -- error per LRM 1.1.1.2
        InBVInit(6 downto 0)    => TestBV_s(6 downto 0)
    );
end OpenTB_A;
```

Figure 6.2.3.1.1 Port Association Examples (ch6_dir\open_ea.vhd)

6.2.3.1.2 Type Conversion

[1] When formal parameters of a port are of different type than their associated actual parameters, function calls or type conversions (for VHDL'93 only) can be used within the association list to convert to correct types (see LRM 4.3.2.2). Thus, conversions are used only to ensure type matching. Table 6.2.3.1.2 summarizes the association list rules. Figure 6.2.3.1.2-1 and 6.2.3.1.2-2 represents examples of the conversion rules in port associations for VHDL'87 and VHDL'93.

Table 6.2.3.1.2 Port Association List

FORMAL mode	FORMAL Type Conversion	ACTUAL Type Conversion	COMMENTS
in	none	FActual_2_formal(ad)	Actual must NOT be **open**
inout	Fformal_2_Actual(fd)	FActual_2_formal(ad)	Actual must NOT be **open**
out	Fformal_2_Actual(fd)	none	Actual must NOT be **open**
buffer	Fformal_2_Actual(fd)	none	Actual must NOT be **open**
in	none	Tformal(ad)	Actual must NOT be **open**
inout	TActual(fd)	Tformal(ad)	Actual must NOT be **open**
out	TActual(fd)	none	Actual must NOT be **open**
buffer	TActual(fd)	none	Actual must NOT be **open**

Fformal_2_Actual(fd) = Type conversion function from the formal type to the actual type with the **f**ormal **d**esignator (fd) as the parameter -- VHDL'87 & VHDL'93
Factual_2_formal(ad) = Type conversion function from the actual type to the formal type with the **a**ctual **d**esignator (ad) as the parameter -- VHDL'87 & VHDL'93
Tactual(fd) = VHDL Type conversion using type mark of actual with the **f**ormal **d**esignator (fd) as the parameter **-- FOR VHDL'93 ONLY**
Tformal(ad) = VHDL Type conversion using type mark of formal with the **a**ctual **d**esignator (ad) as the parameter **-- FOR VHDL'93 ONLY**

```
entity Alist is
  port (InReal_p  : in     real;
        OutInt_p  : out    integer);
end Alist;

architecture Alist_a of Alist is

begin  -- Alist_a
  -- Concurrent signal assignments
  OutInt_p   <= integer(InReal_p); -- Type conversion, legal VHDL'87 &'93
end Alist_a;


-------------------------------------------------------------------------------
--   File name   :  alisttb.vhd
--   Title       :  Testbench for list association test
--   Description :  Tests of various association methods
entity ListTB is
end ListTB;
```

```
architecture ListTB_A of ListTB is
  function ToReal(Integer_v : integer) return real is
    begin
      return real(integer_v);
    end ToReal;

  component Alist
    port( InReal_p              : in         real;
          OutInt_p              : out        integer);
  end component;

  signal Real_s   : real := 3.14;
  signal Real2_s  : real := 0.0;
  signal Int1_s   : integer := 1;
  signal Int2_s   : integer := 2;

begin
  U1_Alist: Alist
    port map(
       InReal_p          => Real_s,     -- formal real gets actual real
       OutInt_p          => Int1_s);    -- actual integer  gets formal integer

  U2_Alist: Alist
    port map(                        -- formal real gets converted actual integer
      InReal_p          => ToReal(Int2_s),
      ToReal(OutInt_p) => Real2_s); -- actual real gets converted formal integer

  U3_Alist: Alist
    port map(                        -- Formal real gets converted actual integer
      InReal_p          => ToReal(Int2_s),

      --ToReal(OutInt_p) => open  -- Illegal, convertion function with open
      OutInt_p          => open);
end ListTB_A;
```

Figure 6.2.3.1.2-1 Examples of Port Association Conversion Rules, VHDL'87 & VHDL'93 (ch6_dir\alist_ea.vhd)

```
...
architecture ListTB_A of ListTB is
...
begin
...
  U2_Alist: Alist
    port map (
      InReal_p    => Real(Int2_s), -- formal real gets converted actual integer
      Real(OutInt_p) => Real2_s    -- actual real gets converted formal integer
    );

  U3_Alist: Alist
    port map (
      InReal_p    => Real(Int2_s),  -- Formal real gets converted actual integer
      --ToReal(OutInt_p)  => open --  Illegal, convertion function with open
      OutInt_p    => open
    );
end ListTB_A;
```

Figure 6.2.3.1.2-2 Examples of Port Association Conversion Rules, VHDL'93 only (ch6_dir\alisttb2.vhd)

6.2.4 Concurrent Procedure Call

A concurrent procedure(LRM 9.3) is equivalent in concept to a component instantiation. Like the component instantiation, the procedure used in the concurrent procedure call is first declared (chapter 7 describes procedures). The procedure could be declared in a package, in the entity declarative section, or in the architecture declarative section. In addition, like a component, a procedure has interfaces with signals, variables (VHDL'93), and constant connections. When a concurrent procedure is instantiated in an architecture it becomes one of the concurrent statements. The connections or associations to the ports, signals and constants of the architecture are made to the parameters of the procedure in a manner similar to the component association list. Thus, when making the association list, it is imperative that the procedure has visibility over the variables, signals, and files being associated.

In addition, just like the reusability of components with multiple component instantiations of the same component, **there can be multiple occurrences of a concurrent procedure with different association lists**. For example, if a TIMING CHECK procedure is written to verify timing violations of a formal signals, then this timing check procedure can be instantiated multiple times. Each instantiation makes a different association between the formal parameters and the actual parameters. Figure 6.2.4 represents a package declaration (described in chapter 8) for two concurrent procedures and an entity/architecture that calls (or instantiates) the concurrent procedures. The package body is not required in the compilation of architectures that use those procedures. The package body, which describes the behavior, is needed though for simulation.

```
library IEEE;
  use IEEE.Std_Logic_1164.all;

package Concurrent_Pkg is                    --   package declaration
                                             --   body described in chptr 6 & 7
  procedure SetupCheck (constant Source_c          : in string;
                        signal   Clock_s           : in Std_Logic;
                        signal   Data_s            : in Std_Logic);

  procedure  HoldCheck (constant Source_c          : in string;
                        signal   Clock_s           : in Std_Logic;
                        signal   Data_s            : in Std_Logic);
end Concurrent_Pkg;

library IEEE;
 use IEEE.Std_Logic_1164.all;

entity ConcProc is
  generic (Setup_g:               time := 30 ns;
           Hold_g:                time := 1 ns);
  port (Ain1     :        in      Std_Logic;
        Ain2     :        in      Std_logic_Vector(31 downto 0);
        ClK      :        in      Std_Logic;
        Aout1    :        buffer  Std_logic;
        Aout2    :        buffer  Std_Logic_Vector(31 downto 0));
end ConcProc;
```

```
library ATEP_Lib;
architecture ConcProc_a of ConcProc is

  use ATEP_Lib.Concurrent_Pkg.all;
  use ATEP_Lib.Concurrent_Pkg;

begin  --  ConcProc_a
  ------------------------------------------------------------------------------
  --  Setup timing checks for all inputs
  ------------------------------------------------------------------------------
  Concurrent_Pkg.SetupCheck
            (Source_c => "Ain1",
             Clock_s  => Clk,
             Data_s   => Ain1);

  Concurrent_Pkg.SetupCheck
            (Source_c => "Ain2",
             Clock_s  => Clk,
             Data_s   => Ain2(0));

  ------------------------------------------------------------------------------
  --  Hold timing checks for all inputs
  ------------------------------------------------------------------------------
  Concurrent_Pkg.HoldCheck
          (Source_c => "Aout1",
           Clock_s  => Clk,
           Data_s   => Aout1);

  Concurrent_Pkg.HoldCheck
          (Source_c => "Aout2",
           Clock_s  => Clk,
           Data_s   => Aout2(31));

  ------------------------------------------------------------------------------
  --  Process:   SetOutputs_Lbl
  --  Purpose:   Sets the outputs through a register
  ------------------------------------------------------------------------------
  SetOutputs_Lbl : process
  begin  --  process SetOutputs_Lbl
    wait until Clk'event and Clk = '1';
    Aout1 <= Ain1 after 10 ns;
    Aout2 <= Ain2;
  end process SetOutputs_Lbl;
end ConcProc_a;
```

Multiple instantiations of the concurrent procedure SetupCheck with different actual parameters

Multiple instantiations of the concurrent procedure HoldCheck with different actual parameters

Figure 6.2.4 Concurrent Procedures Example (ch6_dir\concproc.vhd)

6.2.5 Generate Statement

*[1] A generate statement (LRM 9.7) provides a mechanism for **iterative** or **conditional elaboration** of a portion of a description.*

The **iterative** elaboration of a description is a convenient mechanism to instantiate and replicate concurrent statements. The replication index is either a constant or a generic. This is often used to instantiate and connect components. Some typical applications include the instantiation and connections of multiple identical components (such as half adders to make up a full adder, or exclusive or gates to create a parity tree).

The **conditional** elaboration enables the conditional instantiation of a concurrent statement usually based on a constant or a generic. This is often used to conditionally instantiate a component or a concurrent procedure. For example, if timing check is not needed, then the generate statement can prevent the elaboration of the concurrent timing check procedures. The syntax for the generate is as follows:

generate_statement ::=
  generate_label: generation_scheme
  generate
    {concurrent_statement }
  end generate [generate_label];

generation_scheme ::=
  for generate_parameter_specification
  | **if** condition

Figure 6.2.5 is an example of the use of the generate statement.

```
library IEEE;
  use IEEE.Std_Logic_1164.all;

entity XOR is
  port (A          :  in      Std_Logic;
        B          :  in      Std_Logic ;
        Z          :  out     Std_Logic);
end XOR;

architecture XOR_a of XOR is
begin  --  XOR_a
  Z <= A xor B;
end XOR_a;

--------------------------------------------------------------------------------
--  Entity/architecture with a generate example
--------------------------------------------------------------------------------
library IEEE;
  use IEEE.Std_Logic_1164.all;

entity Generate is
  generic (Setup_g            :     time := 30 ns;
           Hold_g             :     time := 1 ns;
           EnableCheck_g      :     boolean := false);

  port (Ain1      :      in       Std_Logic_Vector(31 downto 0);
        Ain2      :      in       Std_logic_Vector(31 downto 0);
        ClK       :      in       Std_Logic;
        Aout1     :      buffer   Std_logic;
        Aout2     :      buffer   Std_Logic_Vector(31 downto 0));

end Generate;

--------------------------------------------------------------------------------
--  Architecture that demonstrates the use of the generate statement
--------------------------------------------------------------------------------
architecture Generate_a of Generate is

  component XOR                               --   Component declaration
    port (A          :  in      Std_Logic;
          B          :  in      Std_Logic;
          Z          :  out     Std_Logic);
  end component;

  signal  Temp_s     :  Std_Logic_vector(31 downto 0);
```

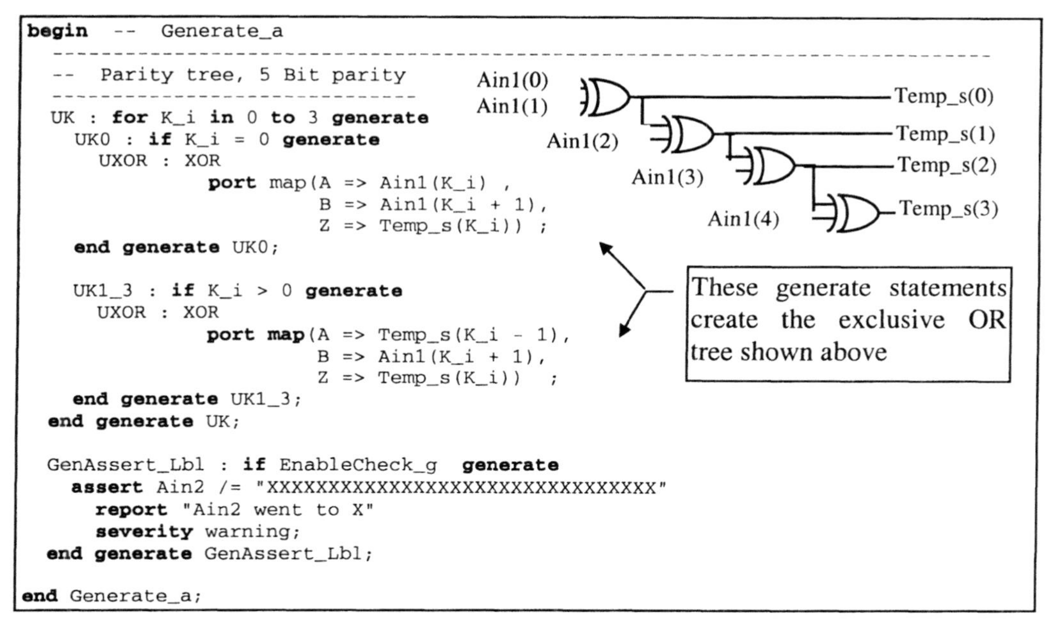

```
begin  -- Generate_a
  -----------------------------------------------------------------------
  -- Parity tree, 5 Bit parity
  -----------------------------
  UK : for K_i in 0 to 3 generate
    UK0 : if K_i = 0 generate
      UXOR : XOR
              port map(A => Ain1(K_i) ,
                       B => Ain1(K_i + 1),
                       Z => Temp_s(K_i)) ;
    end generate UK0;

    UK1_3 : if K_i > 0 generate
      UXOR : XOR
              port map(A => Temp_s(K_i - 1),
                       B => Ain1(K_i + 1),
                       Z => Temp_s(K_i))  ;
    end generate UK1_3;
  end generate UK;

  GenAssert_Lbl : if EnableCheck_g  generate
    assert Ain2 /= "XXXXXXXXXXXXXXXXXXXXXXXXXXXXXXXX"
      report "Ain2 went to X"
      severity warning;
  end generate GenAssert_Lbl;

end Generate_a;
```


Figure 6.2.5 Generate Statement (ch6_dir\generate.vhd)

6.2.6 Concurrent Assertion Statement

[1] (LRM 9.4) A concurrent assertion statement represents a passive process statement (i.e., no signal assignment) containing the sequential assertion statement. Assertions are generally used to report model errors, timing violations, and signal with erroneous values. Assertions are also used as a debugging aid tool to alert the user of a particular situation, and the time occurrence of this situation.

M 👍👍 The following severity levels are recommended:

FAILURE	Error in the model itself
ERROR	Timing violations and invalid data affecting the state of the model.
WARNING	Timing violations and invalid data not affecting the state, but which could affect the simulation behavior of the model (e.g., if data to be sent out from an interface is invalid).
NOTE	Message for debugging use.

Rationale: *Guidelines provided by European Space Agency, and provide consistency in their meaning*

As a general observation, the sequential assertion statement defined in a process or in a concurrent procedure call is more often used than the concurrent assertion statement. This is because more complex synchronization or processing is necessary to determine the assert condition (e.g., condition evaluated at rising edge of clock). The syntax is as follows:

concurrent_assertion_statement ::= -- VHDL'87
    [label :] assertion_statement

assertion_statement ::=
    assert condition -- in VHDL'93, the "assert false " is optional
    [**report** expression] -- of predefined type string
    [**severity** expression]; -- of predefined type severity_level

The assertion statement basically tests the "condition" for normal or good operation, and as long as it is true (i.e., things are normal or OK), no message is reported. However, if the expected or normal condition is false, the report expression is reported to the standard output. The following example demonstrates the point:

```
assert not OpCode_p = "11111111"        -- if OpCode_p = "11111111" the report
    report "Illegal Operation Code"     --    message is displayed
    severity warning;
```

It is important to note that the report expression must be of type string. However, it is possible to report signals or variables by using type conversions, function calls (e.g., the image function in chapter 8), or table lookup conversions as demonstrated in Figure 6.2.6.

```
architecture Assert_a of Assert is
  type B2S_Typ is array(bit) of character;
  type Color_Typ is (Red, Green, Blue);
  type C2S_Typ is array(color_Typ) of string(1 to 5);

  constant Bit2Strg_c  : B2S_Typ :=
      ('0'  => '0',
       '1'  => '1');
  constant Color2Strg_c : C2S_Typ :=
      (Red    => "Red  ",
       Green  => "Green",
       Blue   => "Blue ");

  signal A_s : bit;
  signal Color_s :Color_Typ := Green;

begin  --  Assert_a
  assert Color_s = Red
    report "Color is not Red, it is " & Color2Strg_c(Color_s)
    severity note;

  assert A_s = '1'
    report "A_s /= '1', it is " & Bit2Strg_c(A_s)
    severity note;

end Assert_a;
```

Figure 6.2.6 Use of the Assert Statement (ch6_dir\assert.vhd)

6.2.7 Block Statement

[1] (LRM 9.1) A block is a representation of a portion of the hierarchy of the design. One of the major purpose of a block is to **disable signals** (i.e., the signal drivers) by using a guard expression. *[1] A guard expression is a Boolean-valued expression associated with a block statement that controls assignments to guarded signals within a block. A guard expression defines an implicit signal GUARD that may be used to control the operation of certain statements within the block.* Figure 6.2.7-1 shows the block statement syntax.

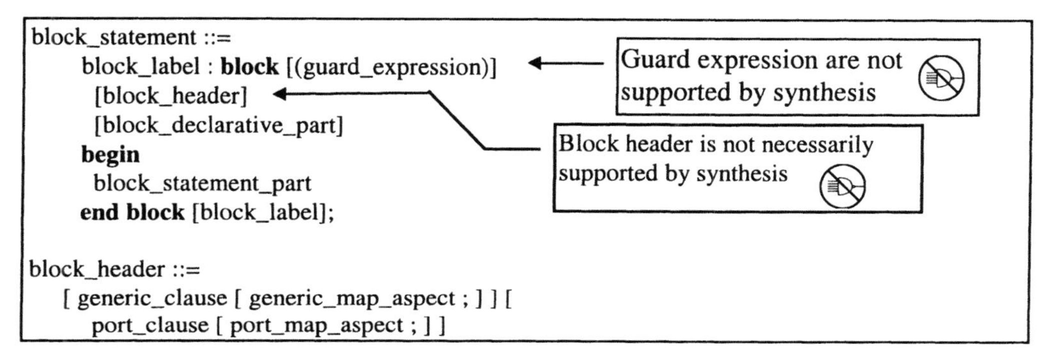

Figure 6.2.7-1 Block Statement Syntax and Example

- The block statement is used in VHDL Initiative Toward ASIC Libraries (*VITAL).* See section 9.7 for more information on *VITAL.*
- Blocks are synthesizable provided signal kind *register* and *bus* and block guards are not used. Use type *Std_Logic* or *Std_Logic_Vector* instead of block guards. In addition, block headers are generally not supported.
- In synthesis, blocks can be used as a separate level of hierarchy to force the synthesizer to compile and optimize sections of the design separately. A complex architecture may present problems in synthesis, layout, and code readability. Partitioning an architecture into blocks provides the following advantages:
  1. Faster synthesis compilation time. Synopsys® provides a *group -hdl_block* directive that groups design partitions and creates new levels of hierarchies. This approach can significantly reduce the compilation time.
  2. Information hiding. Within a block, local type and signal declarations can be defined. The signals within the block are local to the block, and are not visible by the architecture.
  3. Declaration of partition interfaces. The block header enables the definition of local generics, ports, and port maps. All synthesizer vendors do not necessarily support Block headers.
  4. Visibility into architecture. Since a block is a concurrent statement of an architecture, a block has visibility into the ports and signals of the architecture and into the declarations of the component within the architecture. A block also has visibility into packages, constants, and types declared in the entity and architecture. Figure 6.2.7-2 provides an example of a block for a simple design.

Figure 6.2.7-3 represents the same example, but includes a block header. Both examples synthesize in *Synplify-Lite* with *QuickLogic*[8].

```
entity BlkExmpl is
  port(A    : in    Bit_Vector(2 downto 0);
       B    : out   Bit;
       C    : out   Bit);
end BlkExmpl;

architecture BlkExmpl_a of BlkExmpl is
  signal T_s : Bit;            <-- Architectural signal visible by ALL
begin                              concurrent statements of
  Logic_Lbl: block
    signal Local_s : Bit;      <-- Local block signal visible by All concurrent
  begin                            statements of the BLOCK
    -- Concurrent statements
    Local_s <= A(2) and A(1);
    T_s <= not Local_s;
    B <= not A(0);
  end block Logic_Lbl;

  -- Concurrent statement
  C   <= T_s or A(0);
end BlkExmpl_a;
```

Figure 6.2.7-2 Block Example in Synthesis (element\block.vhd)

```
architecture BlkExmpl_a of BlkExmpl is
  signal T_s : Bit;
begin
  Logic_Lbl: block
    port(Data    : in   Bit_Vector(2 downto 0);
         OutB    : out  Bit;
         Temp    : out  Bit);

    port map                   <-- Port header with port declaration
         (Data => A,               and port map. The port header may
          OutB => B,               not be synthesizable.
          Temp => T_s);
    signal Local_s : Bit;
  begin
    -- Concurrent statements
    Local_s <= A(2) and A(1);
    Temp <= not Local_s;
    OutB <= not Data(0);
  end block Logic_Lbl;

  -- Concurrent statement
  C   <= T_s or A(0);
end BlkExmpl_a;
```

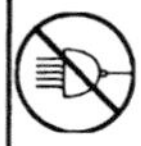

Figure 6.2.7-3 Block Example with Port Header (element\blockh.vhd)

Guarded blocks are not synthesizable. However, guarded blocks enable the partitioning of a complex problem into simple blocks, where each block provides a contribution to the signals when the conditions are ripe (i.e., the guard expression). Since in guarded blocks, the drivers are disconnected if the guard expression is false, the contributions onto signals are only from those blocks whose guard expression is true. If the code is well written, there should only be one connected driver, with all the other drivers disconnected. Conditional signal assignments, with multiple drivers, can be used to emulate blocks by asserting a tri-state values for the emulation of disconnect. If a single driver is desired, then a process is necessary. Figure 6.2.7-4 provides an example of equivalent code using blocks, conditional signal assignments, and a process.

```
...
architecture BlocK_a of Block is
  type Opr_Typ is (AndOP, OrOP, XorOP);
  signal Opr_s   : Opr_Typ;
  signal R1b_s   : Std_Logic register;
  signal R1_s    : Std_Logic;
  signal R1c_s   : Std_Logic;
  signal R2b_s   : Std_Logic register;
  signal R2_s    : Std_Logic;
  signal R2c_s   : Std_Logic;
  signal Clk_s   : Std_Logic;
  signal Data_s  : Std_Logic_Vector(1 downto 0) := "10";

begin  -- BlocK_a
  -- One driver contributes to R2b_s because only one
  -- block statement contributes to the value of R2b_s
  -- (the other bblock statement is disconnected)
  R1_Blk: block(Clk_s = '1' and Opr_s = AndOP)
  begin  -- block R1_Blk
    R1b_s  <= guarded Data_s(1) and Data_s(0);
    R2b_s  <= guarded not(Data_s(1) and Data_s(0));
  end block R1_Blk;

  R2_Blk: block(Clk_s = '1' and Opr_s = OrOp)
  begin  -- block R2_Blk
    R2b_s  <= guarded Data_s(1) or Data_s(0);
  end block R2_Blk;

  ------------------------------------------------------------------------------
  --  Process:   BlockEquiv_Lbl
  --  Purpose:   Equivalent logic to use of blocks
  --          :  Only one driver contributes to R2_s because it is in
  --          :  the same process
  ------------------------------------------------------------------------------
  BlockEquiv_Lbl : process (Clk_s, Opr_s)

  begin  -- process BlockEquiv_Lbl
    if Clk_s = '1' then
      case Opr_s is
        when AndOP  =>
           R1_s  <= Data_s(1) and Data_s(0);
           R2_s  <= not(Data_s(1) and Data_s(0));

        when OrOP   =>
           R2_s  <= Data_s(1) or Data_s(0);

        when XorOP   => null;
      end case;                            -- Opr_s
    end if;
  end process BlockEquiv_Lbl;
```

```
  -- Conditional assignment statement
  R1c_s   <= Data_s(1) and Data_s(0) when
                 (Clk_s = '1' and Opr_s = AndOP)
             else 'Z';
  R2c_s   <= not(Data_s(1) and Data_s(0)) when
                 (Clk_s = '1' and Opr_s = AndOP)
             else 'Z';
  -- 2 drivers for R2c_s
  R2c_s   <= Data_s(1) or Data_s(0) when
                 (Clk_s = '1' and Opr_s = OrOp)
             else 'Z';
end BlocK_a;
```

Figure 6.2.7-4 Application of Block statement (ch6_dir\block.vhd)

1. DO **NOT USE** GUARDED EXPRESSIONS.
2. Use IEEE.Std_Logic_1164 package instead to emulate a Tri-state.

Rationale: *Guarded blocks are not synthesizable. The IEEE-1164 package includes the type Std_Logic and Std_Logic_Vector that are resolved types with the high impedance or disconnect state 'Z'. As a result of this package, guarded expressions, signals and assignments including the reserved words **bus**, **disconnect**, **guarded** and **register** are considered **obsolescent**.*

6.2.7.1 Guarded Signal Assignments

Per LRM 4.3.1.2 and 9.1, *[1] if a signal kind (e.g., register, bus) appears in a signal declaration, then the signals so declared are guarded signals of the kind indicated. Null can be assigned to guarded signals.* Guarded signals can be used in applications where multiple concurrent statements need to either contribute a value onto a signal, or be excluded from such contributions. When a guarded signal is assigned a *null*, that signal assignment is not considered in the resolution function. This contrasts with non-guarded signals where, in *Std_Logic*, the assignment of 'Z' mimics the open contribution or tri-State. However, it does contribute to the resolution function. For example, a CPU with an arithmetic unit, a logic unit, and a floating-point unit needs to send results to a local bus when its computations are valid. With guarded signals, each computational unit (modeled as a concurrent statement), can assign a *null* onto the local bus until it is eligible to assert the computed result. Thus, each computational unit is excluded from contributing a value until a non-*null* is assigned. With *Std_Logic*, each computational unit always contributes a value to the resolution function.

Do not **use guarded signals**

Rationale:
1. *Guarded signals are not synthesizable*
2. *Std_Logic type eliminated the need for guarded statements.*
3. *Guarded signals are harder to read because the guards and the drivers are not collocated.*
4. *Users lack familiarity with guarded signals.*

However, if guards must be used, then the following summarizes rules regarding guarded statements. *[1]*

- *For a guarded signal that is a composite type, each subelement is likewise a guarded signal.*
- *A guarded signal is assigned values under the control of Boolean-valued guard expressions (or guards).*
- Guards can be implicitly or explicitly defined. *If a guard expression appears after the reserved word block, then a signal with the simple name GUARD of predefined type BOOLEAN is implicitly declared at the beginning of the declarative part of the block, and the guard expression must be type BOOLEAN. The implicit signal GUARD must not have a source.* Alternatively, a signal called *GUARD* of type Boolean can be explicitly be declared. That explicit signal can have a source.
- *When a given guard becomes False, the drivers of the corresponding guarded signals are implicitly assigned a null transaction to cause those drivers to turn off.*
- *A disconnect specification is used to specify the time required for those drivers to turn off.*

The value of signal GUARD is always defined within the scope of a given block, and does not implicitly extend to design entities bound to components instantiated within the given block. However, the signal GUARD may be explicitly passed as an actual signal in a component instantiation in order to extend its value to lower-level components.

- *In a concurrent signal assignment, the option GUARDED specifies that the signal assignment statement is executed when a signal GUARD changes from FALSE to TRUE, or when that signal has been TRUE and an event occurs on one of its inputs.*
- *A **register** is a kind of guarded signal that retains its last driven value when all of its drivers are turned off.*

Figure 6.2.7.1-1 represents an application of guarded statements and its testbench. Figure 6.2.7.1-2 represents the simulation results.

```
library IEEE;
  use IEEE.Std_Logic_1164.all;
  use IEEE.Std_Logic_Unsigned.all;

entity TGuard is
  generic(Width_g : Integer := 32);
  port (In1    : in   Std_Logic_Vector(Width_g - 1 downto 0);
        In2    : in   Std_Logic_Vector(Width_g - 1 downto 0);
        Opr    : in   Std_Logic_Vector(1 downto 0);
        Result : out  Std_Logic_Vector(Width_g - 1 downto 0));
end TGuard;

architecture TGuard_a of TGuard is
  constant Plus_c    : Std_Logic_Vector(1 downto 0) := "00";
  constant And_c     : Std_Logic_Vector(1 downto 0) := "01";
  constant MultHi_c  : Std_Logic_Vector(1 downto 0) := "10";
  constant MultLo_c  : Std_Logic_Vector(1 downto 0) := "11";
                                                                 Guarded signal
  signal Result_s  : Std_Logic_Vector(Width_g - 1 downto 0) bus;
```

```
begin  --  TGuard_a
  ALU_Blk: block(Opr = Plus_c)          <-- Guard signal assigned TRUE if Opr = Plus_c
  begin  --  block ALU_Blk                  (implicit use Guard signal)
    Result_s  <= guarded In1 + In2;
  end block ALU_Blk;

  Logical_Blk: block
    signal Guard : Boolean;             <-- Explicit Guard
  begin  --  block Logical_Blk
    Guard <= Opr = And_c;
    Result_s <= guarded In1 and In2;
  end block Logical_Blk;

  --  Process:   Multiply_Lbl
  Multiply_Lbl : process(Opr)
    variable Product_r : Std_Logic_Vector((2 * Width_g) - 1 downto 0);
  begin  --  process Multiply_Lbl
    Product_r := In1 * in2;
    if Opr = MultHi_c then
      Result_s <= Product_r((2 * Width_g) - 1 downto Width_g);

    elsif Opr = MultLo_c then
      Result_s <= Product_r(Width_g - 1 downto 0);

    else                                 <-- Assigning null to
      Result_s <= null;                      guarded signal
    end if;
  end process Multiply_Lbl;

  Result <= Result_s;
end TGuard_a;

library IEEE;                           <-- Testbench
  use IEEE.Std_Logic_1164.all;

entity TGuardTB is
    generic(Width_g : Integer := 32);
end TGuardTB;
                                                        <-- Constants to
architecture TGuardTB_a of TGuardTB is                      enhance readability
  constant Plus_c     : Std_Logic_Vector(1 downto 0)  := "00";
  constant And_c      : Std_Logic_Vector(1 downto 0)  := "01";
  constant MultHi_c   : Std_Logic_Vector(1 downto 0)  := "10";
  constant MultLo_c   : Std_Logic_Vector(1 downto 0)  := "11";

  component TGuard
    generic(Width_g : Integer := 32);
    port (In1    : in   Std_Logic_Vector(Width_g - 1 downto 0);
          In2    : in   Std_Logic_Vector(Width_g - 1 downto 0);
          Opr    : in   Std_Logic_Vector(1 downto 0);
          Result : out  Std_Logic_Vector(Width_g - 1 downto 0));
  end component;

  signal In1    : Std_Logic_Vector(Width_g - 1 downto 0);
  signal In2    : Std_Logic_Vector(Width_g - 1 downto 0);
  signal Opr    : Std_Logic_Vector(1 downto 0);
  signal Result : Std_Logic_Vector(Width_g - 1 downto 0);
```

```
begin
  In1 <= "10101010000010101010001111110100 1" after 100 ns,
         "11111111111111110000000000000000" after 150 ns,
         "00000000111111100000000001111000" after 200 ns,
         "11111110000000100011100100001111" after 250 ns;

  In2 <= "10000000000000000000001111101001" after 100 ns,
         "10000000000000000000001110000000" after 150 ns,
         "00111111110000000000000001111000" after 200 ns,
         "11000000000000111111111111001111" after 250 ns;

  process
  begin
    Opr <= "00" after 75 ns,
           "01" after 85 ns,
           "10" after 100 ns,
           "11" after 120 ns;
    wait for 150 ns;
  end process;


  U1: TGuard
    generic map(Width_g => 32)
    port map(In1    => In1,
             In2    => In2,
             Opr    => Opr,
             Result => Result);
end TGuardTB_a;
```

Figure 6.2.7.1-1 Application of Guarded Statements and Testbench (element\tguard.vhd)

ns	delta	in1	In2	opr	result	Insruction
0	+2	XXXXXXXX	XXXXXXXX	X	ZZZZZZZZ	
75	+2	XXXXXXXX	XXXXXXXX	0	XXXXXXXX	Plus
85	+3	XXXXXXXX	XXXXXXXX	1	XXXXXXXX	And
100	+3	AA1547E9	800003E9	2	550AA68D	MultHi
120	+2	AA1547E9	800003E9	3	8D362E11	MultLo
150	+0	FFFF0000	80000380	3	8D362E11	
200	+0	00FE0078	3FC00078	3	8D362E11	
225	+2	00FE0078	3FC00078	0	40BE00F0	Plus
235	+3	00FE0078	3FC00078	1	00C00078	And
250	+3	FE02390F	C007FFCF	2	BE899AAC	MultHi
270	+2	FE02390F	C007FFCF	3	6A0B1421	MultLo

Figure 6.2.7.1-2 Simulation Results for Model shown in Figure 6.2.7.1-1

Figures 6.2.7.1-3 and 6.2.7.1-4 represent another example for the potential use of guarded statements where three functional blocks perform separate operations in data. If an instruction is for a specific block, then that block executes the operation and supplies the result onto a resolved signal of type *integer*. The resolution function for this resolved *integer* is relatively simple since only one driver will drive a value onto the guarded signal (*Rint_s*) while the other drivers will drive a *null*. That resolution function detects the condition of multiple drivers and reports an error. If there are no drivers, then it returns *Integer'low*. If there is one driver, it then returns the value of that driver. The simulation results are shown in Figure 6.2.7.1-5.

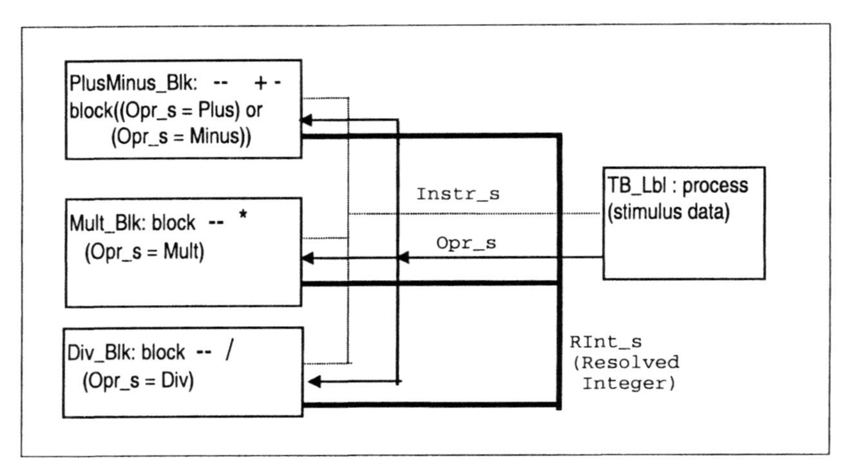

Figure 6.2.7.1-3 Functional Diagram for T4Guard Model

```
entity T4Guard is
end T4Guard;

architecture T4Guard_a  of T4Guard is
  type Opr_Typ is (Plus, Minus, Mult, Div, None);

  type Instr_typ is record
   Int1    : Integer;
   Int2    : Integer;
  end record;

  type AInteger_Typ is array(Natural range <>) of Integer;

  function Resolve(D : AInteger_Typ) return Integer is
  begin
    assert (D'length = 1 or  D'length = 0)
      report "More than one driver of Integer type"
      severity Warning;

    if D'length = 0 then
      return Integer'low;
    else
      return D(D'low);
    end if;
  end Resolve;

  subtype RInteger_Typ is Resolve Integer;

  signal Instr_s   : Instr_typ :=
             (Int1     => 5,
              Int2     => 6);
  signal RInt_s : RInteger_Typ bus;

  signal Opr_s     : Opr_Typ;
```

```
begin  --  T4Guard_a
  PlusMinus_Blk: block((Opr_s = Plus) or (Opr_s = Minus))
  begin  --  block Int_Blk
    Int_Lbl : process(Opr_s)
    begin  --  process Flt_Lbl
      if Guard then
        if Opr_s = Plus  then
          RInt_s <= Instr_s.Int1 + Instr_s.Int2;
        elsif Opr_s = Minus then
          RInt_s <= Instr_s.Int1 - Instr_s.Int2;
        end if;
      else
        RInt_s <= null;
      end if;
    end process Int_Lbl;
  end block PlusMinus_Blk;

  Mult_Blk: block(Opr_s = Mult)
  begin  --  block Int_Blk
    RInt_s <= guarded (Instr_s.Int1 * Instr_s.Int2);
  end block Mult_Blk;

  Div_Blk: block(Opr_s = Div)
  begin  --  block Int_Blk
    RInt_s <= guarded (Instr_s.Int1 / Instr_s.Int2);
  end block Div_Blk;

  ---------------------------------------------------------------------------
  --  Process:   TB_Lbl
  --  Purpose:   Provide stimulus
  ---------------------------------------------------------------------------
  TB_Lbl : process
  begin  --  process TB_Lbl
    TestOpr_Lbl: for I in Opr_Typ loop
      Opr_s  <=  I;
      Instr_s.Int1    <= Instr_s.Int1 + 1;
      Instr_s.Int2    <= Instr_s.Int2 + 1;
      wait for 100 ns;
    end loop TestOpr_Lbl;
  end process TB_Lbl;
end T4Guard_a ;
```

In a process, one must assign a null if Guard is false

In a concurrent signal assignment , the *null* is automatically assigned if the Guard is false

Figure 6.2.7.1-4 Application of Guards (element\T4Guard.vhd)

```
-- Simulation results
--          ns   delta                      instr_s       rint_s opr_s
--           0      +1                      (6, 7)            11  plus
--         100      +2                      (7, 8)            -1 minus
--         200      +2                      (8, 9)            72  mult
--         300      +2                     (9, 10)             0   div
--         400      +2                    (10, 11) -2147483648  none
--         500      +2                    (11, 12)            23  plus
--         600      +2                    (12, 13)            -1 minus
--         700      +2                    (13, 14)           182  mult
--         800      +2                    (14, 15)             0   div
--         900      +2                    (15, 16) -2147483648  none
```

Figure 6.2.7.1-5 Simulation Results for T4Guard Model shown

EXERCISES

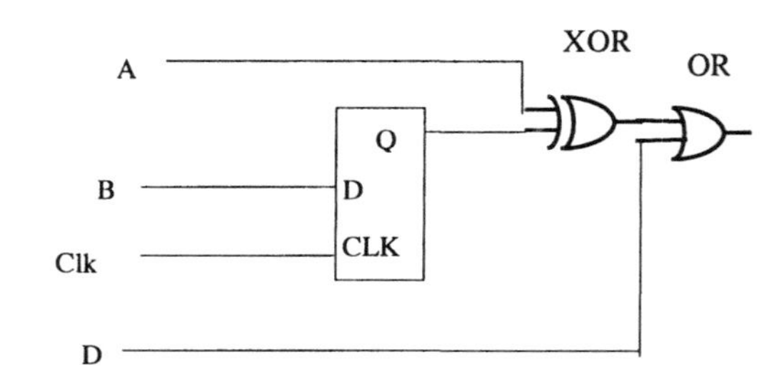

1. Given the following circuit, write a VHDL description that describes its behavior. Use the following:
- 1 process to describe the positive edge triggered flip-flop
- 1 process to describe the random logic
- 1 component instantiation of the circuit. The component includes the "in" *ports A, B, D, Clk* and the "out" port *Z*. All ports are of type Std_Logic.
- 1 concurrent statement to simulate the clock (50 ns '0', 50 ns '1').
- 3 concurrent statements to generate waveforms on A, B, and D. Try to get a good mix of combinations.
- 1 concurrent assertion statement that notes that 'Z' equals '0'. Use the "generate" based on the value of a generic to generate this statement.
- Assume 0 delay for everything.

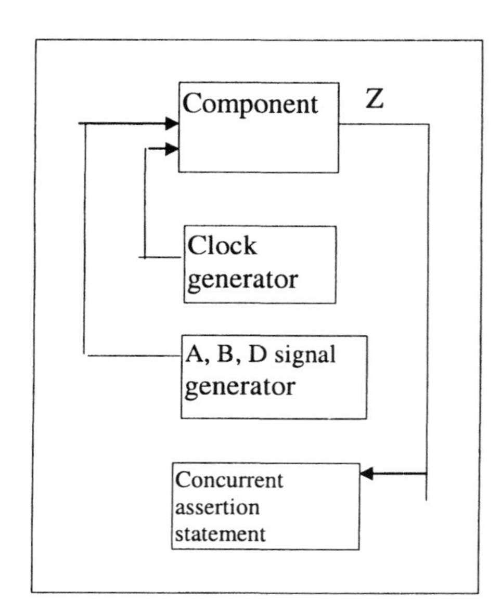

2. Write a model for a 16 bit pseudo-random (PN) number generator whose output is of type IEEE.Std_Logic_Vector. The PN generator shall consist of a 16 bit register (R (15 downto 0)). This register shifts left at every clock. The data into bit 15 is equal to R(0) xor R(4) xor R(13) xor (R15).

7. SUBPROGRAMS

This chapter defines procedures, functions, concurrent procedures, and overloaded operators. It also defines the rules and methodologies in the application of subprograms. Techniques are presented for separating high level tasks from low level bus protocols using subprograms and modular design approaches.

7.1 SUBPROGRAM DEFINITION

(LRM 2.1) [1] Subprograms define algorithms for computing values or exhibiting behavior. Subprograms include procedures and functions. *[1] They may be used as computational resources to define a portion of a sequential operation, convert between values of different types, to define the resolution output values driving a common signal, to define portion of a process,* or to define a concurrent procedure statement.

[1] There are two forms of subprograms:

1. **Procedure:** A procedure is a subroutine that performs operations using all the visible parameters and objects, and can modify one or more of the visible parameters and objects in accordance to certain rules. A procedure may include **formal parameters** that define the objects used in the execution of a procedure. A procedure call is a call to the procedure with the **actual parameters** being mapped onto the **formal parameters.** A procedure may include ***wait*** statements. There are two types of procedures:
    - Sequential procedures called from within processes.
    - Concurrent procedures instantiated as concurrent statements.
2. **Function:** A function is a routine that returns a value. The function defines how the return value is computed based on the values of the **formal parameters**. A function

call is a call to the function with the **actual parameters** being mapped onto the **formal parameters.**

Figure 7.1-1 represents the relationship between a process and a sequential procedure call.

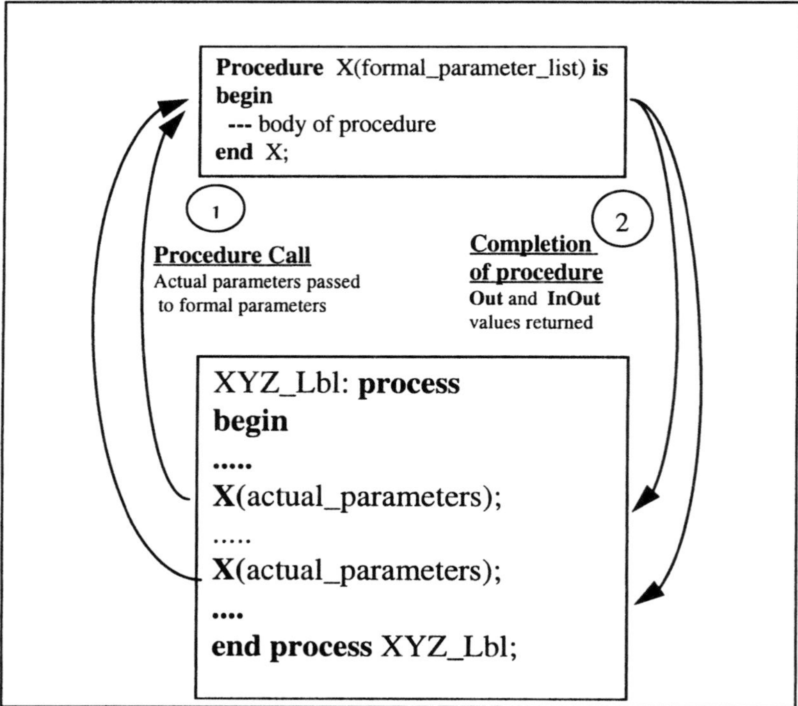

Figure 7.1-1 Relationship between a Process and a Sequential Procedure Call

A sequential procedure call is equivalent to inserting the contents of the procedure body in-line with the code. Subprograms are used in two different classes:

1. Commonly used subprograms that are highly reusable, such as *Std.TextIO.Write*, convert functions from *IEEE.Std_Logic_Vector* to *integer*, and user defined type conversions.

2. Subprograms that provide information hiding to increase readability, such as sending data. Those subprograms tend not to be reusable.

Use procedures instead of in-line code to "can" or group operations performing a particular purpose (e.g., send data using a specific protocol). Note that procedures have many restrictions in synthesis (see chapter 13).
Rationale: *Procedures tend to modularize the code, and create more compact code for the body of processes. In addition, procedures are reusable, thus increasing flexibility and minimization of errors because there is less code duplication.*

The subprogram syntax is shown below.

```
subprogram_body ::=
    subprogram_specification is
        subprogram_declarative_part
    begin
        subprogram_statement_part
    end [ subprogram_kind ] [designator] ;

subprogram_statement_part ::=
    { sequential_statement }

formal_parameter_list ::=
    parameter_interface_list

subprogram_kind ::= procedure | function

parameter_interface_list ::=
    interface_element { ; interface_element }

interface_element ::=
    interface_declaration

interface_declaration ::=
     interface_constant_declaration
    | interface_signal_declaration
    | interface_variable_declaration
    | interface_file_declaration     -- '93 only

interface_variable_declaration ::=
    [variable] identifier_list : [mode]
        subtype_indication [:= static_expression]

interface_file_declaration ::= -- '93 only
    file identifier_list : subtype_indication
     -- NO MODE for interface_file_declaration

subprogram_specification ::=
        procedure designator
          [(formal_parameter_list)]
       | [ pure | impure ] function designator
          [(formal_parameter_list)]
          return type_mark;

subprogram_declarative_part ::=
       { subprogram_declarative_item }

subprogram_declarative_item ::=
        subprogram_declaration
       | subprogram_body
       | type_declaration
       | subtype_declaration
       | constant_declaration
       | variable_declaration
       | file_declaration
       | alias_declaration
       | attribute_declaration
       | attribute_specification
       | use_clause

interface_constant_declaration ::=
       [constant] identifier_list : [in]
           subtype_indication [:= static_expression]

interface_signal_declaration ::=
       [signal] identifier_list : [mode]
           subtype_indication
              [:= static_expression]

mode ::= in -- default
       | out
       | inout
```

The subprogram syntax shows that the formal parameter interface declaration consists of five parts:

1. A class definition (i.e., constant, variable, signal, file).
2. An interface identifier or the name of the formal parameter.
3. A mode for the parameter. These mode can be
   - **in** -- Specifies that the parameter is an input or READ only.
   - **out** -- Specifies that the parameter is an output or a WRITE
   - **Inout** -- Specifies that the parameter is READ and WRITTEN.
4. A subtype indication.
5. An optional static expression (i.e., the initialized value of the object).

Figure 7.1-2 is a subprogram example used to demonstrate several subprogram concepts.

7.2 SUBPROGRAM RULES AND GUIDELINES

7.2.1 Unconstrained Arrays in Subprograms

[1] If a formal parameter is of unconstrained array (e.g., string, Bit_Vector*) then the subtype of the formal in any call to the subprogram is taken from the actual associated parameter.* Thus, the size and direction of the formal is taken from the actual parameter. This is demonstrated in the subprogram ShiftRight in Figure 7.1-2.

M 👍👍 Whenever possible, use unconstrained arrays, instead of constrained arrays, to define the type of the subprogram formal parameters that are arrays.

Rationale: *Use of the unconstrained array causes the procedure to be more useful or "generic" because the sizes of the arrays are determined by the size of the actual parameters. Thus, the subprogram can be reused for different array lengths.*

M 👍👍 Do not rely on the direction and range of the actual parameter. Instead, if direction and range are needed, normalize the range and direction with the use of either aliases or variables. Example:

```
procedure ShiftRight(variable Data_v      : inout Bit_Vector) is
  alias    Data1_v : Bit_Vector(Data_v'length - 1 downto 0) is Data_v;
  variable Data2_v : Bit_Vector(Data_v'length - 1 downto 0) := Data_v;
begin -
```

Warning: User defined aliases are currently not supported in synthesis.

Rationale: *Range and direction of actual parameters are not predictable. To create a reusable subprogram, normalization is necessary. The use of attributes with aliases or variable can bind the size and direction of the formal parameters of the arrays to enable access to individual array elements in a controlled and readable manner. Otherwise, a subprogram might become invalid.*

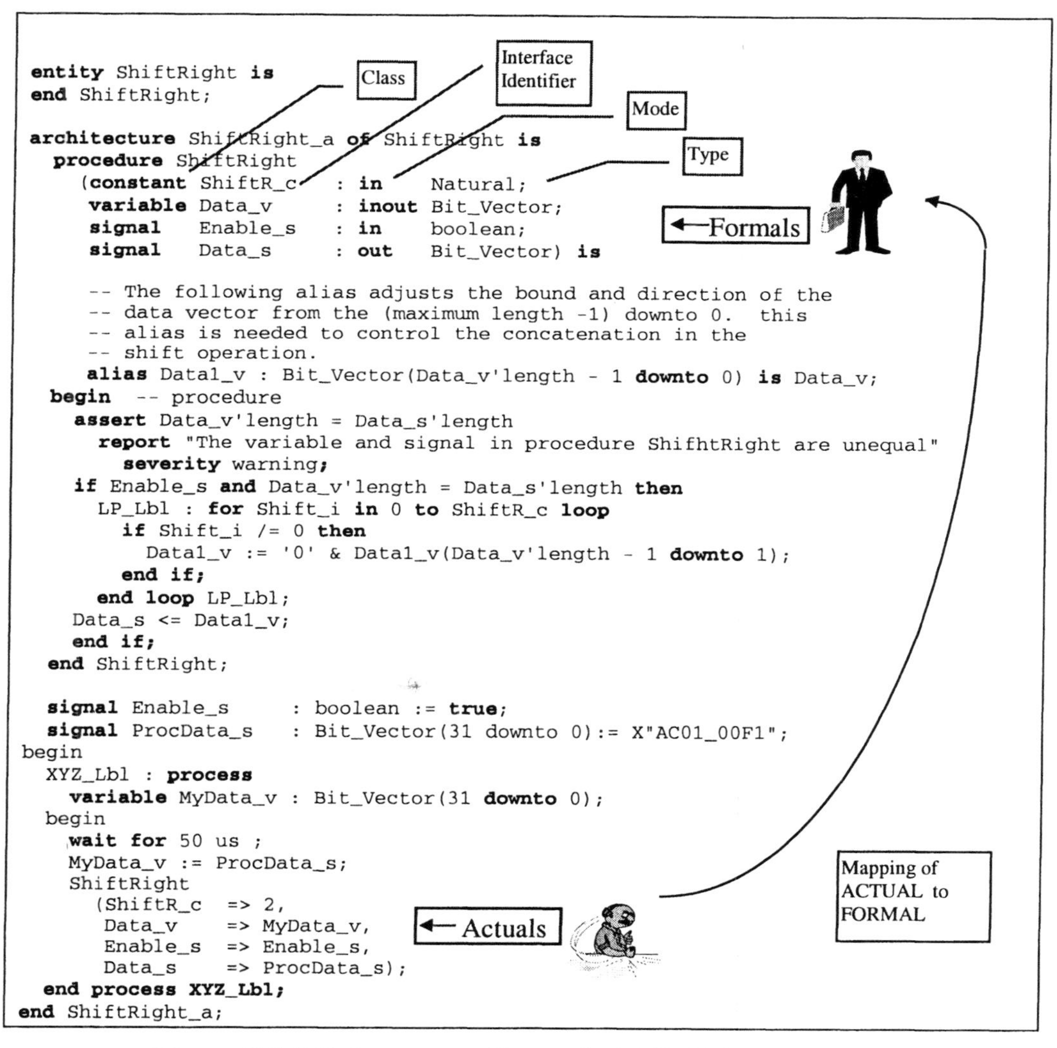


```
entity ShiftRight is
end ShiftRight;

architecture ShiftRight_a of ShiftRight is
  procedure ShiftRight
    (constant ShiftR_c   : in    Natural;
     variable Data_v     : inout Bit_Vector;
     signal   Enable_s   : in    boolean;
     signal   Data_s     : out   Bit_Vector) is

     -- The following alias adjusts the bound and direction of the
     -- data vector from the (maximum length -1) downto 0.  this
     -- alias is needed to control the concatenation in the
     -- shift operation.
     alias Data1_v : Bit_Vector(Data_v'length - 1 downto 0) is Data_v;
  begin  -- procedure
    assert Data_v'length = Data_s'length
      report "The variable and signal in procedure ShifhtRight are unequal"
        severity warning;
    if Enable_s and Data_v'length = Data_s'length then
      LP_Lbl : for Shift_i in 0 to ShiftR_c loop
        if Shift_i /= 0 then
          Data1_v := '0' & Data1_v(Data_v'length - 1 downto 1);
        end if;
      end loop LP_Lbl;
    Data_s <= Data1_v;
    end if;
  end ShiftRight;

  signal Enable_s     : boolean := true;
  signal ProcData_s   : Bit_Vector(31 downto 0):= X"AC01_00F1";
begin
  XYZ_Lbl : process
    variable MyData_v : Bit_Vector(31 downto 0);
  begin
    wait for 50 us ;
    MyData_v := ProcData_s;
    ShiftRight
      (ShiftR_c  => 2,
       Data_v    => MyData_v,
       Enable_s  => Enable_s,
       Data_s    => ProcData_s);
  end process XYZ_Lbl;
end ShiftRight_a;
```

Figure 7.1-2 Subprogram Example (ch7_dir\shftrght.vhd)

7.2.2 Interface class declaration

In a subprogram call, the actual designators associated with the formal parameters are dependent upon the class of the formal and actual parameters. Table 7.2.2 summarizes the class associations in subprograms. Figure 7.2.2-1 and Figure 7.2.2-2 provide examples for the application of those rules.

Table 7.2.2 Class Association in Subprograms

FORMAL DESIGNATOR CLASS	ACTUAL DESIGNATOR CLASS	COMMENTS
signal	**signal**	**Signals** can only be associated with **signals**
variable	**variable** **file** (VHDL'87 only)	**Variables** can only be associated with **variables**. In VHDL'87 a file object is a member of the variable class of objects (LRM'87 4.3.2). This is not the case for VHDL'93. Thus, for VHDL'87 an actual designator of class **file** can be associated with a formal designator of class **variable.**
constant	**signal**, **variable**, **constant** literal or an expression	A formal of class constant can be associated with an actual of class **signal, variable,** or an expression.
file	**file** (VHDL'93 only)	**files** can only be associated with **files**

MI👍👍 If the subprogram mode is **in,** then use the class **variable** if:

1. The type of the identifier is of access type (e.g., *Std.TextIO.line*).
2. The type of the identifier is of file type for VHDL'87 (e.g., *Std.TextIO.Text*).

If signal attributes of the formal parameters are used (e.g., *S'last_value*) then use the class **signal.** Otherwise, use the class **constant.**

Rationale: *When the mode is* ***in*** *and the class is* ***constant*** *then the actual parameter can then be of the class:* ***constant*** *literal, an expression, a* ***variable****, or a* ***signal.*** *This provides an enhanced flexibility in the procedure calls. If the class of the formal is defined as a* ***variable****, then ONLY a* ***variable*** *actual can be passed. If the class is defined as a* ***signal****, then ONLY a* ***signal*** *actual can be passed.*

Class ***Variable*** *must be used for identifiers of type access because the actual can only be of class* ***variable*** *(an object of type* ***access*** *cannot be a* ***signal*** *or a* ***constant*** *or an expression). In addition, in VHDL'87 a* ***file*** *object is a member of the* ***variable*** *class, and thus the actual can only a variable. Class* ***file*** *is not allowed in subprograms for VHDL'87.*

```
architecture FileIO_a of FileIO is
begin  --  FileIO_a
  File_Lbl : process
    use Std.TextIO.all;
    use Std.TextIO;
    file DataIn_f        : TextIO.text is in  "datain.txt";

    procedure Test
      (variable Line_v : in TextIO.Line;          -- ☺
       variable File_v : in TextIO.Text) is       -- ☺

    begin  --  Test
    end Test;

    procedure TestError
      (constant Line_v : in TextIO.Line;          --
       constant File_v : in TextIO.Text) is       --
    begin  --  TestError
    end TestError;

  begin  --  process ReadWriteFile_Lbl
    wait;
  end process File_Lbl ;
end FileIO_a;
```

Object with access type must be a variable

Object with file type must be a **variable** -- VHDL'87 only --Must be **file** for -- VHDL'93

Figure 7.2.2-1 Restrictions on Class Variable and File (VHDL'87) (ch7_dir\subp_ea.vhd)

```
architecture ModeIn_a of ModeIn is
  procedure ProcOr
    (constant Data_c     : in    Bit_Vector;
     variable Data_v     : in    Bit_Vector;
     signal   Enable_s   : in    boolean;
     constant Enable_c   : in    boolean;
     signal   Data_s     : out   Bit_Vector) is

  begin  -- procedure
    if Enable_s and Enable_c and
        Data_s'length = Data_v'length and
        Data_s'length = Data_c'length then
      Data_s <= Data_v or Data_c;
    end if;
  end ProcOr;

  signal   Enable_s      : Boolean := true;
  signal   ProcData_s    : Bit_Vector(31 downto 0)  := X"AC01_00F1";
  constant ProcData_c    : Bit_Vector(31 downto 0)  := X"FFAA_BCD0";
```

Procedure declaration

```
begin
  XYZ_Lbl : process
    variable MyData_v : Bit_Vector(31 downto 0);
    variable IsGood_v : boolean := true;
    constant Good_c   : boolean := true;
  begin
    wait for 50 us;
    MyData_v := not ProcData_s;
    ProcOr                                 -- Association:
      (Data_c     => X"FFAA_BCD0",         --   Literal is OK
       Data_v     => MyData_v,             --   Must be a variable, OK
       Enable_s   => Enable_s,             --   Must be a signal, OK
       Enable_c   => Enable_s,             --   Signal, literal, variable OK
       Data_s     => ProcData_s);          --   Must be a signal, OK
    wait for 1 ns;

    ProcOr
      (Data_c     => MyData_v,             --   Variable is OK
       Data_v     => X"FFAA_BCD0",         --   Illegal, must be a variable
       Enable_s   => IsGood_v,             --   illegal, must be a signal
       Enable_c   => IsGood_v,             --   Variable is OK
       Data_s     => ProcData_s);          --   line 58
    wait for 10 ns;

    ProcOr
      (Data_c     => ProcData_s,           --   Signal is OK
       Data_v     => ProcData_c,           --   Illegal, must be a variable
       Enable_s   => Good_c,               --   Illegal, must be a signal
       Enable_c   => Good_c,               --   OK
       Data_s     => ProcData_s);          --   OK

    ProcOr
      (Data_c     => ProcData_s,           --   OK
       Data_v     => ProcData_s,           --   Illegal, must be a signal
       Enable_s   => true,                 --   Illegal, must be a signal
       Enable_c   => true,                 --   OK
       Data_s     => ProcData_s);          --   OK
  end process XYZ_Lbl;
end ModeIn_a;
```

Figure 7.2.2-2 Restrictions on Class Variable and File (VHDL'87) (ch7_dir\modein.vhd)

If an interface class is not specified, the class **constant** is assumed for a mode **in.** If the mode is **out**, then the default class is **variable.**

Always define the interface class in procedures. For functions use the default class (i.e., **constant)** if no other class is used.

Rationale: *This method enhances readability and constancy when writing procedures and functions.*

If no mode is explicitly given in the interface declaration then mode **in** is assumed.

Always declare the **mode of the formal procedure parameters**. For **functions**, the formal parameters are always of mode **in**, and thus, the mode should **NOT be specified**.

Rationale: *This method enhances readability and constancy when writing procedures and functions.*

7.2.3 Subprogram Initialization

[1] If an interface declaration contains a ":=" symbol, the expression is said to be the default expression of the interface object. The type of the default expression must be that of the corresponding interface object. It is an error if the default expression appears in an interface declaration and any of the following conditions hold:

1. *The interface object is a formal signal parameter*
2. *The interface object is a formal variable parameter of mode other than **in**.*

Figure 7.2.3 demonstrates initialization rules of subprogram formal parameters.

```
architecture SubError_a of SubError is
  procedure ErrorConditions
       (constant Count_c     : in    natural  := 10;
        constant Sting_c     : in    string; -- size determined by actual
        constant CountErr_c  : out   bit;    -- ERROR, can't return a constant 💣
        constant CountErr2_c : inout bit;   -- ERROR, can't return a constant  💣

        variable Value1_v    : in    positive := 5;
        variable Value2_v    : in    positive;
        variable Value3_v    : out   Bit_Vector;
        variable Value4_v    : inout natural;
        --    can't initialize an OUT or INOUT
        variable Value5_v    : inout natural := 10;     -- ERROR 💣
        variable Value6_v    : out   natural := 10;     -- ERROR 💣
        signal   Clk_s       : in    bit;
        signal   Data1_s     : in    natural := 100; -- ERROR, Init. on signal 💣

                                --    can't initialize an OUT or INOUT
        signal   Data2_s     : out   natural := 100;  -- ERROR 💣
        signal   Data3_s     : inout bit_Vector := "1011";--    ERROR 💣

        signal   Data4_s     : out   natural;
        signal   Data5_s     : in    bit_Vector) is

      -- Procedure declarations go here
      variable Result_v   : natural;
    begin
      Result_v := Data1_s;
      Value6_v := 2 * Result_v;
    end ErrorConditions;

begin  --   SubError_a
end SubError_a;
```

Figure 7.2.3 Initialization Rules of Subprogram Formal Parameters (ch7_dir\suberror.vhd)

[1] If a procedure or function initializes its ***in*** *parameters to a default value, then when the procedure or function is called, the actual parameter can override the default value if it is passed. If the actual parameter is omitted, the value of the actual parameter used by the subprogram is the default value.*

M When desired, initialize the **constant in** formal parameters to default values, and position those formal parameters as the last parameters in the list.

Rationale: *Initialization of constants is useful when defining a default value for that constant. Positioning the initialized constants as the last parameters in the list enables a call to the subprogram without explicitly specifying the actual parameters for which the default values are used.*

For example, in package *Std.TextIO* the following procedure is declared.

```
procedure Write(L          : inout LINE;
                VALUE      : in    bit;
                JUSTIFIED  : in    SIDE := right;   -- default specified
                FIELD      : in    WIDTH := 0);     -- default specified
```

A call to the write procedure can be written as follows:

```
Write(OutLine_v, '1'); -- default are used for Justified and Field
Write(Outline_v, '1', left); -- default used for field
Write(L         => OutLine_v,
      VALUE     => '1',
      JUSTIFIED => right,
      FIELD     => 2);
Write(L         => OutLine_v,
      VALUE     => '1',   -- default used for Justified
      FIELD     => 2);
```

7.2.4 Subprogram Implicit Signal Attributes

[1] It is an error if signal-valued attributes 'stable, 'quiet, 'transaction, and 'delayed of formal signal parameters of any mode are read within a subprogram. This restriction is necessary because subprograms are elaborated at call time, and cease to exist when they are completed. However the signals valued attributes *'stable, 'quiet, 'transaction, and 'delayed are* signals, and signals are only created at elaboration time, and not at runtime.

M If it is necessary to use the implicit signals, then declare a formal signal of the implicit type, and pass as the actual the implicit signal.

Rationale: *signal-valued attributes are signals, and cannot be accessed from within the subprogram. However, passing signal-valued attributes as actual parameters is legal.*

Figure 7.2.4 is an example of setup procedures that demonstrates this concept.

```
architecture Set_a of Set is
  signal         D_s                : Std_Logic := '0';
  signal         Clk_s              : Std_Logic := '0';

-------------------------------------------------------------------------------
-- Procedure with error in using implicit signals inside procedure
-------------------------------------------------------------------------------
  procedure SetUpError
    (signal    Clk_s         : in    Std_Logic;
     signal    D_s           : in    Std_Logic;
     constant  SetUpTime_c   : in    time;
     variable  Violation_v   : out   boolean) is

  begin
    --                             ->|   |<- hold
    --         --+    <-- setup ---->+------
    --   CLK     |___________________|
    --
    --         -----+                  +---
    --   D_S        |__________________|
    --
    --   Check rising edge of clock
    if (Clk_s'event and Clk_s ='1') then
      --   Check setup time
      Violation_v := not D_s'stable(SetUpTime_c) < SetUpTime_c; --  #46
    end if;
  end SetUpError;
```

Attribute stable may not be accessed from parameter d_s.

```
-------------------------------------------------------------------------------
-- If Implicit signals are used, they must be passed
-------------------------------------------------------------------------------
  procedure SetUpFix
    (signal    Clk_s         : in    Std_Logic;
     signal    Dstable       : in    boolean;  --   pass D_s'stable(T)
     variable  Violation_v   : out   boolean) is

  begin
    --   Check rising edge of clock
    if (Clk_s'event and Clk_s ='1') then
      Violation_v := not Dstable;     --    (no implicit signals)

    end if;
  end SetUpFix;


-------------------------------------------------------------------------------
-- Implicit signals can sometimes be avoided,
-------------------------------------------------------------------------------
  procedure SetUp
     (signal    Clk_s         : in    Std_Logic;
      signal    D_s           : in    Std_Logic;
      constant  SetUpTime_c   : in    time;
      variable  Violation_v   : out   boolean) is


  begin
    --   Check rising edge of clock
    if (Clk_s'event and Clk_s ='1') then --    (no implicit signals)
      --   Check setup time
      Violation_v := D_s'last_event < SetUpTime_c;
    end if;
  end SetUp;
```

```
begin  --  Set_a
...
-------------------------------------------------------------------------------
--  Process:   SetCheck_Lbl
-------------------------------------------------------------------------------
  SetCheck_Lbl : process (D_s, Clk_s)
    constant SetUpTime_c    :  time := 10 ns;
    variable Violation_v    :  boolean;
  begin
    SetUpError(Clk_s        => Clk_s,       --   error is in the procedure
               D_s          => D_s,
               SetUpTime_c => SetUpTime_c,
               Violation_v => Violation_v);

    SetUpFix(Clk_s        => Clk_s,
             Dstable      => D_s'stable(SetUpTime_c),
             Violation_v => Violation_v);

    SetUp(Clk_s        => Clk_s,
          D_s          => D_s,
          SetUpTime_c => SetUpTime_c,
          Violation_v => Violation_v);
  end process SetCheck_Lbl;
end Set_a;
```

Figure 7.2.4 Example of Setup Procedures (ch7_dir\sethsubp.vhd)

7.2.5 Passing Subtypes

If a formal parameter is of a parent type (e.g., *Std_uLogic, integer*) **then the actual parameter can be of either the parent type or a subtype of the parent type** (e.g., *Std_Logic, Intger0to5_Typ*). Figure 7.2.5 demonstrates this concept.

```
Architecture Test_a of Test is
  subtype Intger0to5_Typ is integer range 0 to 5;

  procedure Increment(variable D_v   : inout  integer) is
  begin
    D_v := D_v + 1;
  end Increment;

begin
  Test_Lbl: process
    variable T_v : Intger0to5_Typ := 0;
  begin
    Increment(T_v);
    wait;
  end process Test_Lbl;
end Test_a;
```

Figure 7.2.5 Actual Parameters as Subtypes of Formal Parameters (ch7_dir\test_ea.vhd)

7.2.6 Drivers in Subprograms

*[1] A process statement contains a driver for each actual signal associated with a formal signal parameter of mode **out or inout** in a subprogram call. A subprogram contains a driver for each formal signal parameter of mode out or inout declared in its subprogram specification. For a signal parameter of mode **inout** or **out**, the driver of an actual signal is associated with the corresponding driver of the formal signal parameter at the START of each call. Thereafter, during the execution of the subprogram, an assignment to the driver of the formal signal parameter is equivalent to an assignment to the driver of the actual.* This concept is demonstrated in figure 7.2.6-1, where the procedure *Invert* is called from process *A_Lbl*. At the call and during the execution of the procedure, signal drivers *Data_s* becomes associated with formal parameter *A_s*. At termination of the procedure, formal parameter *A_s* has no more significance. Driver *Data_s* was created during elaboration because it is a signal of the architecture.

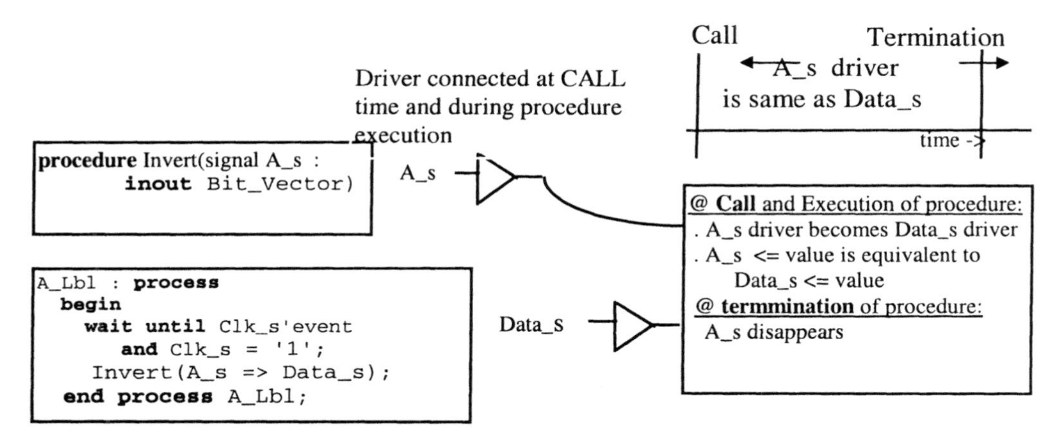

Figure 7.2.6-1 Drivers in Procedure Calls

Figure 7.2.6-2 represents a model to demonstrate drivers in procedure calls.

Every signal assignment in a process (or equivalent process) creates one (and only one) driver for that signal. If a procedure, declared in the <u>declarative portion of a process</u>, attempts to assign a value onto a signal or port that is not in the formal parameter list, then that transaction is legal for the following reasons:

1. Signals of an architecture or ports of the entity are visible by the process.
2. The process owns the driver associated with that signal or port.

If a procedure, declared in the <u>declarative portion of architecture</u>, attempts to assign a value onto a signal or port <u>that is not in the formal parameter list</u>, then that transaction is in <u>error</u> because there is NO owner for the signal driver. To remedy this problem a formal parameter list of class **signal** is needed. See section 7.2.8 on recommendations for side effects and means to separate high level tasks from low level protocols.

```
architecture Driver_a of Driver is
  procedure Invert
    (signal A_s : inout Bit_Vector) is
  begin
    A_s <= not A_s;  -- A_s has a driver  @ call time
    wait for 100 ns;
    A_s <= not A_s;
    wait for 200 ns;
    A_s <= not A_s;
  end Invert;

  signal Clk_s   : Bit;
  signal Data_s  : Bit_Vector(7 downto 0) := X"AB";

begin  --  Driver_a

  Clk_s <= not Clk_s after 50 us;  -- One driver for Clk

  A_Lbl : process
  begin
    wait until Clk_s'event and Clk_s = '1';
    Invert(A_s => Data_s);  -- Data_s has a driver
  end process A_Lbl;
  -- during the call, assigment to A_S driver
  -- is equivalent to driver on Data_s
end Driver_a;
```

Figure 7.2.6-2 Model for Drivers in Procedure Calls (ch7_dir\driver.vhd)

7.2.7 Signal Characteristics in Procedure Calls

[1] ***The actual signal associated with a signal parameter must be static.*** *No conversion functions or type conversion must appear in either the formal part or the actual part of an association element that associates an actual signal with a formal signal parameter.* In other words, the range of the actual cannot be dynamic or change during the simulation. Figure 7.2.7-1 demonstrates an example of this requirement.

```
entity Static is
  generic (Numb_g   :  natural := 4);
end Static;

architecture Static_a of Static is
  procedure Invert(signal Data_s : inout Bit_Vector) is
    begin
      Data_s <= not Data_s after 10 ns;
    end Invert;

  function Set2One(constant Data_s : Bit_Vector)
      return  Bit_Vector is
    begin
      return Data_s or not Data_s;
    end Set2One;

  signal D_s      : Bit_Vector(7 downto 0) := X"A5";
  signal Numb_s   : natural := 5;
```

```
begin  --  Static_a
                                        The actual for parameter data_s must
  XYZ_Lbl : process                     denote a static signal name.
  begin
    Invert(Data_s => D_s(7 downto Numb_s));    --  Not static
    wait for 100 ns;

    Invert(Data_s => D_s(7 downto Numb_g));    -- ☺
    wait;

    Invert(Data_s => Set2One(D_s));   --  No conversion function allowed
  end process XYZ_Lbl;
end Static_a;
```

Figure 7.2.7-1 Actual Parameter must be Static (ch7_dir\static.vhd)

[1] If an actual signal is associated with a formal signal parameter, and if the formal parameter is a constrained array subtype, then it is an error if the actual does not contain a matching element of the formal. Figure 7.2.7-2 demonstrates this rule.

```
architecture More_a of More is
  subtype BV10 is Bit_vector(9 downto 0);
  signal  Bus9_s   : Bit_Vector(8 downto 0);
  signal  Bus10_s  : Bit_Vector(9 downto 0);
  signal  Bus1_s   : Bit;

  procedure Invert(signal In_s : inout Bit_Vector(9 downto 0))  is
   begin
    In_s <= not In_s;
  end Invert;

begin  --  More_a

  Test_Lbl : b
  begin
    Invert(Bus10_s);          -- OK
    wait for 1 ns;

    Invert(Bus1_s & Bus9_s);  --
          -- The result of "&" is an expression and not a signal
    wait for 1 ns;

    Invert(Bus9_s);  -- Actual length is 9, expected 10
  end process Test_Lbl;
end More_a;
```

Figure 7.2.7-2 Matching elements in procedure calls for constrained array signals (ch7_dir\morerule.vhd)

Subelement association can be used to associate multiple actual signals to formal parameters. This is equivalent in hardware to the "joining" of several busses or signals into a component. The subelement association can be used for actual parameters of procedures or components. Figure 7.2.7-3 represents an example of an *invert* procedure in which two signals are associated with the input formal parameter.

```
architecture More_a of More is
  signal  Bus1_s   : Bit := '1';
  signal  Bus9_s   : Bit_Vector(8 downto 0) := "101000101";
  signal  Bus10_s : Bit_Vector(9 downto 0) := "1111000010";

  procedure Invert(signal In_s  : in   Bit_Vector;
                   signal Out_s : out  Bit_Vector)  is
   begin
    assert In_s'length = Out_s'length
      report "In_s /= Out_s"
      severity warning;
    Out_s <= not In_s;
  end Invert;

begin  --  More_a
  Test_Lbl : process
  begin
    -- sequential procedure call
    -- NO Implied "wait on" statements.
    Invert(In_s(9)            => Bus1_s,
           In_s(8 downto 0) => Bus9_s,
           Out_s              => Bus10_s);
    wait for 10 ns;
  end process Test_Lbl;
end More_a;
```

This is equivalent to Bus1_s & Bus9_s except that the concatenation is **ILLEGAL** because the "&" is an expression

Figure 7.2.7-3 Subelement Association (ch7_dir\ morerul2.vhd)

7.2.8 Side Effects

A subprogram is said to have a side effect if its behavior is dependent not only on the formal parameters, but also on other parameters that the subprogram has visibility or access. Thus, a subprogram has side effects if it affects or changes something (variable, signal, or file) not declared in its parameter list. Procedures and impure functions (VHDL'93) declared in an architecture have **read** visibility to the ports of the entity and signals declared in the architecture. They have **read** visibility over variables and loop counter declared in the calling process. They also have **read** access to global signals and shared variables (VHDL'93) declared in packages made visible to the architecture. All this visibility is made without explicitly declaring the objects in the formal parameter list of procedures. Procedures and impure functions (VHDL'93) declared in the declarative section of a process have in addition to the **read** visibility, **write** access to signals and ports of the architecture because the drivers are associated with the process.

An example of a procedure with side effect is the WRITE procedure for a Bus Functional Model (BFM). The formal parameters of the procedure include the address and the data, whereas the procedure body accesses the ports and other signals of the architecture. This can be used to stimulate the control signals necessary to emulate the WRITE algorithm. The procedure with side effects omits from the procedure the low level control signals. Thus, the procedure call would be something like:

```
-- with Side effects
WRITE(Address_s  => "ABCD_0000",
      Data_s      => "1234_5678");
```

control signals omitted because of visibility rules. 👎👎

```
-- Without Side Effects
WRITE(Address_s  => "ABCD_0000",
      Data_s      => "1234_5678",
      RwF_s       => RwF,
      Parity_s    => Prty,
      Cntrlx_s    =>  Cntrlx1,
      ....
```

```
                            Cntrlz_s    =>  Cntrlz1);
```

Avoid side effects on procedures. Avoid impure functions with side effects. If the side effects are unavoidable or desirable for some explicit reasons, then document the side effects in the header of the procedure declaration.

Rationale: *This loose visibility rule can cause unexpected results when not used correctly.*

7.2.8.1 Separating High Level Tasks From Low Level Protocols

Tasks deal with high level jobs such as READ, WRITE, AND SEND DATA. Tasks represent the system concept of the desired transactions. Protocol deals with manipulation of port interfaces to achieve the system tasks. Handling complex tasks at the protocol level hinders readability, maintainability and is more prone to errors because of the complexity (see also chapter 10).

When writing a procedure that accesses a large number of signals to achieve the protocol function (e.g., a microprocessor bus), the following options are available:

1. **Write a procedure within the process declaration section**, but do not use the signal names in the formal parameter list. Rely instead on the visibility rules and on the fact that the process will own the drivers for those signals if an assignment is made. This approach is **NOT RECOMMENDED** because it is poor software practice since procedure **has side effects** and is not well documented.

2. **Write a procedure** (in a package, architectural declaration section, or process declaration section) that includes ALL the signals used in the formal parameter list. This approach requires the association of the actual signals to the formal signals when the procedure is called. This approach does not provide information hiding and supports good design practices.

3. **Declare task control signals in the architecture of the design.** Let a *client* processs or component define the sequence of tasks to be executed. Those tasks are assigned onto the task signals and are sensed by a *server* process that handles the low level protocol interface. This *server* process translates the high level command from the requesting *client* process into low level control signals at the interface ports. Figure 7.2.8.1-1 demonstrates the concept of task control. Figure 7.2.8.1-2 represents a VHDL example implementing this concept with procedures. Chapter 10 expends upon these notions.

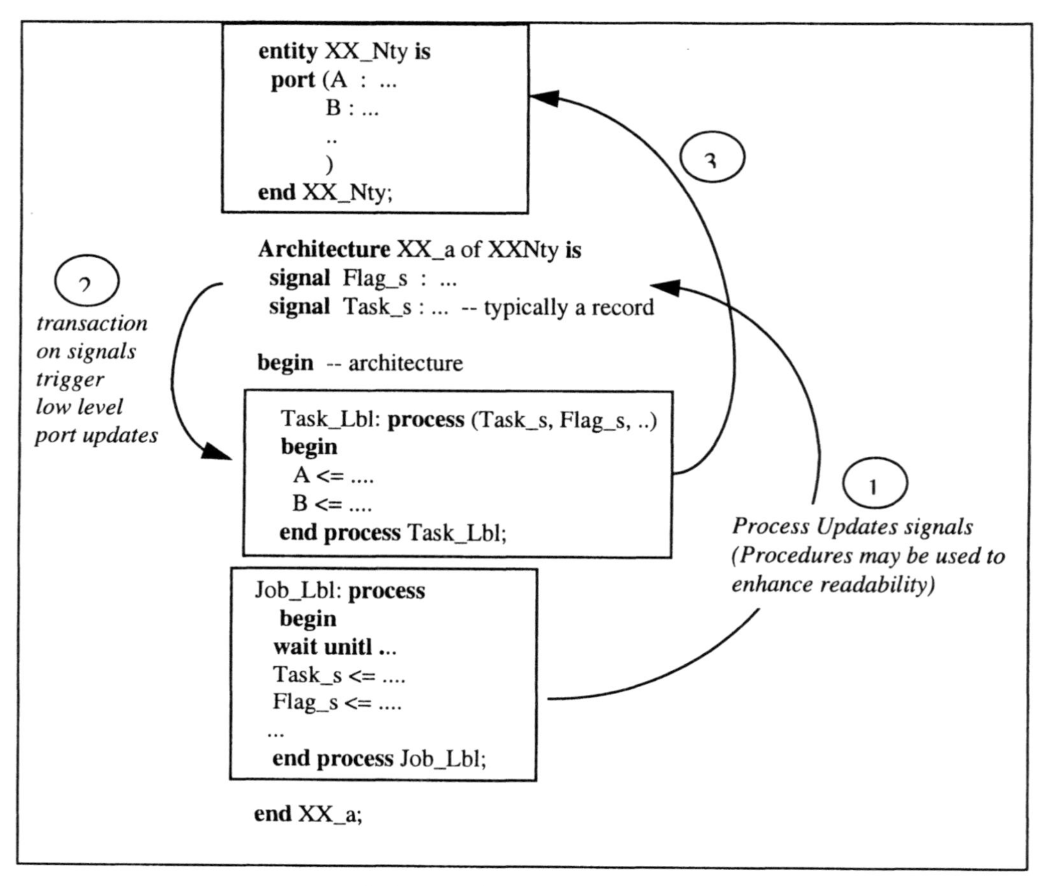

Figure 7.2.8.1-1 Task Control Concept

```
library IEEE;
  use IEEE.Std_Logic_1164.all;
entity Hide is
  port (Data      : inout   Std_Logic_Vector(7 downto 0);
        Address   : out     Std_logic_Vector(7 downto 0);
        RdF       : out     Std_Logic;
        DataRdy   : out     Std_Logic;
        Clk       : in      Std_Logic
       );
end Hide;

architecture Hide_a of Hide is
  type Action_typ is (Read, Write);
  type Task_Typ is record
    Address   : Std_Logic_Vector(7 downto 0);
    Data      : Std_Logic_Vector(7 downto 0);
    Action    : Action_typ;
  end record;

  signal Task_s : Task_Typ;
```

```
  ------------------------------------------------------------------------
  --  Procedure: GetData
  --  Purpose:   Request to send a READ request onto the bus
  --  Inputs:    The request address (no control)
  --  Outputs:   The task to be handle by the process Task_Lbl
  ------------------------------------------------------------------------
  procedure   GetData(constant Address_c : in  Std_Logic_Vector;
                      signal   Task_s    : out Task_Typ) is
    variable Task_v : Task_Typ;
  begin  --  GetData
    Task_v.Address  :=  Address_c;
    Task_v.Action   :=  Read;
    Task_s  <= Task_v;  -- send task
  end GetData;


begin  --  Hide_a
  ------------------------------------------------------------------------
  --  Process:   Task_Lbl
  --  Purpose:   Accepts a high level task and converts it
  --         :   to low level control signals
  ------------------------------------------------------------------------
  Task_Lbl : process
  begin  --  process Task_Lbl
    wait on Task_s'transaction;
    case Task_s.Action is
      when Read    =>
        Data       <= (others  => 'Z');
        Address    <= Task_s.Address;
        RdF        <= '1';
        DataRdy    <= '1';
        wait until Clk'event and Clk = '1';
        Data       <= (others  => 'Z');
        Address    <= (others  => 'H');
        RdF        <= '1';
        DataRdy    <= '0';

      when Write   =>
        Data       <= Task_s.Data;
        Address    <= Task_s.Address;
        RdF        <= '0';
        DataRdy    <= '1';
        wait until Clk'event and Clk = '1';
        Data       <= (others  => 'Z');
        Address    <= (others  => 'H');
        RdF        <= '1';
        DataRdy    <= '0';
    end case;                           --  Task_s.Action
  end process Task_Lbl;

  ------------------------------------------------------------------------
  --  Process:   HighLevelJob_Lbl
  --  Purpose:   Provides high level jobs to be executed on the bus
  ------------------------------------------------------------------------
  HighLevelJob_Lbl : process

  begin  --  process HighLevelJob_Lbl
    wait until Clk'event and Clk = '1';
    GetData(Address_c  =>  "00001010",
            Task_s     =>  Task_s);
    wait;
  end process HighLevelJob_Lbl;
end Hide_a;
```

Low level protocol handles by a separate process, not involved with the task generation.

Figure 7.2.8.1-2 Task Control Concept (ch7_dir\drvproc.vhd)

7.2.9 Positional and Named Notation

[1] Positional association may not follow named association, and vice versa.

M 👍👎 Just like aggregates for arrays and records, parameters for procedures and functions should be passed by named notation, unless the number of parameters is less than two.

Rationale: *Named notation enhances readability. However, when the number of parameters is few, Positional notation is superior.*

7.3 SUBPROGRAM OVERLOADING

[1] (LRM 2.3) A given subprogram designator can be used in several subprogram specifications. The subprogram designator) is then said to be overloaded; the designated subprograms are also said to be overloaded and *to overload each other. If two subprograms overload each other, one of them can hide the other only if both subprograms have the same parameter and result type profile. A call to an overloaded subprogram is ambiguous (and in error) if the name of the subprogram, the number of parameter associations, the types and order of the actual parameters, the names of the formal parameters (when named association is used), and the result type (for functions) are not sufficient to identify exactly one overloaded subprogram specification.* Thus, it is an error if the compiler can't figure it out. Examples:

```
procedure Check(constant Setup_c : in time;       -- Setup check
                signal   D_s     : in Bit_Vector;
                signal   Clk_s   : in Bit);

procedure Check(constant Hold_c  : in time;       -- Hold check
                signal   D_s     : in Bit_Vector;
                signal   Clk_s     : in Bit);
```

```
-- Procedure calls
-- Non Ambiguous calls
Check (Setup_c => 10 ns,
       D_s     => DataBus,
       clk_s   => Clk);

Check (Hold_c  => 4 ns,
       D_s     => DataBus,
       clk_s   => Clk);
```

```
-- Ambiguous calls
Check (10 ns, DataBus, Clk); -- Setup?
Check (4 ns, DataBus, Clk);  -- Hold?
-- Which check should be used ??
-- Can't tell.
```

7.4 FUNCTIONS

Functions were defined in section 7.1, and the rules defined about subprograms are applicable to functions since functions are subprograms. *[1] An impure function (in VHDL'93 only) may return a different value each time it is called, even when different calls have the same actual parameter values. A pure function returns the same value each time it is called with the same values as actual parameters.*

Avoid the definition of impure functions.

Rationale: *Impure functions have side effects. If they must be used, clearly identify the rationale for using it, and the parameters that the impure function depends upon. Impure functions are not compatible with VHDL'87.*

*The default class and mode for subprograms is **constant** and **in**. Formal parameters of functions can only be of mode **in**.*

when the class of all the formal parameters is **constant**, do NOT identify the class in the function declarations. Also, do NOT identify the mode of the formal parameters for function declarations.

Rationale: *Enhances code readability. The mode of functions is always **in**, and thus it is redundant to identify the mode. Since the class and mode of subprograms default to **constant** it is not necessary to specify the class when all the formal parameters are of class **constant.** However, when classes other than **constant** are also used in functions (e.g., signal), the identification of the class for all the formal parameters enhances readability.*

*[1] A **RETURN** statement is used to complete the completion of the innermost enclosing function or procedure body. A return statement appearing in a procedure body must not have an expression. A return statement appearing in a function body must have an expression. The value of the expression defines the result returned by the function.*

Figure 7.4 represents examples of functions. More examples are provided in chapter 8, 10 and 12.

```
...
architecture Function_a of Function is
-------------------------------------------------------------------------
--  Function: IsAtoZ determines if a character is in the range of 'A' to 'Z'
-------------------------------------------------------------------------
  function IsAtoZ(Cin_c    : Character) return boolean is
    begin
      return (Cin_c > '@' and Cin_c < '[');
    end IsAtoZ;

-------------------------------------------------------------------------
--  Function: TimingError
-------------------------------------------------------------------------
  function TimingError
      (signal    D_s           : Std_logic;
       constant SetUpTime_c   : time) return boolean is
    begin
      return D_s'last_event < SetUpTime_c;
    end TimingError;
```

Default class **constant** and mode **in** used for clarity

class is specified for all parameters because class of all parameters is not the default **constant. Default** mode in is used for clarity

```
---------------------------------------------
--  Function: Bit2Std converts a bit type to Std_Logic
---------------------------------------------
  function Bit2Std(Bit_c  : Bit) return Std_Logic is
    begin
      case Bit_c is
        when '0'  => return '0';
        when '1'  => return '1';
      end case;                           --  Bit_c
    end Bit2Std;
```

```
--   Overloaded function operating on a bit vector
```

```
  function Bit2Std(Bit_c  : Bit_Vector) return Std_Logic_Vector is
    alias Bits_c : Bit_Vector(Bit_c'length - 1 downto 0) is Bit_c;
    variable Bits_v : Std_Logic_Vector(Bits_c'range);
    begin
      Convert_Lbl : for Idx_i in Bits_c'range loop
        Bits_v(Idx_i) := Bit2Std(Bits_c(Idx_i)); --   call to above function
      end loop Convert_Lbl;
      return Bits_v;
    end Bit2Std;
-----------------------------------------------------------------------------
--  Function: Std2Natural  converts a Std_Logic_Vector to natural number
--            Std_Logic_Vector number is unsigned
--            Number of bits cannot exceed 31 since this is the
--            largest natural number = 2**31 -1
-----------------------------------------------------------------------------
  function Std2Natural(Bits_c  : Std_Logic_Vector)  return natural is
    alias ABits_c : Std_Logic_Vector(Bits_c'length - 1 downto 0) is Bits_c;
    variable  Result_v : natural := 0;
    type Powers_Typ is array(0 to 30) of natural;

    constant Powers_c : Powers_Typ :=    --   2**nth power weights
      (0  => 1,
       1  => 2,
       2  => 4,
       3  => 8,
       4  => 16,
       5  => 32,
       6  => 64,
       7  => 128,
       8  => 256,
       9  => 512,
       10 => 1024,
       11 => 2 ** 11,
       12 => 2 ** 12,
       13 => 2 ** 13,
       14 => 2 ** 14,
       15 => 2 ** 15,
       16 => 2 ** 16,
       17 => 2 ** 17,
       18 => 2 ** 18,
       19 => 2 ** 19,
       20 => 2 ** 20,
       21 => 2 ** 21,
       22 => 2 ** 22,
       23 => 2 ** 23,
       24 => 2 ** 24,
       25 => 2 ** 25,
       26 => 2 ** 26,
       27 => 2 ** 27,
       28 => 2 ** 28,
       29 => 2 ** 29,
       30 => 2 ** 30);
      begin
      if Abits_c'length > 31 then
```

Efficient use of constant for table lookup

```
        return 0;
        assert false
          report " Input number is out of range"
          severity warning;
      end if;
      Weights_Lbl : for Ix_i in ABits_c'range loop
        if Abits_c(Ix_i) = '1' then
          Result_v := Result_v + Powers_c(Ix_i);
        end if;
      end loop Weights_Lbl;
      return Result_v;
    end Std2Natural;

    signal Data_s  : Std_logic := '0';

begin  --  Function_a
  --  Process:   Test_Lbl
  --  Purpose:   Test the functions

  Test_Lbl : process
    variable Char_v    : character;
    variable Std_v     : Std_Logic_Vector(15 downto 0) := "0000100000001011";
    variable Bits_v    : Bit_Vector(15 downto 0) := X"00A4";
    variable Bit_v     : Bit  := '1';
    variable IsTrue_v  : boolean;
    variable Number_v  : natural;
    constant SetUp_c   : time := 20 ns;
  begin  --  process Test_Lbl
    Char_v := 'V';
    IsTrue_v := IsAtoZ(Char_v);          <---- Function calls
    Char_v := '1';
    IsTrue_v := IsAtoZ(Char_v);
    Char_v := 'z';
    IsTrue_v := IsAtoZ(Char_v);

    Std_v(0)  := Bit2Std(Bit_v);
    Std_v     := Bit2Std(Bits_v);

    Number_v := Std2Natural(Std_v);
    wait for 110 ns;                          <---- Function calls
    if TimingError(D_s          => Data_s,
                   SetUpTime_c  => SetUp_c) then
      assert false
        report "Timing error on Data_s"
          severity warning;
    end if;
    wait; -- suspend process
  end process Test_Lbl;

  --Concurrent statement
  Data_s <= '1' after 100 ns,
            '0' after 200 ns,
            'H' after 300 ns;
end Function_a;
```

Figure 7.4 Examples of Functions (ch7_dir\morefnct.vhd)

7.5 RESOLUTION FUNCTION

[1] (LRM 2.4) A resolution function is a function that defines how the values of multiple sources of a given signal are to be resolved into a single value for that signal. Resolution functions are associated with signals that require resolution by including the name of the resolution function in the declaration of the signal or in the declaration of the subtype of the signal (the preferred approach*). A signal with an associated resolution function is called a resolved signal.*

Resolution functions are ***implicitly*** *invoked during each simulation cycle in which corresponding resolved signals are active* (i.e., more than on driver onto the same signal). *Each time a resolution function is invoked, it is passed an array value, each element of which is determined by a corresponding source of the resolved signal.*

Resolution functions must be **associative** or order independent. Thus, the order of the calculation must not modify the result.

Figure 7.5-1 demonstrates the declarations of resolved types as defined in package Std_logic_1164.

```
-- std_logic_1164 Package declarations, the following is declared:
type Std_uLogic is ( 'U',   -- Uninitialized
                     'X',   -- Forcing  Unknown
                     '0',   -- Forcing  0
                     '1',   -- Forcing  1
                     'Z',   -- High Impedance
                     'W',   -- Weak     Unknown
                     'L',   -- Weak     0
                     'H',   -- Weak     1
                     '-'    -- Don't care
                   );
type Std_uLogic_vector is array ( natural range <> ) of Std_uLogic;
-- resolved  is the name of the resolution function
function Resolved( s : Std_uLogic_Vector ) return Std_uLogic;

-- Subtype Std_Logic and type Std_Logic_Vectors are resolved type.
-- Thus, the resolution function "resolved" is implicitly called
-- if there are more than 1 driver sharing a common signal.
subtype Std_Logic is resolved Std_uLogic;
type std_logic_vector is array ( natural range <>) of Std_Logic;

type Stdlogic_Table is array(Std_uLogic, Std_uLogic) of Std_uLogic;
-- In PACKAGE std_logic_1164  package body, the following is defined:
constant Resolution_Table : Stdlogic_Table := (
    --      ---------------------------------------------------------
    --      | U    X    0    1    Z    W    L    H    -          |   |
    --      ---------------------------------------------------------
            ( 'U', 'U', 'U', 'U', 'U', 'U', 'U', 'U', 'U' ), -- | U |
            ( 'U', 'X', 'X', 'X', 'X', 'X', 'X', 'X', 'X' ), -- | X |
            ( 'U', 'X', '0', 'X', '0', '0', '0', '0', 'X' ), -- | 0 |
            ( 'U', 'X', 'X', '1', '1', '1', '1', '1', 'X' ), -- | 1 |
            ( 'U', 'X', '0', '1', 'Z', 'W', 'L', 'H', 'X' ), -- | Z |
            ( 'U', 'X', '0', '1', 'W', 'W', 'W', 'W', 'X' ), -- | W |
            ( 'U', 'X', '0', '1', 'L', 'W', 'L', 'W', 'X' ), -- | L |
            ( 'U', 'X', '0', '1', 'H', 'W', 'W', 'H', 'X' ), -- | H |
            ( 'U', 'X', 'X', 'X', 'X', 'X', 'X', 'X', 'X' )  -- | - |
       );
```

```
function Resolved(S : Std_uLogic_Vector ) return Std_uLogic is
      variable Result : Std_uLogic := 'Z';  -- weakest state default
   begin
      -- the test for a single driver is essential otherwise the
      -- loop would return 'X' for a single driver of '-' and that
      -- would conflict with the value of a single driver unresolved
      -- signal.
      if    (S'length = 1) then    return S(S'low);
      else
          for I in S'range loop
              Result := Resolution_Table(Result, S(i));
          end loop;
      end if;
      return result;
   end resolved;
```

Resolution function effectively scans all the sources of the signals and resolves the final value.

Figure 7.5-1 Declarations of Resolved Types in Package Std_Logic_1164

Another example of a resolution function for signals of type boolean where TRUE dominates is shown in figure 7.5-2.

```
-- TRUE wins over FALSE
  function fb_Resolve (DRIVERS : Bool_Array) return boolean is
    variable Found_True : boolean := false;
  begin
    LoopThruAllDrivers: for Idx_i in DRIVERS'range loop
      if DRIVERS(Idx_i) then
        Found_True := true;
        exit LoopThruAllDrivers;
      end if;
    end loop LoopThruAllDrivers;
    return Found_True;
  end fb_Resolve;

  subtype Rbool_type is fb_Resolve boolean;
```

Figure 7.5-2 resolution function for type boolean where True dominates (ch7_dir\frslvbol.vhd)

Figure 7.5-3 demonstrates an example of a poorly written resolved function that violates the order dependency rule. A correction to this function is also provided in this figure.

```
-- this Function is order dependent.
-- Thus 3 drivers with value '1', '0', and '1' would return
-- nand('1', '0') nand '1' = '1' nand '1' = '0' which is
-- incorrect.  If the order of evaluation is '1', '1', '0' the
-- result would have been a '1'

function BadResolveNand(Drivers : Bit_Vector) return Bit is  --
  alias Drivers_c : Bit_Vector(1 to Drivers'length) is Drivers;
  variable Result_v : Bit := Drivers_c(1);
begin
  Lp_Lbl : for Ix_i in 2 to Drivers_c'right loop
    Result_v := result_v nand Drivers_c(Ix_i);
  end loop Lp_Lbl;
  return result_v;
end BadResolveNand;
```

```
function OKResolveNand(Drivers : Bit_Vector) return Bit is --
  alias Drivers_c : Bit_Vector(1 to Drivers'length) is Drivers;
  variable Found0_v  : boolean := false;
begin
  Lp_Lbl : for Ix_i in Drivers_c'range loop
    if Drivers_c(Ix_i) = '0' then  -- search for a '0'
      Found0_v := true;
      exit Lp_Lbl;
    end if;
   end loop Lp_Lbl;
  If Found0_v then
     return '1'
  else
    return '0';
  end if;
end OKResolveNand;
```

Figure 7.5-3 Example of a Resolved Function (ch7_dir\BadGood.vhd)

Resolution function for a record type is demonstrated in section 8.1.6 because it also represents an application of a record.

7.6 OPERATOR OVERLOADING

VHDL operators (e.g., "*", "/", "+", "-", "and") are written as infix operators, but are really functions that operate on left and right operands, and return a value of a certain type. These operators are predefined for certain types of operands. For example the "*" (multiplication) function is predefined for left operands of type **integer**, and right operand of type **integer**. However, it is not defined for a left operand of type **integer** and a right operand of type **real**. To enhance code readability, VHDL (like *Ada)* allows the operators to be overloaded, thus operating on types defined by the user. Figure 7.6-1 is an example of overloading the operator "*" for operations between integer and real objects. Figure 7.6-1 represents the overloaded operator "+" for Std_Logic_Vector types. A Two's complement function that uses the "+" operator is also demonstrated.

```
--   Title       :  Operations on Real with overloaded operators
--   Description :  Sample operations on Real numbers
architecture Overload_Beh of Overload is
  subtype Real5to10_Typ is real range 5.0 to 10.0;

  function "*" (L_v  : integer;
                R_v  : real) return real is
  begin
    return real(L_v) * R_v;
  end "*";

  function "*" (L_v  : real;
                R_v  : integer) return real is
  begin
    return L_v * real(R_v);
  end "*";
```

Overloaded operator "*" functions

```
begin
  Demo2_Lbl: process
     variable Real_v  : real := 30.0;
     variable Real2_v : Real5to10_Typ;
     variable Int_v   : integer := 3;
  begin
     Real_v  :=  3.0 * Real_v;
     Real_v  :=  Real_v * real(Int_v);   --   type conversion
     Real_v  := 2 * 3.0;    -- Overloaded "*"
     Real2_v := 2 * 3.5; -- Overloaded "*"
     Real_v  := 2.0 * Int_v; -- Overloaded "*"
     Real_v  := 2 * Real_v; -- Overloaded "*"
     wait;
  end process Demo2_lbl;
end Overload_Beh;
```

Figure 7.6-1 Overloading Operator "*" Between Integers and Reals (ch7_dir\overld1.vhd)

```
architecture StdPlus_a of StdPlus is
   function "+" (L_c : Std_Logic_Vector;
                 R_c : Std_Logic_Vector)
      return STD_Logic_Vector is
      variable Carry_v : Std_Logic := '0';
      variable L_v     : Std_Logic_Vector ((L_c'length -1) downto 0) := L_c;
      variable R_v     : Std_Logic_Vector ((R_c'length -1) downto 0) := R_c;
      variable Sum_v   : Std_Logic_Vector ((R_c'length -1) downto 0)
                          := (others => '0');
      variable Three_v : Std_Logic_Vector (2 downto 0);
   begin
      if L_c'length /= R_c'length then
        assert false
          report "Left and right length of + operator are unequal"
          severity warning;
        return L_c;
      else
        DO_ALL_BITS_Lbl:
          for I in 0 to (L_c'length - 1) loop
            Three_v := Std_Logic_Vector'(L_v (I) & R_v (I) & Carry_v);
            case Three_v is
              when "000" =>
                Sum_v (I)   := '0';
                Carry_v     := '0';

              when "001" =>
                Sum_v (I)   := '1';
                Carry_v     := '0';

              when "010" =>
                Sum_v (I)   := '1';
                Carry_v     := '0';

              when "011" =>
                Sum_v (I)   := '0';
                Carry_v     := '1';

              when "100" =>
                Sum_v (I)   := '1';
                Carry_v     := '0';

              when "101" =>
                Sum_v (I)   := '0';
                Carry_v     := '1';
              when "110" =>
                Sum_v (I)   := '0';
```

Bounds and direction are redefined. Variables are also initialized.

Addition of 3 bits of type Std_Logic. States 'H', 'L', 'Z', 'X','W', '-','U' would yield an 'X' in this model.

```
                Carry_v      := '1';

              when "111" =>
                Sum_v (I)    := '1';
                Carry_v      := '1';

              when others =>
                Sum_v (I)    := 'X';
                Carry_v      := 'X';
            end case;
          end loop DO_ALL_BITS_Lbl;
        return Sum_v;
      end if;
    end  "+";


    function TwosC (L_c: Std_Logic_Vector )
      return Std_Logic_Vector is
        variable One_v : Std_Logic_Vector
          ((L_c'length - 1) downto 0) := (0       => '1',
                                          others => '0');
        variable Result_v : Std_Logic_Vector
          ((L_c'length -1) downto 0) := L_c;
    begin
      Result_v := not Result_v; -- 1's Complement
      return (Result_v + One_v); -- use of overloaded "+"
    end TwosC;

    signal A_s : Std_Logic_Vector(7 downto 0)  := "01011101";
    signal B_s : Std_Logic_Vector(7 downto 0)  := "01101110";
    signal C_s : Std_Logic_Vector(7 downto 0)  := "00000000";
    signal D_s : Std_Logic_Vector(7 downto 0)  := "01011101";
begin  --   StdPlus_a
    C_s <= A_s + B_s after 10 ns;
    D_s <= TwosC(B_s) after 20 ns;
    B_s <= TwosC(C_s) after 50 ns;
end StdPlus_a;
```

Use of the "+" overloaded operator.

Figure 7.6-2 Overloading Operator "+" for Std_Logic_Vector types (ch7_dir\stdplus.vhd)

7.7 CONCURRENT PROCEDURE

Concurrent procedures were described in section 6.2.4. A concurrent procedure represents a procedure that is instantiated as a concurrent statement. It is equivalent to a process with a procedure call followed by a **wait** on sensitivity list extracted from all the actual signals whose mode in the formal parameter list is of **in** or **inout.** The syntax for a concurrent procedure call is:

```
concurrent_procedure_call ::=
    [ label : ] [postponed] procedure_call_statement

procedure_call_statement ::=
    procedure_name [ ( actual_parameter_part ) ] ;
```

1. If the procedure is called as a SEQUENTIAL procedure from WITHIN A PROCESS, then there is **NO IMPLIED WAIT** at the END of the procedure.

2. If the procedure is called as a CONCURRENT procedure within an architecture, then there **IS an IMPLIED WAIT** at the **END of the procedure** on the signals declared in the formal parameter list whose mode is IN or INOUT.

3. **Variable** parameters are not allowed in concurrent procedures (VHDL'87) since variables cannot be declared outside of a process. In VHDL'93 shared variables can be declared in an architecture declaration and in a package declaration part.

Avoid the use of **wait** statements in concurrent procedures. Use the implied **wait** statement.

Rationale: *If user defined **wait** statements are included in the concurrent procedures, the user **must be aware** of the implied **wait** statement at the end of the procedures. This may be confusing. Concurrent procedures are usually used to fire an equivalent process when signals connected to the formal parameters have an event.*

Figure 7.7-1 represents an example of a concurrent procedure to verify setup and hold time, and report the errors on the transcript output and also onto a file. Figure 7.7-2 represents the timing waveforms and the transcript output. The code uses VHDL'93. File cshld87.vhd represents the same file for VHDL'87, and is available on disk only.

```
--   Title       :  Concurrent Setup and Hold procedures with VHDL'93
--   Description :  Concurrent procedures and instantiation
library IEEE;                                  --  used because Std_Logic
  use IEEE.std_logic_1164.all;                   --   signals are resolved

use Std.TextIO.all;
use Std.TextIO;

entity SetHold is
  generic (SetupTime_g:          time := 15 ns;
           HoldTime_g:           time := 5 ns);
end SetHold;


architecture SetHold_a of SetHold is
  -- file          Error_f          : TextIO.Text is out "error.rpt"; -- VHDL'87
  file          Error_f          : TextIO.Text
                                      open Write_Mode is "error.rpt";
  signal        D1_s             : Std_Logic := '0';
  signal        D2_s             : Std_Logic := '1';
  signal        Clk_s            : Std_Logic := '0';
  signal        Violation_s      : Std_Logic := 'L';
```

```
---------------------------------------------------------------------------
--  Setup concurrent procedure
--  Wakes up when an event occurs on in or inout signals
--  This procedure has a side effect in that it writes to a file
--  declared in the architecture.  The file is not passed to the procedure.
--  Upon detection of a violation a short pulse appears on the
--  violation_s signal
---------------------------------------------------------------------------

    --                          ->|  |<- hold
    --       --+   <-- setup ---->+----------------------
    --   CLK    |__________________|
    --
    --       -----+                 +--------------
    --   D_S       |__________________|
    --

  procedure SetUp
      (signal    Clk_s          : in    Std_Logic;
       signal    D_s            : in    Std_Logic;
       constant  SetUpTime_c    : in    time;
       constant  SignalName_c   : in    string;
       file      Error_f        :       TextIO.Text; -- file interface declaration
                                                      -- must not contain a mode
       signal    Violation_s    : out   Std_Logic) is
     variable Outline_v   : Std.TextIO.line;  -- line for transcript
     variable Outlinef_v  : Std.TextIO.line;  -- line for file output
  begin
    --   Check rising edge of clock
    if(Clk_s'event and Clk_s ='1') then
    --   Check setup time
      if D_s'last_event < SetUpTime_c then
        Violation_s <= '1', 'L' after 1 ns;
        TextIO.Write(OutLine_v, string'("time "));
        TextIO.Write(OutLine_v, now);
        TextIO.Write(OutLine_v, string'(" Setup time violation on "));
        TextIO.Write(OutLine_v, SignalName_c);
        TextIO.Write(Outlinef_v, OutLine_v.all); -- copy contents of line
        TextIO.Writeline(Output, OutLine_v);    -- write to transcript
        TextIO.Writeline(Error_f, OutLinef_v);  -- write to file
      end if;
    end if;


    -- The following commented code is implied ONLY if the procedure
    -- is called as a CONCURRENT PROCEDURE.   Thus it is not implied
    -- if the procedure is called from within a process.
    -- wait on Clk_s, D_s;  -- all Formal signals of mode IN or INOUT
  end SetUp;
---------------------------------------------------------------------------
--  Concurrent Hold Procedure.
--  Wakes up when an event occurs on in or inout signals
--  This procedure has a side effect in that it writes to a file
--  declared in the architecture.  The file is not passed to the procedure.
--  Upon detection of a violation a short pulse appears on the
--  violation_s signal
---------------------------------------------------------------------------
  procedure Hold
    (signal    Clk_s          : in    Std_Logic;
     signal    D_s            : in    Std_Logic;
     constant  HoldTime_c     : in    time;
     constant  SignalName_c   : in    string;
     file      Error_f        :       TextIO.Text;
     signal    Violation_s    : out   Std_Logic) is

    variable Outline_v : Std.TextIO.line;
    variable Outlinef_v : Std.TextIO.line;
```

File interface declaration is illegal in VHDL'87. Mode is illegal in VHDL'93.

```
  begin
    --   Check Hold violation
    if (D_s'event and Clk_s'last_value = '0') then
      if Clk_s'last_event < holdTime_c then
        Violation_s <= '1', 'L' after 1 ns;
        TextIO.Write(OutLine_v, string'("time "));
        TextIO.Write(OutLine_v, now);
        TextIO.Write(OutLine_v, string'(" Hold time violation on "));
        TextIO.Write(OutLine_v, SignalName_c);
        TextIO.Write(Outlinef_v, OutLine_v.all); -- copy contents of line
        TextIO.Writeline(Output, OutLine_v);    -- write to transcript
        TextIO.Writeline(Error_f, OutLinef_v);  -- write to file
      end if;
    end if;

    -- The following commented code is implied ONLY if the procedure
    -- is called as a CONCURRENT PROCEDURE.
    -- wait on Clk_s, D_s;  -- all Formal signals of mode IN or INOUT
  end Hold;

begin  --  Set_a
-------------------------------------------------------------------------
--  Process:   ClockGen_Lbl
--  Purpose:   Generates clock pulses
-------------------------------------------------------------------------
  ClockGen_Lbl : process
  begin  --  process ClockGen_Lbl
    Clk_s <= not Clk_s;
    wait for 50 ns;
  end process ClockGen_Lbl;

-------------------------------------------------------------------------
--  Process:   MakeD_Change_Lbl
--  Purpose:   Cause D_s to change in value to test setup
-------------------------------------------------------------------------
  MakeD_Change_Lbl : process

  begin  --  process MakeD_Change_Lbl
    D1_s  <= '0' after   1 ns,
             '1' after  80 ns,
             '0' after 120 ns,
             '1' after 195 ns,
             '0' after 202 ns;

    wait for 300 ns;

    D2_s  <= '1' after   1 ns,
             '0' after  70 ns,
             '1' after 110 ns,
             '0' after 197 ns,
             '1' after 204 ns;
    wait for 300 ns;

  end process MakeD_Change_Lbl;


-------------------------------------------------------------------------
--  Instantiation of concurrent procedures
-------------------------------------------------------------------------
  SetUp(Clk_s         => Clk_s,
        D_s           => D1_s,
        SetUpTime_c   => SetUpTime_g,
        SignalName_c  => "D1_s",
        Error_f       => Error_f,
        Violation_s   => Violation_s);
```

```
  SetUp(Clk_s         => Clk_s,
        D_s           => D2_s,
        SetUpTime_c   => SetUpTime_g,
        SignalName_c  => "D2_s",
        Error_f       => Error_f,
        Violation_s   => Violation_s);

  Hold (Clk_s         => Clk_s,
        D_s           => D1_s,
        HoldTime_c    => HoldTime_g,
        SignalName_c  => "D1_s",
        Error_f       => Error_f,
        Violation_s   => Violation_s);

  Hold (Clk_s         => Clk_s,
        D_s           => D2_s,
        HoldTime_c    => HoldTime_g,
        SignalName_c  => "D2_s",
        Error_f       => Error_f,
        Violation_s   => Violation_s);

end SetHold_a;
```

Figure 7.7-1 Concurrent Procedures to Verify Setup and Hold Time (ch7_dir\concshld.vhd)

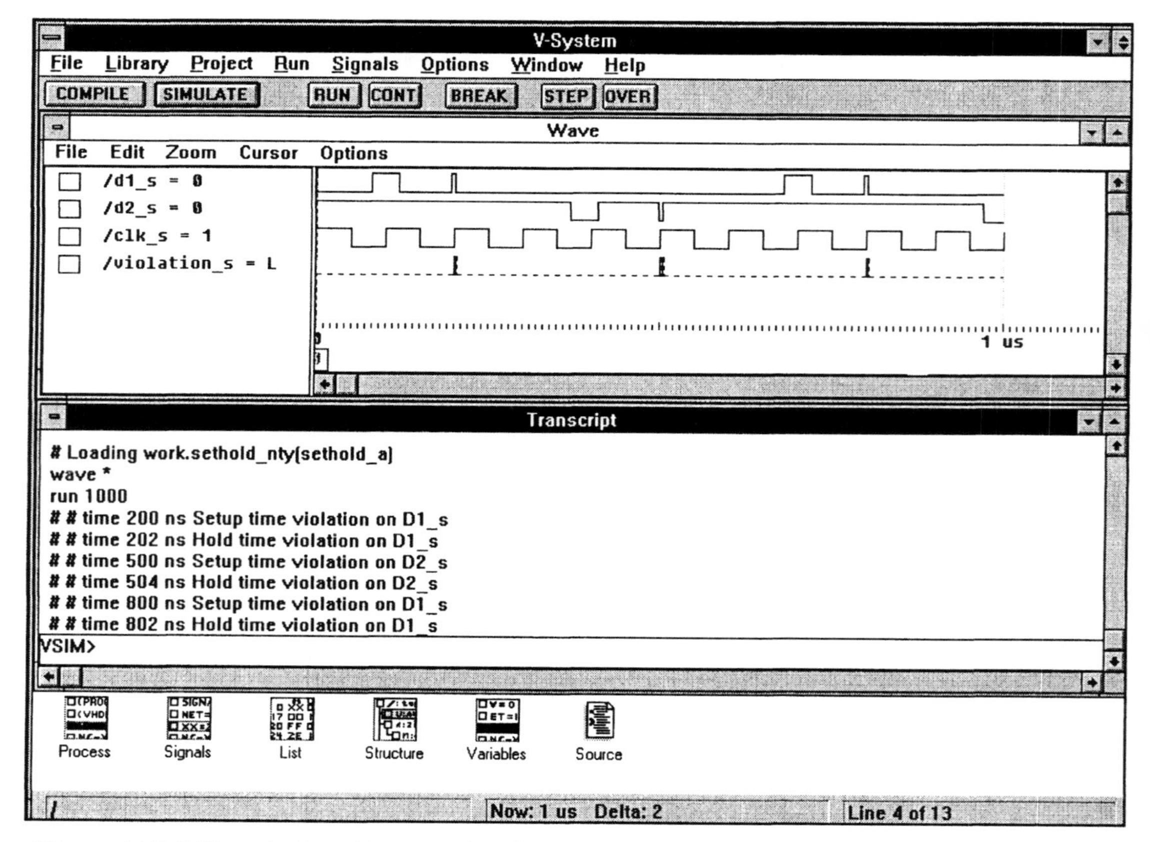

Figure 7.7-2 Simulation Outputs for Concurrent Procedures to Verify Setup and Hold Time ***(with Vsystem from Model Technology)***

EXERCISES

1. Write a function that performs the parity tree (exclusive OR) on the Bit_Vector being passed to it. This function returns a Bit.

2. Given the following definition:
type ArrNatural_Typ **is array** (natural <>) **of** integer;
Write an overloaded operator "+" that adds elements of two formal parameters of type ArrNatural_Typ.
Thus if the following signals are defined:
signal S1_s : ArrNatural_Typ(1 **to** 3) := (1 => 2, 2 => 7, 3 => 5);
signal S2_s : ArrNatural_Typ(1 **to** 3) := (1 => 1, 2 => 2, 3 => 3);
signal S3_s : ArrNatural_Typ(1 **to** 3);

then the following concurrent statement should yield the following result
S3_s <= S1_s + S2_s;
-- 1 => 3, 2 => 9, 3 => 8

3. Write a resolution function for the natural type that resolves to the LARGEST value Thus if driver1 drives a 3, and driver 2 drives a 100, and driver 3 drives a 50, the resolution function would yield 100.

4. Write a concurrent procedure that generates a pulse on a signal if the MSB of the input signal is ever a '1'. The input signal is of type *Std_Logic_Vector*.

5. Write a function that converts a string to a string of 80 characters. The function accepts as input a parameter of type string (unconstrained array of characters). It returns a string of 80 characters (constrained array) where any unused portion is a blank. Thus if the string "Hello" is passed, it returns "Hello ... " (where " ..." represents blank characters to fill the 80 character line). Test this function by reading data from a file into a variable of type line, and pass the contents of that line to the function. Display the output string on the transcript window (Output).

8. PACKAGES

This chapter explains the concept of packages and includes examples and coding methodologies in using packages. This chapter includes a linear feedback shift register (*LFSR*) package and an *Image* package that provides flexibility in converting various data types, including bit vector arrays, into binary decimal, and hexadecimal strings. It also explains the TextIO package and provides several examples in manipulating files.

8.1 PACKAGE

(LRM 2.5) A package represents a program unit that allows the specification of groups of logically related declarations. Packages include pools of type declarations, constant declarations, global signal declarations, and (for VHDL'93) global variables. They could also represent component declarations, attribute declarations, attribute specification, and subprograms. The subprograms (i.e., procedures and functions) declared in packages provide information hiding. These subprograms can be called from outside the package, while the inner workings remain hidden and protected from users.

M 👍👍 A user-defined package should contain CLOSELY related items that support an application function. These include:

1. Common items of a design (Types, procedures, functions, constants)
2. Common functionality (Convert to types or to IO utilities)
3. Global signals and global variables (VHDL'93).

Rationale: *The contents of a package should be cohesive.*

Packages are defined in two parts as represented in Figure 8.1

1. **Package declaration** defines the visible contents of a package. A design unit can access these declarations when the library name and the "***use*** *library.Package_Name.**all***" are specified in the design unit code (i.e., entity, architecture, and configuration). A user can compile code that makes use of a package when the package declaration is compiled. It is not necessary to compile the package body to compile the design unit. The package body is necessary to execute the simulation.
2. **Package body** provides the implementation details of subprograms, and the actual values of deferred constants (constant defined in the package declaration, but whose value is defined in the package body).

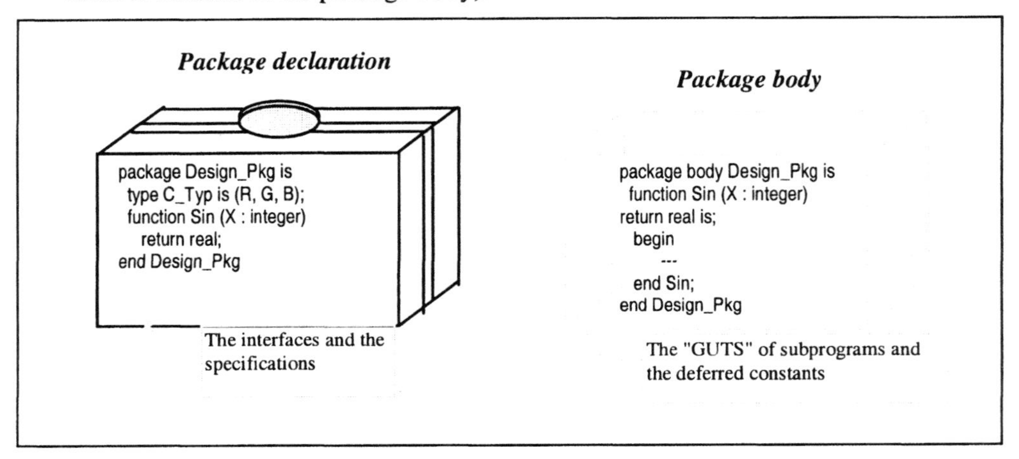

Figure 8.1 Parts of a Package

8.1.1 Package Declaration

A package declaration defines the interface to a package. The syntax is as follows:

```
package_declaration ::=
    package identifier is
     package_declarative_part
    end [package] [package_simple_name];

package_declarative_part ::=
    { package_declarative_item }

package_declarative_item ::=
      subprogram_declaration
    | type_declaration
    | subtype_declaration
    | constant_declaration
    | signal_declaration
    | file_declaration
    | alias_declaration
    | component_declaration
    | attribute_declaration
    | attribute_specification
    | disconnection_specification
    | use_clause
```

8.1.2 Package Body

A package body defines the bodies of subprograms and the values of deferred constants declared in the interface to the package. The syntax is as follows.

```
package_body ::=
   package body package_simple_name is
     package_body_declarative_part
   end [package body] package_simple_name ];

package_body_declarative_part ::=
     { package_body_declarative_item }

package_body_declarative_item ::=
       subprogram_declaration
     | subprogram_body
     | type_declaration
     | subtype_declaration
     | constant_declaration
     | file_declaration
     | alias_declaration
     | use_clause
```

The package body declarative items not declared in the package declarative part of a package declaration are NOT visible by declarations that "use" the package. They are only accessible to body of the package to facilitate the definition of the bodies of the subprograms.

Declare in the package declaration, only those items be used by users of the package. Hide items used only by the package body in the body of the package.

Rationale: *This method provides information hiding, thus making information unnecessary to users unavailable.*

Figure 8.1.2-1 represents a package declaration, a package body and a testbench for the package. Note that the package declaration includes a subtype definition, a function *ToChar* that converts a string of any length to a sized string, where the size is an input parameter. Function *To80Char* converts an unconstrained string to an 80-character string, and uses the function *ToChar.* In conformance to the methodology, the architecture identifies the package paths when accessing objects, types, and subprograms declared in other packages. This enhances readability. The process within the testbench architecture reads a line from a file and converts that line to a string of 80 characters. This clips the line if it is longer than 80 characters. It also pads the line with spaces if it is shorter than 80 character. The process then displays the line on the output transcript.

This package is useful when a user needs to read a line and to transfer it to another component via a signal. In VHDL, a signal cannot be of access type. Thus, a signal cannot be of type *Std.TextIO.line.* In addition, a signal must be static and its size cannot dynamically change. Therefore, a user can declare a signal of type *String(1 to 80),* read a line from file and convert that line to 80 characters. The 80-character string can then be assigned to the signal for transfer to another component. The assumption here is that the file will contain information only within the first 80 characters of a line.

```
package Char_Pkg is
  subtype String80_Typ is string(1 to 80);
  function ToChar(Size_c   : positive;
                  string_c : string) return string;
  function To80Char(string_c : string) return String80_Typ;
end Char_Pkg;

package body Char_Pkg is
  function ToChar(Size_c   : positive;
                  string_c : string) return string is
    variable Char_v : string(1 to Size_c) := (others => nul);
    begin
      if String_c'length <= Size_c then -- string is short
        LPshort_Lbl : for Index_i in 1 to String_c'length  loop
          Char_v(Index_i) := String_c(Index_i);
        end loop LPshort_Lbl;
        return Char_v;
      else  -- long string.  Clip to size
        LPlong_Lbl  : for Index_i in 1 to Size_c loop
          Char_v(Index_i) := String_c(Index_i);
        end loop LPlong_Lbl;
        return Char_v;
       end if;
    end ToChar;

  function To80Char(String_c : string) return String80_Typ is
    begin
      return ToChar(Size_c   => 80,
                    String_c => String_c);
    end To80Char;
end Char_Pkg;

library ATEP_Lib;
entity TestPkg is
  generic(FileId_g  : string :=
     "h:\atepvhdl\doc_ver2\src_dir\ch8_dir\top10.txt");
end TestPkg;

architecture TestPkg_a of TestPkg is
  use Std.TextIO.all;   use Std.TextIO;

  use ATEP_Lib.Char_Pkg.all;    use ATEP_Lib.Char_Pkg;
  file    Top10_f   : TextIO.Text is in FileId_g;
begin  --  TestPkg_a
  Test_Lbl : process
    variable InLine_v  : Std.TextIO.line;
    variable OutLine_v : Std.TextIO.line;
    variable Char80_v  : Char_Pkg.String80_Typ;
  begin  --  process Test_Lbl
    ReadFile_Lbl: while not TextIO.Endfile(Top10_f) loop
      -- Read 1 line from the input file
      TextIO.ReadLine(top10_f, InLine_v);
      Char80_V := Char_Pkg.To80Char(InLine_v.all);
      TextIO.Write(OutLine_v, Char80_v);
      TextIO.WriteLine(Output, OutLine_v);
    end loop ReadFile_Lbl;
    wait;
  end process Test_Lbl;
end TestPkg_a;
```


Figure 8.1.2-1 Package Declaration, Package Body, and Package Testbench (ch8_dir\char_pb.vhd)

Figure 8.1.2-2 shows the declaration of types, subtypes, and global signal declarations in a package. This package is used in the next section to demonstrate the application of the **use** clause.

```
package Design_Pkg is
  type       State_Typ is (Idle, On1, Off);
  subtype    Int03_Typ is integer range 0 to 3;
  type       StateArray_Typ  is array(State_Typ) of integer;
  constant StateToInt_c : StateArray_Typ :=
      (Idle   => 0,
       On1    => 1,
       Off    => 2);
  signal    Count1_s  : natural;
  signal    Count2_s  : natural;
  signal    Count3_s  : natural;
  signal    Count4_s  : natural;
  signal    Count5_s  : natural;
  signal    Count6_s  : natural;
  signal    State_s   : State_Typ;

  function ToNextState(State_c : State_Typ) return State_Typ;
end Design_Pkg;

package body Design_Pkg is
Subtype XInt_Typ is integer range 0 to 10;

  function ToNextState(State_c : State_Typ) return State_Typ is
    begin  --  ToNextState
      case State_c is
        when Idle => return On1;
        when On1  => return Off;
        when Off  => return Idle;
      end case;                              --  State_c
    end ToNextState;
end Design_Pkg;
```

DEMONSTRATION: Type conversion lookup constant. "ON1" used instead of "ON" because "ON" is a reserved word.

Gobal signals. Architectures with visibility into this package can access these signals without using ports.

Declaration NOT visible outside this package

Figure 8.1.2-2 Package with Type, Function, and Global Signal Declarations (ch8_dir\desgn_pb.vhd)

8.1.3 Deferred Constant

[1] (LRM 2.6) A deferred constant is a constant that is declared without an assignment symbol (:=) and represents an expression in a package declaration. A corresponding full declaration of the constant must exist in the package body to define the value of the constant. Figure 8.1.3 represents a package with a deferred constant. Deferred constants are currently not synthesizable (see chapter 13).

```
package Deferred_Pkg is
    constant MaxCount_c : natural;
end Deferred_Pkg;

package body Deferred_Pkg is
    constant MaxCount_c : natural := 10;
end Deferred_Pkg;
```

Deferred constants require minimum recompilation (see section 8.1.6, compilation order)

Figure 8.1.3 Package with a Deferred Constant (ch8_dir\const_pb.vhd)

8.1.4 The "use" Clause

(LRM 10.4) Items declared within a package declaration become accessible by a design unit using the following methods:

1. **By selection**. If the design unit is given access to the library, then an item can be made visible by providing the full path name of that item. For example:

   signal XYZ_s : LibraryName.PackageName.Item_Typ;
   MyVar_v := LibraryName.PackageName.ToBit(XYZ_s);

2. By **use** clause. If the design unit is given access to the library, and a "**use** library.pPackageName.**all;**" statement is made, then ALL items declared in that package are visible by the design unit.

A **significant advantage** in using the clause "**use** LibraryName.PackageName.**all**;" is the visibility and direct access of all the implicit operators of enumeration data types (i.e., the "<", ">", and "="). Without the "**.all**" clause, the implicit operators must be used as function calls with the direct path explicitly defined (e.g., "**if** LibraryName.PackageName."="(A_v, B_v) **then**").

Another **controversial advantage** of the "**.all**" is the direct usage of the objects (e.g., global signals and global variables (for VHDL'93)) and types and subprograms declarations defined in the packages without the explicit naming of the path for those objects and declarations. Several VHDL programmers argue that since the package is visible through the **use clause**, it is clear as to which object or declaration the author intended, and readability is not compromised. However, when several packages are declared using the **use clause**, it becomes difficult to keep track of the sources of those objects and declarations. For example, signals can be declared in entities (as signals or ports), architectures, in blocks, and in packages (as global signals). Thus, when a signal assignment is made, it could be difficult to identify its specification source. Good code can be defined as *"code that can be read and understood by others, or by the code's author after a period of more than 6 months"*. Thus, to **enhance readability and code maintainability** the following recommendations are made:

1. When using a package:
   - **FOR VHDL'87 or VHDL'93,** use the following **use** clauses

```
-- To access all objects and operators
use LibraryName.PackageName.all;
-- To eliminates need to identify library name
use LibraryName.PackageName;
```

- **FOR VHDL'93 only,** one may use aliases:

```
-- To Access all objects and operators alias the package name
-- (e.g., SPN_Pkg represents a user defined short alias for the package name)
alias SPN_Pkg is LibraryName.PackageName;
use SPN_Pkg.all;
```

2. When accessing an object (signal, constant, or variable (VHDL'93)) or a declaration that is defined in a package **provide the package name in-line with the object declaration or access** (this enhances readability and debugging). Example:

```
variable State_v  : PackageName.State_Typ; --VHDL'87 or 93
variable State93_v : SPN_Pkg.State_Typ;  -- alias:  VHDL'93
PackageName.SignalName_s <= Value; --VHDL'87 or 93
SPN_Pkg.SignalName_s <= Value; -- VHDL'93
```

3. When accessing a subprogram defined in a package, use one of the following options:

   a) Provide either the **package name** or the alias of the package name in the access of the subprogram. This is the recommended method because the path is in-line with the code, making it more readable and maintainable. For example:

```
State_v := PackageName.FunctionName(ObjectIdentifier);
if PackageName.IsInA2Z(CharIn_v) then
  PackageName.SendData(Data_s    => DataBus);
end if;
State_v := SPN_Pkg.FunctionName(ObjectIdentifier); -- 93 only
```

   b) In the declaration section of the architecture, **declare the access to the subprogram with the "use" statement.** This documents the source of the subprogram. The subprogram can then be called without the package name. This solution does not provide the path for the subprogram in-line with the code. However, this solution is often preferred because the path need not be added. Note that VHDL'93 helped with that respect with aliases, where the path can be a short name. For example:

```
architecture XYZ_a of XYZ is
  use LibraryName.PackageName;
  use PackageName.FunctionName;
begin
                                   Y_s <= FunctionName(X_s);
end XYZ_a;
```

4. When **accessing enumeration** of objects, do **NOT use the package name** since it can be cumbersome, and does not contribute to the readability since the type of the object identifies its source. Note that the "**.all**" is required here. For example:

```
    variable State_v  : PackageName.State_Typ  := Idle;
      -- ...
  if State_v = Idle then
```

Figure 8.1.4 represents an example of the application and restrictions of the **use** clause for accessing objects, types, and subprograms declared in packages.

```
library ATEP_Lib;

entity PackTest is
end PackTest;

architecture PackTest_a of PackTest is
begin
  P1_Lbl : process
   -- To have access to ALL objects, implied operators of enumerated types
   -- and subprograms
   use ATEP_Lib.Design_Pkg.all;

   -- To have access to objects by providing a path without naming the
   -- library, but naming the package
   use ATEP_Lib.Design_Pkg;
   use Design_Pkg.ToNextState;  -- makes the function visible  -- 
                                -- Path is well documented
   variable State1_v   : State_Typ := Idle;
   variable State2_v   : Design_Pkg.State_Typ := Idle; --
   variable State3_v   : ATEP_Lib.Design_Pkg.State_Typ := Idle;
   variable IsEqual_v  : boolean;

  begin
    if State1_v < ATEP_Lib.Design_Pkg.State_Typ'high then
      State1_v := State_Typ'succ(State1_v);
    end if;
    State_s <= ToNextState(State1_v);
    -- State_s <= Design_Pkg.ToNextState(State1_v);
    State3_v := State1_v;
    IsEqual_v := State1_v = State3_v;
    Count1_s  <= 3;

    Design_Pkg.Count2_s  <= Design_Pkg.StateToInt_c(State1_v) ;

    ATEP_Lib.Design_Pkg.Count3_s  <= Design_Pkg.StateToInt_c(State2_v);
    wait on ATEP_Lib.Design_Pkg.Count4_s;
  end process P1_Lbl;

  P2_Lbl : process
   use ATEP_Lib.Design_Pkg;  -- library name
   variable State1_v  : Design_Pkg.State_Typ := Design_Pkg.Idle;
   variable State2_v  : Design_Pkg.State_Typ := Design_Pkg.Idle;
   variable State3_v  : ATEP_Lib.Design_Pkg.State_Typ :=  Design_Pkg.Idle;
   variable IsEqual_v : boolean;

  begin
    if State1_v <  Design_Pkg.State_Typ'high then  --

      -- No feasible entries for infix op: "<"
      -- Need the "use Atep_Lib.design_Pkg.all" to gain access to
      -- implied operators
```

Path is declared here, thus the path need not be repeated.

Alternate method because path is in-line with the code.

Path of object provides GOOD documentation --
Providing path when using **objects** enhances readability.

Alternate method that includes the library name and package name.

```
        State1_v := off;  \_  -- Unknown identifier: off
                              --needed "use LibraryName.PackageNmae.all; -- or
                              -- State1_v := Design_Pkg.State_Typ.off;

        State1_v :=  Design_Pkg.State_Typ'succ(State1_v);
      end if;
      State2_v := Design_Pkg.ToNextState(State1_v);
      IsEqual_v := State1_v = State2_v;  --

      -- No feasible entries for infix op: "="
      -- Need the "use Atep_Lib.design_Pkg.all" to gain access to
      -- implied operators

      Count4_s  <= 3; -- Unknown identifier: Count4_s  --

      -- signal is defined in package, and no visibility is provided

      Design_Pkg.Count5_s  <= 5;
      ATEP_Lib.Design_Pkg.Count6_s  <= 6;
      wait on ATEP_Lib.Design_Pkg.Count1_s;
    end process P2_Lbl;
end PackTest_a;
```

Figure 8.1.4 Application and Restrictions of the Use **Clause (ch8_dir\packtest.vhd)**

8.1.5 Signals in Packages 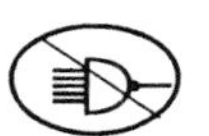

Signals declared in packages are useful for applications where information is transferred among components in an abstract manner such as control and data between BFMs (Bus Functional Models – see chapter 10). Global signals have the following characteristics:

1. **Global signals are not allowed in synthesizable** descriptions.

2. **Global signals** represent an **abstraction** useful **in inter-component communications** including the following:

   a) Handshake control signals to synchronize operations between components.

   b) Data and control transfer information to control the mode or data transfer of a *server* (or slave) component under the control of a *client* (or master) component.

   c) Transfer of actual data sent and received by models to a verifier component for error detection. For example, the comparison of expected data sent or received by a model against data transferred through the interfaces of the unit under test.

3. **Global signals can represent quick fixes** for high level or Bus Functional Models to add functionality. The use of global signals allows the addition of inter-component communication without the need to modify the entity and testbench architecture (because there are no changes to the ports of the components).

An example of an application that makes use of global signals is the *UART* project provided in the chapter 11. In that example, the *UART* transmitter sends characters over the RS232 bus using the RS232 protocol. It also sends a copy of the transmitted character sent over a global signal defined in a package. The *UART* receiver copies the received character onto another global signal. The verifier process examines both global signals to determine if an error in the link has occurred.

8.1.6 Resolution Function in Packages

Figure 8.1.6-1 demonstrates the definition of a resolution function for a record type with two fields, and where one of the fields is of type *Std_uLogic*, and the other filed is of type *Std_Logic_Vector*(31 downto 0). The *IEEE.Std_Logic_1164* package handles resolved signals of type *Std_Logic* and *Std_Logic_Vector*, but not records. Figure 8.1.6-2 demonstrates a testbench to verify this package.

```
library IEEE;
  use IEEE.Std_Logic_1164.all;

package ResolvedRec_pkg is
  -- note: Std_Logic is a subtype and thus is type compatible
  -- with Std_ulogic (the parent type)
  type uRec_Typ is record    -- unresolved
    A  : Std_uLogic;
    B  : Std_uLogic_Vector(31 downto 0);
  end record;

  type uRecVect_Typ is array (natural range <>) of uRec_Typ;   -- -- unresolved
  type Stdlogic_Table is array(Std_uLogic, Std_uLogic) of Std_uLogic;
    constant Resolution_Table_c : stdlogic_table := (
    --        ---------------------------------------------------------
    --        | U    X    0    1    Z    W    L    H    -        |  |
    --        ---------------------------------------------------------
              ( 'U', 'U', 'U', 'U', 'U', 'U', 'U', 'U', 'U' ), -- | U |
              ( 'U', 'X', 'X', 'X', 'X', 'X', 'X', 'X', 'X' ), -- | X |
              ( 'U', 'X', '0', 'X', '0', '0', '0', '0', 'X' ), -- | 0 |
              ( 'U', 'X', 'X', '1', '1', '1', '1', '1', 'X' ), -- | 1 |
              ( 'U', 'X', '0', '1', 'Z', 'W', 'L', 'H', 'X' ), -- | Z |
              ( 'U', 'X', '0', '1', 'W', 'W', 'W', 'W', 'X' ), -- | W |
              ( 'U', 'X', '0', '1', 'L', 'W', 'L', 'W', 'X' ), -- | L |
              ( 'U', 'X', '0', '1', 'H', 'W', 'W', 'H', 'X' ), -- | H |
              ( 'U', 'X', 'X', 'X', 'X', 'X', 'X', 'X', 'X' )  -- | - |
       );

  function ResolvedRec(S_c : uRecVect_Typ) return uRec_Typ;
  subtype Rec_Typ is ResolvedRec uRec_Typ;
end ResolvedRec_pkg;
-------------------------------------------------------------------------------
```

```
package body ResolvedRec_pkg is
  function ResolvedRec (S_c : uRecVect_Typ) return uRec_Typ is
    variable Result_v  : uRec_Typ := S_c(S_c'low);
  begin
    if S_c'length = 1 then -- Only one driver on bus
      return Result_v;
    else
      Result_v.A := 'Z';
      Result_v.B := (others  => 'Z');
      Drivers_Lbl : for Driver_i in S_c'range loop
        -- Compute Result_v.A since it is not a vector
        Result_v.A := Resolution_Table_c(Result_v.A, S_c(Driver_i).A);
        -- COmpute Result_B for each bit in the vector B
        BResolve_Lbl : for Bfield_i in S_c(Driver_i).B'range loop
          Result_v.B(Bfield_i) :=
          Resolution_Table_c(Result_v.B(Bfield_i),
                             S_c(Driver_i).B(Bfield_i));
        end loop BResolve_Lbl;
      end loop Drivers_Lbl;
      return Result_v;
    end if;

  end ResolvedRec;
end ResolvedRec_pkg;
```

Figure 8.1.6-1 Definition of a Resolution Function for a Record Type (ch8_dir\rslvrec.vhd)

```
library IEEE;
  use IEEE.Std_Logic_1164.all;
library ATEP_Lib;
  use ATEP_Lib.ResolvedRec_pkg.all;
  use ATEP_Lib.ResolvedRec_pkg;

entity RecTB is
end RecTB;

architecture RecTB_a of RecTB is
  signal Data_s : Rec_Typ;

begin  --  RecTB_a
  ---------------------------------------------------------------------
  --  Process:   P1_lbl
  --  Purpose:   Put data onto Data_s
  ---------------------------------------------------------------------
  P1_lbl : process
    variable Data_v : Rec_Typ :=
      (A   =>  '1',
       B   =>  "11110000111111110000000010101010");
  begin  --  process P1_lbl
    Data_v.A   := not Data_v.A;
    Data_v.B   := not Data_v.B;
    Data_s   <= Data_v;
    wait for 100 ns;
  end process P1_lbl;
```

```
---------------------------------------------------------------------------
-- Process:   P2_lbl
-- Purpose:   Put data onto Data_s
---------------------------------------------------------------------------
P2_lbl : process
  variable Data_v : Rec_Typ :=
    (A   =>  '1',
     B   =>  "00111111110000000010101010111100");
begin  --  process P1_lbl
  Data_v.A  := not Data_v.A;
  Data_v.B  := not Data_v.B;
  Data_s  <= Data_v;
  wait for 75 ns;
end process P2_lbl;

end RecTB_a;
```

Figure 8.1.6-2 Testbench to Verify Resolution Function for a Record Type (ch8_dir\rslvtb.vhd)

8.1.7 Subprograms in Packages

Unlike subprograms declared in architectures, **subprograms declared in packages have limited side effects** because they do not have access to entity ports, signals of architectures, or variables of processes. Subprograms in packages have READ visibility of objects declared in packages, including other packages if access is provided with the "**use**" statement. Thus, formal parameters of subprograms must include ALL the non-visible interfaces required for the operation of the subprograms. Subprograms declared in packages can be used by all architectures that access the package through the **library** and **use** clauses. **Procedures** can be called by processes as **sequential statements**. They can also be called as **concurrent statements**. **Concurrent procedures** are sensitive to all signals declared in the parameter list of the procedure of mode **in** or **inout. Concurrent procedures** also have an implied "**wait**" statement at the end of the procedure on all signals declared in the parameter list of mode **in** or **inout**.

Functions can be called within subprograms, within architectures, or in concurrent statements. NO conversion functions or type conversion function must appear in either the formal part or the actual part of an association element that associates an actual signal with a formal signal parameter (LRM 7.4). See section 6.2.3.1 for the port association rules.

8.2 CONVERTING TYPED OBJECTS TO STRINGS

There is often a need to convert objects of type *Bit, Bit_Vector, Std_uLogic, Std_Logic_Vector, Signed, Unsigned, Integer, Real,* or *Time* to a string. In VHDL'93, the *'image* attribute provides this conversion for scalar prefixes (e.g., *Bit, Std_Logic, Integer).* However, the *'image* attribute does not operate on vectors, such as *Bit_Vector, Std_Logic_Vector, Signed, and Unsigned.* A conversion function is required for the array types. In VHDL'87, a conversion function is required for any type. Figure 8.2 provides a package for the overloaded *image, HexImage,* and *DecImage* functions. The function is called *image* because of the close connotation to the attribute *'image.*

```
library IEEE;
  use IEEE.Std_Logic_1164.all;
  use IEEE.Std_Logic_TextIO.all;
  use IEEE.Std_Logic_Arith.all;

library Std;
  use STD.TextIO.all;

package Image_Pkg is
  function Image(In_Image : Time) return String;
  function Image(In_Image : Bit) return String;
  function Image(In_Image : Bit_Vector) return String;
  function Image(In_Image : Integer) return String;
  function Image(In_Image : Real) return String;
  function Image(In_Image : Std_uLogic) return String;
  function Image(In_Image : Std_uLogic_Vector) return String;
  function Image(In_Image : Std_Logic_Vector) return String;
  function Image(In_Image : Signed) return String;
  function Image(In_Image : UnSigned) return String;

  function HexImage(InStrg   : String) return String;
  function HexImage(In_Image : Bit_Vector) return String;
  function HexImage(In_Image : Std_uLogic_Vector) return String;
  function HexImage(In_Image : Std_Logic_Vector) return String;
  function HexImage(In_Image : Signed) return String;
  function HexImage(In_Image : UnSigned) return String;

  function DecImage(In_Image : Bit_Vector) return String;
  function DecImage(In_Image : Std_uLogic_Vector) return String;
  function DecImage(In_Image : Std_Logic_Vector) return String;
  function DecImage(In_Image : Signed) return String;
  function DecImage(In_Image : UnSigned) return String;
end Image_Pkg;

package body Image_Pkg is
  function Image(In_Image : Time) return String is
    variable L : Line;  -- access type
    variable W : String(1 to 25) := (others => ' ');
      -- Long enough to hold a time string
  begin
    -- the WRITE procedure creates an object with "NEW".
    -- L is passed as an output of the procedure.
    Std.TextIO.WRITE(L, in_image);
    -- Copy L.all onto W
    W(L.all'range) := L.all;
    Deallocate(L);
    return W;
  end Image;
```

An implementation must allow a physical type to include the range -2147483647 to 2147483647 (or 10 characters). It is safe to assume that L.all'range will be less than 25 characters.

```
  function Image(In_Image : Bit) return String is
    variable L : Line;  -- access type
    variable W : String(1 to 3) := (others => ' ');
  begin
    Std.TextIO.WRITE(L, in_image);
    W(L.all'range) := L.all;
    Deallocate(L);
    return W;
  end Image;
  function Image(In_Image : Bit_Vector) return String is
```

```
    variable L : Line;  -- access type
    variable W : String(1 to In_Image'length) := (others => ' ');
  begin
    Std.TextIO.WRITE(L, in_image);
    W(L.all'range) := L.all;
    Deallocate(L);
    return W;
  end Image;

  function Image(In_Image : Integer) return String is
    variable L : Line;  -- access type
    variable W : String(1 to 32) := (others => ' ');
     -- Long enough to hold a time string
  begin
    Std.TextIO.WRITE(L, in_image);
    W(L.all'range) := L.all;
    Deallocate(L);
    return W;
  end Image;

  function Image(In_Image : Real) return String is
    variable L : Line;  -- access type
    variable W : String(1 to 32) := (others => ' ');
      -- Long enough to hold a time string
  begin
    Std.TextIO.WRITE(L, in_image);
    W(L.all'range) := L.all;
    Deallocate(L);
    return W;
  end Image;

  function Image(In_Image : Std_uLogic) return String is
    variable L : Line;  -- access type
    variable W : String(1 to 3) := (others => ' ');
  begin
    IEEE.Std_Logic_Textio.WRITE(L, in_image);
    W(L.all'range) := L.all;
    Deallocate(L);
    return W;
  end Image;

  function Image(In_Image : Std_uLogic_Vector) return String is
    variable L : Line;  -- access type
    variable W : String(1 to In_Image'length) := (others => ' ');
  begin
    IEEE.Std_Logic_Textio.WRITE(L, in_image);
    W(L.all'range) := L.all;
    Deallocate(L);
    return W;
  end Image;

  function Image(In_Image : Std_Logic_Vector) return String is
    variable L : Line;  -- access type
    variable W : String(1 to In_Image'length) := (others => ' ');
  begin
     IEEE.Std_Logic_TextIO.WRITE(L, In_Image);
     W(L.all'range) := L.all;
     Deallocate(L);
     return W;
  end Image;
  function Image(In_Image : Signed) return String is
  begin
    return Image(Std_Logic_Vector(In_Image));
  end Image;
  function Image(In_Image : UnSigned) return String is
  begin
    return Image(Std_Logic_Vector(In_Image));
```

```
end Image;

function HexImage(InStrg  : String) return String is
  subtype Int03_Typ is Integer range 0 to 3;
  variable Result : string(1 to ((InStrg'length - 1)/4)+1) :=
      (others => '0');    -- length of result string
  variable StrTo4 : string(1 to Result'length * 4) :=
      (others => '0');    -- length of extended bit vector
  variable MTspace : Int03_Typ;  --  Empty space to fill in
  variable Str4    : String(1 to 4);
  variable Group   : Natural := 0;
begin
  MTspace := Result'length * 4  - InStrg'length;
  StrTo4(MTspace + 1 to StrTo4'length) := InStrg; -- padded with '0'
  Cnvrt_Lbl : for I in Result'range loop
    Group := Group + 4;  -- identifies end of bit # in a group of 4
    Str4 := StrTo4(Group - 3 to Group); -- get next 4 characters
    case Str4 is
      when "0000"  => Result(I) := '0';
      when "0001"  => Result(I) := '1';
      when "0010"  => Result(I) := '2';
      when "0011"  => Result(I) := '3';
      when "0100"  => Result(I) := '4';
      when "0101"  => Result(I) := '5';
      when "0110"  => Result(I) := '6';
      when "0111"  => Result(I) := '7';
      when "1000"  => Result(I) := '8';
      when "1001"  => Result(I) := '9';
      when "1010"  => Result(I) := 'A';
      when "1011"  => Result(I) := 'B';
      when "1100"  => Result(I) := 'C';
      when "1101"  => Result(I) := 'D';
      when "1110"  => Result(I) := 'E';
      when "1111"  => Result(I) := 'F';
      when others  => Result(I) := 'X';
    end case;                          --  Str4
  end loop Cnvrt_Lbl;
  return Result;
end HexImage;
function HexImage(In_Image : Bit_Vector) return String is
begin
  return HexImage(Image(In_Image));
end HexImage;

function HexImage(In_Image : Std_uLogic_Vector) return String is
begin
  return HexImage(Image(In_Image));
end HexImage;

function HexImage(In_Image : Std_Logic_Vector) return String is
begin
  return HexImage(Image(In_Image));
end HexImage;

function HexImage(In_Image : Signed) return String is
begin
  return HexImage(Image(In_Image));
end HexImage;
```

```
function HexImage(In_Image : UnSigned) return String is
begin
  return HexImage(Image(In_Image));
end HexImage;

function DecImage(In_Image : Bit_Vector) return String is
  variable In_Image_v : Bit_Vector(In_Image'length downto 1) := In_Image;
begin
  if In_Image'length > 31 then
    assert False
      report "Number too large for Integer, clipping to 31 bits"
      severity Warning;
    return Image(Conv_Integer
                  (Unsigned(To_StdLogicVector
                      (In_Image_v(31 downto 1)))));
  else
    return Image(Conv_Integer(Unsigned(To_StdLogicVector(In_Image))));
  end if;
end DecImage;

function DecImage(In_Image : Std_uLogic_Vector) return String is
  variable In_Image_v : Std_uLogic_Vector(In_Image'length downto 1) := In_Image;
begin
  if In_Image'length > 31 then
    assert False
      report "Number too large for Integer, clipping to 31 bits"
      severity Warning;
     return Image(Conv_Integer(Unsigned(In_Image_v(31 downto 1))));
  else
      return Image(Conv_Integer(Unsigned(In_Image)));
  end if;
end DecImage;

function DecImage(In_Image : Std_Logic_Vector) return String is
  variable In_Image_v : Std_Logic_Vector(In_Image'length downto 1) := In_Image;
begin
  if In_Image'length > 31 then
    assert False
      report "Number too large for Integer, clipping to 31 bits"
      severity Warning;
     return Image(Conv_Integer(Unsigned(In_Image_v(31 downto 1))));
  else
      return Image(Conv_Integer(Unsigned(In_Image)));
  end if;
end DecImage;
```

```
  function DecImage(In_Image : Signed) return String is
    variable In_Image_v : Signed(In_Image'length downto 1) := In_Image;
  begin
    if In_Image'length > 31 then
      assert False
        report "Number too large for Integer, clipping to 31 bits"
        severity Warning;
       return Image(Conv_Integer(In_Image_v(31 downto 1)));
    else
        return Image(Conv_Integer(In_Image));
    end if;
  end DecImage;

  function DecImage(In_Image : UnSigned) return String is
    variable In_Image_v : UnSigned(In_Image'length downto 1) := In_Image;
  begin
    if In_Image'length > 31 then
      assert False
        report "Number too large for Integer, clipping to 31 bits"
        severity Warning;
       return Image(Conv_Integer(In_Image_v(31 downto 1)));
    else
        return Image(Conv_Integer(In_Image));
    end if;
  end DecImage;

end Image_Pkg;
```

Figure 8.2 Image Package (ch8_dir\image_pb.vhd)

8.3 PACKAGE TEXTIO

Section 2.4 defines the file declarations and the implicit file procedures as the result of the file declarations, including the *File_Open* and the *File_Close* procedures available in VHDL'93. *TextIO* package supports human readable IO. *TextIO* is one of the packages defined in the *IEEE Std 1076-1987* and *IEEE Std 1076-1993* Language Reference Manuals. *[1] Package TextIO contains declarations of types and subprograms that support formatted ASCII I/O operations.*

Package TextIO defines the following types and subtypes:

type LINE **is access** string;
type TEXT **is file** of string;
type SIDE **is** (right, left);
subtype WIDTH **is** natural;

TextIO defines the following file declarations:

-- VHDL'93 syntax:
file input : TEXT **open** read_mode **is** "STD_INPUT";
file output : TEXT **open** write_mode **is** "STD_OUTPUT";

-- VHDL'87
file input : TEXT **is in** "STD_INPUT";
file output : TEXT **is out** "STD_OUTPUT";

TextIO defines the following overloaded procedures:

```
procedure READLINE(file f: TEXT; L: out LINE);
procedure READ(L:inout LINE; VALUE: out bit; GOOD : out BOOLEAN);
procedure READ(L:inout LINE; VALUE: out bit);
procedure READ(L:inout LINE; VALUE: out bit_vector; GOOD : out BOOLEAN);
procedure READ(L:inout LINE; VALUE: out bit_vector);
procedure READ(L:inout LINE; VALUE: out BOOLEAN; GOOD : out BOOLEAN);
procedure READ(L:inout LINE; VALUE: out BOOLEAN);
procedure READ(L:inout LINE; VALUE: out character; GOOD : out BOOLEAN);
procedure READ(L:inout LINE; VALUE: out character);
procedure READ(L:inout LINE; VALUE: out integer; GOOD : out BOOLEAN);
procedure READ(L:inout LINE; VALUE: out integer);
procedure READ(L:inout LINE; VALUE: out real; GOOD : out BOOLEAN);
procedure READ(L:inout LINE; VALUE: out real);
procedure READ(L:inout LINE; VALUE: out string; GOOD : out BOOLEAN);
procedure READ(L:inout LINE; VALUE: out string);
procedure READ(L:inout LINE; VALUE: out time; GOOD : out BOOLEAN);
procedure READ(L:inout LINE; VALUE: out time);

procedure WRITELINE(file f : TEXT; L : inout LINE);
procedure WRITE(L : inout LINE; VALUE : in bit;
      JUSTIFIED: in SIDE := right;  FIELD: in WIDTH := 0);
procedure WRITE(L : inout LINE; VALUE : in bit_vector;
      JUSTIFIED: in SIDE := right;  FIELD: in WIDTH := 0);
procedure WRITE(L : inout LINE; VALUE : in BOOLEAN;
      JUSTIFIED: in SIDE := right;  FIELD: in WIDTH := 0);
procedure WRITE(L : inout LINE; VALUE : in character;
      JUSTIFIED: in SIDE := right;  FIELD: in WIDTH := 0);
procedure WRITE(L : inout LINE; VALUE : in integer;
      JUSTIFIED: in SIDE := right;  FIELD: in WIDTH := 0);
procedure WRITE(L : inout LINE; VALUE : in real;
      JUSTIFIED: in SIDE := right;  FIELD: in WIDTH := 0;
      DIGITS: in NATURAL := 0);
procedure WRITE(L : inout LINE; VALUE : in string;
      JUSTIFIED: in SIDE := right;  FIELD: in WIDTH := 0);
procedure WRITE(L : inout LINE; VALUE : in time;
      JUSTIFIED: in SIDE := right;  FIELD: in WIDTH := 0;
      UNIT: in TIME := ns);

-- function ENDFILE(file F : TEXT) return boolean;, -- is implicit built-in:
```

Default value for Justified is right. Field defines width of characters. "0" width means the minimum width that can accommodate the data.

To make those type, file, and procedure declarations visible, a **use** clause must be provided. Those overloaded procedures operate on objects of type *Bit*, *Bit_Vector*, *Boolean, Character, Integer, Real, String*, and *Time*. Note that the *Read* and *Write* procedures use a variable of type *Line* defined as an *access* to a *string*. An examination of the body of this package reveals that a *WRITELINE* will deallocate the data pointed by the *access* object. Thus, at the conclusion of the *WRITELINE* call, the line pointer shall point to null, and the data is lost. If it is desired to maintain the data pointed by the line pointer then a new pointer must be created to point to the data. Figure 8.3-1 demonstrates the use of each of these overloaded procedures, and the method used to maintain the data pointed by the line pointer.

```
entity FileIO is
  generic(DataIn_g  : string := "datain2.txt";
          DataOut_g : string := "dataout.txt");
end FileIO;

architecture FileIO_a of FileIO is
begin  --  FileIO_a
  ---------------------------------------------------------------------
```

```
--  Process:   ReadWriteFile_Lbl
--  Purpose:   Read a file, and display its content to the screen.
--             Also, copy content onto an output file.
--             Use All the overloaded Read and Write TextIO procedures
--  Inputs:    file: datain2.txt
--             Data                Type
--             ----------          -----------
--             1                   bit
--             10110011            bit_vector
--             true                boolean
--             A                   character
--             15                  integer
--             3.1416              real
--             String__10          string
--             10 ns               time
--
--  Outputs:   file: dataout.txt
--------------------------------------------------------------------------
ReadWriteFile_Lbl : process
  use Std.TextIO.all;
  use Std.TextIO;
  file DataIn_f          : TextIO.text is in  DataIn_g;
  file DataOut_f         : TextIO.text is out DataOut_g;
  variable InLine_v      : TextIO.line;  -- pointer to string
  variable OutLine_v     : TextIO.line;  -- pointer to string
  variable ScreenLine_v  : TextIO.line;  -- pointer to string
  variable Bit_v         : Bit;
  variable BitV8_v       : Bit_Vector(1 to 8);
  variable Bool_v        : boolean;
  variable Char_v        : character;
  variable Int_v         : integer;
  variable Real_v        : real;
  variable String_v      : string(1 to 10);
  variable time_v        : time;
begin  -- process ReadWriteFile_Lbl
  ReadFile_Lbl: while not TextIO.Endfile(DataIn_f) loop
    -- ******************** BIT ********************
    -- Read 1 line from the input file

    TextIO.ReadLine(DataIn_f, InLine_v);
    -- Create a new pointer to point to a new string
    -- so that the new line can be used for OUTPUT to the monitor
    -- without affecting the read line.  Note that a Read, Write, and
    -- WriteLine  deallocate the data pointed by the access object.
    -- Thus,  after a Read, Write, and WriteLine, the pointer points
    -- to null, and the data is lost.  To maintain it, a new
    -- pointer must be created that points to that data.
    ScreenLine_v := new string(InLine_v'low to InLine_v'high);

    -- Must copy the data, but not the pointers
    ScreenLine_v.all := InLine_v.all;

    -- ScreenLine_v is deallocated after the write of the line
    TextIO.WriteLine(output, ScreenLine_v);       -- to screen

    -- Read value from InLine_v line (a bit).  InLine_v points to
    -- null after the Read operation.
    TextIO.Read(InLine_v, Bit_v);

    -- Write the bit variable
    TextIO.Write(OutLine_v, Bit_v); -- write bit to Outline
    TextIO.WriteLine(DataOut_f, OutLine_v);
    -- ******************** BIT_VECTOR ********************
    TextIO.ReadLine(DataIn_f, InLine_v);  -- read line
    ScreenLine_v := new string(InLine_v'low to InLine_v'high); --new line
    ScreenLine_v.all := InLine_v.all; -- copy new line
    TextIO.Read(InLine_v, BitV8_v); -- Update variable
    TextIO.WriteLine(output, ScreenLine_v); -- Send line to screen
```

```
      TextIO.Write(OutLine_v, BitV8_v); -- write bit vector to Line
      TextIO.WriteLine(DataOut_f, OutLine_v);

      -- ******************** BOOLEAN ********************
      TextIO.ReadLine(DataIn_f, InLine_v);
      ScreenLine_v := new string(InLine_v'low to InLine_v'high);
      ScreenLine_v.all := InLine_v.all; -- copy new line
      TextIO.Read(InLine_v, Bool_v);
      TextIO.WriteLine(output, ScreenLine_v); -- Send line to screen
      TextIO.Write(OutLine_v, Bool_v);  -- send to line
      TextIO.WriteLine(DataOut_f, OutLine_v);

      -- ******************** CHARACTER  ********************
      TextIO.ReadLine(DataIn_f, InLine_v);
      ScreenLine_v := new string(InLine_v'low to InLine_v'high);
      ScreenLine_v.all := InLine_v.all; -- copy new line
      TextIO.Read(InLine_v, Char_v);
      TextIO.WriteLine(output, ScreenLine_v);      -- to screen
      TextIO.Write(OutLine_v, Char_v);  -- send to line
      TextIO.WriteLine(DataOut_f, OutLine_v);

      -- ******************** INTEGER ********************
      TextIO.ReadLine(DataIn_f, InLine_v);
      ScreenLine_v := new string(InLine_v'low to InLine_v'high);
      ScreenLine_v.all := InLine_v.all; -- copy new line
      TextIO.Read(InLine_v, Int_v);
      TextIO.WriteLine(output, ScreenLine_v);      -- to screen
      TextIO.Write(OutLine_v, Int_v);  -- send to line
      TextIO.WriteLine(DataOut_f, OutLine_v);

      -- ******************** REAL ********************
      TextIO.ReadLine(DataIn_f, InLine_v);
      ScreenLine_v := new string(InLine_v'low to InLine_v'high);
      ScreenLine_v.all := InLine_v.all; -- copy new line
      TextIO.Read(InLine_v, Real_v);
      TextIO.WriteLine(output, ScreenLine_v);      -- to screen
      TextIO.Write(OutLine_v, Real_v);  -- send to line
      TextIO.WriteLine(DataOut_f, OutLine_v);

      -- ******************** STRING ********************
      TextIO.ReadLine(DataIn_f, InLine_v);
      ScreenLine_v := new string(InLine_v'low to InLine_v'high);
      ScreenLine_v.all := InLine_v.all; -- copy new line
      TextIO.Read(InLine_v, String_v);
      TextIO.WriteLine(output, ScreenLine_v);      -- to screen
      TextIO.Write(OutLine_v, String_v);  -- send to line
      TextIO.WriteLine(DataOut_f, OutLine_v);
      -- ******************** TIME ********************
      TextIO.ReadLine(DataIn_f, InLine_v);
      ScreenLine_v := new string(InLine_v'low to InLine_v'high);
      ScreenLine_v.all := InLine_v.all; -- copy new line
      TextIO.Read(InLine_v, Time_v);
      TextIO.WriteLine(output, ScreenLine_v);      -- to screen
      TextIO.Write(OutLine_v, Time_v);  -- send to line
      TextIO.WriteLine(DataOut_f, OutLine_v);
    end loop ReadFile_Lbl;
    wait;
  end process ReadWriteFile_Lbl;
end FileIO_a;
```

Figure 8.3-1 Demonstration of file IO with VHDL'87 (ch8_dir\ftxio87.vhd)

Significant changes occurred in the use of files between *VHDL'87* and *VHDL'93*, thus making the file declarations and usage different between the two standards. *VHDL'93* includes explicit file open and close statements. Files can also be passed as parameters to subprograms (see chapter 6). Figure 8.3-2 is an example of file utilization using *TextIO* package, and compiled with *VHDL'87*. Figure 8.3-3 represents the same example compiled with VHDL'93.

```
architecture FileIO_a of FileIO is
begin -- FileIO_a
  ------------------------------------------------------------------------------
  -- Process:   ReadWriteFile_Lbl
  -- Purpose:   Read a file, and display its content to the screen.
  --            Also, copy content onto an output file.
  --            The copy is performed on a character by character basis
  --            instead of a copy of the whole line.
  -- Inputs:    file: datain.txt
  -- Outputs:   file: dataout.txt
  ------------------------------------------------------------------------------
  ReadWriteFile_Lbl : process
    use Std.TextIO.ALL;
    use Std.TextIO;
    file DataIn_f        : TextIO.text is in  "datain.txt";
    file DataOut_f       : TextIO.text is out "dataout.txt";
    variable OutLine_v   : TextIO.line;   -- pointer to string
    variable OutLine2_v  : TextIO.line;   -- pointer to string
    variable Index_v     : natural;
  begin -- process ReadWriteFile_Lbl
    while not TextIO.Endfile(DataIn_f) loop
      -- Read 1 line from the input file
      TextIO.ReadLine(DataIn_f, OutLine_v);

      -- Create a new pointer to point to a new string
      -- so that the new line can be used for OUTPUT to transcript
      -- without affecting the read line.  Note that a WriteLine
      -- deallocates the line, and the data is deleted.
      OutLine2_v := new string(OutLine_v'low to OutLine_v'high);

      -- Must copy the data, but not the pointers
      OutLine2_v.all := OutLine_v.all;

      -- OutLine2_v is deallocated after the write of the line
      TextIO.WriteLine(output, OutLine2_v);      -- to screen

      -- Test if OutLine_v is an empty line, and write it if empty
      if OutLine_v'length = 0 then              -- null line
        TextIO.WriteLine(DataOut_f, OutLine_v);
      else             -- must copy the characters onto OutLine2_v
        -- Scan the characters in the line
        for Index_i in OutLine_v'low to OutLine_v'high loop

          -- Copy each characters from OutLine_v to OutLine2_v
          TextIO.Write(OutLine2_v, OutLine_v(Index_i));
        end loop;
        -- Now write the whole line
        TextIO.WriteLine(DataOut_f, OutLine_v);  -- write the copied line
      end if;
    end loop;
    wait;
  end process ReadWriteFile_Lbl;
end FileIO_a;
```

Figure 8.3-2 File Utilization Using TextIO Package, and compiled with VHDL'87 (ch8_dir\file87.vhd)

```
...
architecture FileIO93_a of FileIO is
begin  --  FileIO_a
  ReadWriteFile_Lbl : process
    use Std.TextIO.ALL;
    use Std.TextIO;
    file DataIn_f        : TextIO.text open Read_Mode  is "datain.txt";
    file DataOut_f       : TextIO.text open Write_Mode is "dataout.txt";
    variable OutLine_v  : TextIO.line;  -- pointer to string
    variable OutLine2_v : TextIO.line;  -- pointer to string
    variable Index_v    : natural;

  begin  --  process ReadWriteFile_Lbl
    while not TextIO.Endfile(DataIn_f) loop
      TextIO.ReadLine(DataIn_f, OutLine_v);
      OutLine2_v := new string(OutLine_v'low to OutLine_v'high);
      OutLine2_v.all := OutLine_v.all;
      TextIO.WriteLine(output, OutLine2_v);      -- to screen
    if OutLine_v'length = 0 then                -- null line
        TextIO.WriteLine(DataOut_f, OutLine_v);
      else                 -- must copy the characters onto OutLine2_v
        for Index_i in OutLine_v'low to OutLine_v'high loop
          TextIO.Write(OutLine2_v, OutLine_v(Index_i));
        end loop;
        TextIO.WriteLine(DataOut_f, OutLine_v);  -- write the copied line
      end if;
    end loop;
    wait;
  end process ReadWriteFile_Lbl;
end FileIO93_a;
```

Figure 8.3-3 File Utilization Using TextIO Package, and Compiled with VHDL'93 (ch8_dir\file93.vhd)

8.3.1 Printing Objects from VHDL

There are two methods to produce outputs on the screen:

1. The *Write* and *WriteLine* procedures.
2. The *Assert* statement

The *Image* function or the <u>*'Image* attribute (VHDL'93) for scalar prefix only</u> can be used to convert vectors to strings. The *'Image* cannot be used for composite type as prefixes (e.g., *Bit_Vector'Image* is illegal). The *TextIO* package in library *STD* and the Synopsys *Std_Logic_TextIO* package provide the overloaded *Write* and *WriteLine* procedures for various types (see packages for details). These packages are used in the definition of the *image* function. Figure 8.3.1-1 demonstrates the application of these concepts.

```
library IEEE;
  use IEEE.Std_logic_1164.all;

library Work;
  use Work.Image_Pkg.all;
  use Std.TextIO.all;

entity StringOut is
end StringOut;
```

```
architecture StringOut_a of StringOut is
  signal S_Bit          : Bit;
  signal S_BitV         : Bit_Vector(15 downto 0);
  signal S_Int          : Integer;
  signal S_Real         : Real;
  signal S_uStd         : Std_uLogic;
  signal S_Std          : Std_Logic;
  signal S_StdV         : Std_Logic_Vector(33 downto 0);

begin
  Test_Lbl: process
    variable T_v : Time;
    variable L_v : Line;
  begin
    S_Bit   <= '1';
    S_BitV  <= "0000111101011010";
    S_Int   <= 25;
    S_Real  <= 3.1416;
    S_uStd  <= 'H';
    S_Std   <= 'L';
    S_StdV  <= "111010HLXU00001111HHHHXXXXHHHHLLL1";
    wait for 152 ns;
    -- Method 1, writing to OUTPUT
    Std.TextIO.Write(L_v, String'("T = "));
    Std.TextIO.Write(L_v, now);
    Std.TextIO.Write(L_v, String'("  S_Bit = "));
    Std.TextIO.Write(L_v, S_Bit);
    Std.TextIO.WriteLine(Output, L_v);

    Std.TextIO.Write(L_v, String'("T = "));
    Std.TextIO.Write(L_v, now);
    Std.TextIO.Write(L_v, String'("  S_BitV = "));
    Std.TextIO.Write(L_v, S_BitV);
    Std.TextIO.WriteLine(Output, L_v);

    Std.TextIO.Write(L_v, String'("Thex = "));
    Std.TextIO.Write(L_v, now);
    Std.TextIO.Write(L_v, String'("  S_BitV = ") & HexImage(S_BitV));
    Std.TextIO.WriteLine(Output, L_v);

    Std.TextIO.Write(L_v, String'("T = "));
    Std.TextIO.Write(L_v, now);
    Std.TextIO.Write(L_v, String'("  S_Int = "));
    Std.TextIO.Write(L_v, S_Int);
    Std.TextIO.WriteLine(Output, L_v);

    Std.TextIO.Write(L_v, String'("T = "));
    Std.TextIO.Write(L_v, now);
    Std.TextIO.Write(L_v, String'("  S_Real = "));
    Std.TextIO.Write(L_v, S_Real);
    Std.TextIO.WriteLine(Output, L_v);

    Std.TextIO.Write(L_v, String'("T = "));
    Std.TextIO.Write(L_v, now);
    Std.TextIO.Write(L_v, String'("  S_uStd = "));
    IEEE.Std_Logic_TextIO.Write(L_v, S_uStd);
    Std.TextIO.WriteLine(Output, L_v);

    Std.TextIO.Write(L_v, String'("T = "));
    Std.TextIO.Write(L_v, now);
    Std.TextIO.Write(L_v, String'("  S_Std = "));
```

Output is a file class defined in package *TextIO*. It is defined as the standard output, usually the screen.

Use of the HexImage function for HEX display

```
    IEEE.Std_Logic_TextIO.Write(L_v, S_Std);
    Std.TextIO.WriteLine(Output, L_v);

    Std.TextIO.Write(L_v, String'("T = "));
    Std.TextIO.Write(L_v, now);
    Std.TextIO.Write(L_v, String'("  S_StdV = "));
    IEEE.Std_Logic_TextIO.Write(L_v, S_StdV);
    Std.TextIO.WriteLine(Output, L_v);

    S_StdV  <= "11101010101010101111000001110001 11";
    wait for 10 ns;

    -- Method 2 -- ASSERT
    assert False
      report "T = " & Image(now) & "  S_BitV = " & Image(S_BitV) &
        HexImage(S_BitV)
      severity Note;

    assert False
      report "S_Bit=" & Image(S_Bit) &
             "; S_Int=" & Image(S_Int) &
             ";S_Real=" & Image(S_Real) &
             ";S_uStd=" & Image(S_uStd) &
             ";S_Std =" & Image(S_Std)  &
             ";S_StdV=" & Image(S_StdV)
      severity Note;

    assert False
      report "S_StdV=" & HexImage(S_StdV)
      severity Note;            \__ Use of the HexImage
                                    function for HEX display
    wait;
  end process Test_Lbl;
end StringOut_a;
```

Figure 8.3.1 Strings to Screen, VHDL'87 and VHDL'93 (ch8_dir\strgnout.vhd)

8.4 DESIGN OF A LINEAR FEEDBACK SHIFT REGISTER (LFSR)

An LFSR is a feedback shift register. It is used to generate pseudo-random numbers for chip Built-In-Self-Test (BIST) and for testbenches. LFSRs are particularly practical in the generation of pseudo-random numbers of type *Bit_Vector, Std_Logic_Vector, Unsigned or Signed* types. They are preferred over the use of *Integer* type because very wide ranges can be generated (e.g., 100+ bits). Figure 8.4-1 demonstrates a hardware implementation of a four-bit LFSR. Figure 8.4-2 shows an LFSR declaration package for the LFSR function. Figure 8.4-3 represents a portion of the package body for this package (the complete body is on the disk). Figure 8.4-4 demonstrates the design of a four-bit LFSR register. Figure 8.4-5 demonstrates the results of synthesis for this LFSR design. Figure 8.4-6 represents a testbench for the LFSR function.

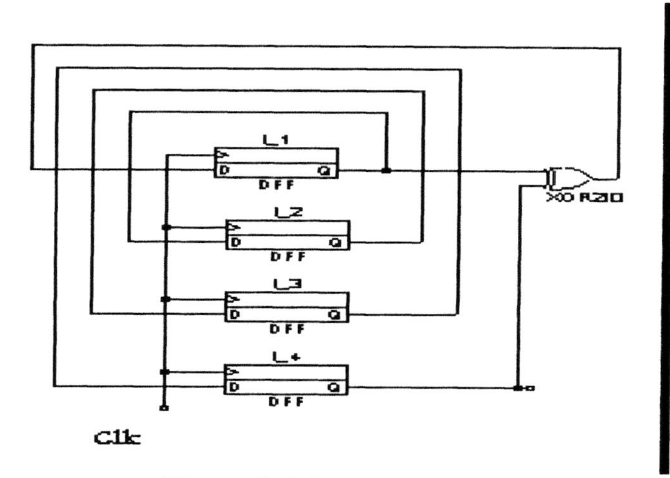

Figure 8.4-1 Four Bit LFSR

Note that other versions of the LFSR package are provided on disk to accommodate the following types:

TYPE	FILE ON DISK	COMMENTS
Std.Standard.Bit_Vector	package\lfsrbit.vhd	
IEEE.Numeric_Std.Signed	package\lfsrnstd.bhd	
IEEE.Numeric_Std.Unsigned	package\lfsrnstd.bhd	
IEEE.Numeric_Bit.Signed	package\lfsrnbit.vhd	
IEEE.Numeric_Bit.Unsigned	package\lfsrnbit.vhd	
Lib.Std_Logic_Arith.Signed	package \lfsrarith.vhd	Lib is any user defined
Lib.Std_Logic_Arith.Unsigned	package \lfsrarith.vhd	library

```
--    File name   :  lfsrstd.vhd
--    Title       :  Linear Feedback Shift Register Package
--    Description :  for registers of lengths 2 through 64
--                :  and  100, 132, 164, 200, 300 bits.
--                :  Initial actual value should be a not zero vector.
--                :  If is it, this function will return the value of 1.
--                :  Data should NOT contain any of the following values
--                :  'U' | 'X' | 'Z' | 'W' | '-'
--                : Actual parameter can be either a variable or a
--                : signal.  Size of vector can be between
--                :     2 to 64, 100. 132. 164. 200, or 300.
--                : Function detects size of actual and returns
--                : a value of the same size as
--                : the actual parameter.  Example of application:
--                :   signal R32_s : Std_Logic_Vector(31 downto 0) :=
--                :   "00101100010101111110001010100110";
--                :         ....
--                :         R32_s <= LFSR(R32_s);
--                :
--    Source of   :  "Built-In Test for VLSI: Pseudorandom Techniques",
--    equations   :  Paul H. Bardell, William H. McAnney, Jacob Savir,
--                :  John Wiley and Sons, 1987.  [10]
--                :  Equations are described in Appendix, and provide
--                :  the polynomials for degrees 2 through 300.
--                :
--    Synthesis   :  This code is synthesizable.  Pragmas are written
--                :  for Synopsys synthesizer to ignore the assert
```

```
--                 :  statement written for error detection
---------------------------------------------------------------
library IEEE;
  use IEEE.Std_Logic_1164.all;

-- pragma translate_off
Library Work;
  -- for conversion of Std_Logic_Vector to a string
  use Work.Image_Pkg.all;

-- pragma translate_on
package LfsrStd_Pkg is
  function LFSR(S : Std_Logic_Vector)  return Std_Logic_Vector;
  function LFSR(S : Std_uLogic_Vector) return Std_uLogic_Vector;
end LfsrStd_Pkg;
```

Figure 8.4-2 LFSR Declaration Package (package\lfsrstd.vhd).

```
package body LfsrStd_Pkg is
  function LFSR(S : Std_Logic_Vector) return Std_Logic_Vector is
    variable S_v   : Std_Logic_Vector(1 to S'Length);
    constant S_c   : Std_Logic_Vector(1 to S'Length)
                       := (others => '0');
  begin
    -- pragma translate_off
    S_v := To_X01(S);    -- function is in Std_Logic_1164 package
    if Is_X(S_v) then
      assert False
        report "Passed parameter contains one of the following " &
               "characters  'U' | 'X' | 'Z' | 'W' | '-' "
        severity Warning;
      assert False
        report "Data passed = " &
               Work.Image_Pkg.Image(S)
        severity Note;
       return S;       -- Return unchanged value
    end if;
    -- pragma translate_on
    S_v := S;
    if  S_v = S_c then
      return (S_c(1 to S_c'length - 1) & '1');
    else
      case S'Length is
        when 2 =>    --  X^2 + X^1 + 1
          return (S_v(2) xor S_v(1)
                 ) & S_v(1 to S'Length - 1);
        when 3 =>    -- X^3 + X^1 + 1
          return (S_v(3) xor S_v(1)
                 ) & S_v(1 to S'Length - 1);

        when 4 =>    -- X^4 + X^1 + 1
          return (S_v(4) xor S_v(1)
                 ) & S_v(1 to S'Length - 1);
        ....

        when 32 =>  -- X^32 + X^28 + X^27 + X^1 + 1
          return (S_v(32) xor S_v(28) xor
                  S_v(27)  xor S_v(1)
                 ) & S_v(1 to S'Length - 1);


        ...
        when 64 =>  -- X^64 + X^4 + X^3 + X^1 + 1
          return (S_v(64) xor S_v(4) xor
                  S_v(3)  xor S_v(1)
                 ) & S_v(1 to S'Length - 1);

        when 100 =>  -- X^100 + X^37 + 1
```

```
            return (S_v(100) xor S_v(37)
                   ) & S_v(1 to S'Length - 1);

          when 132 =>  -- X^132 + X^29 + 1
            return (S_v(132) xor S_v(29)
                   ) & S_v(1 to S'Length - 1);

          when 164 =>  -- X^164 + X^14 + X^13 + X^1 + 1
            return (S_v(164) xor S_v(14) xor
                    S_v(13)  xor S_v(1)
                   ) & S_v(1 to S'Length - 1);
          when 200 =>  -- X^200 + X^163 + X^2 + X^1 + 1
            return (S_v(200) xor S_v(163) xor
                    S_v(2)  xor S_v(1)
                   ) & S_v(1 to S'Length - 1);

          when 300 =>  -- X^300 + X^7 + 1
            return (S_v(300) xor S_v(7)
                   ) & S_v(1 to S'Length - 1);


          when others =>
            -- pragma translate_off
            assert False
              report "Length of vector is NOT in proper range"
              severity Warning;
            -- pragma translate_on
            return S_v;
        end case;
      end if;
    end LFSR;

    function LFSR(S : Std_uLogic_Vector) return Std_uLogic_Vector is
    begin
      return To_StdULogicVector(LFSR(To_StdLogicVector(S)));
    end LFSR;
end LfsrStd_Pkg;
```

Figure 8.4-3 LFSR Package Body (package\lfsrstd.vhd).

```
library IEEE;
  use IEEE.Std_Logic_1164.all;

library Work;
  use Work.LfsrStd_Pkg.all;

entity TestLFSR is
  port(Clk    : in   Std_Logic;
       LFSR4  : out  Std_Logic_Vector(3 downto 0));
end TestLFSR;

architecture TestLFSR_a of TestLFSR is
  signal LFSR4_s : Std_Logic_Vector(3 downto 0);
begin
  LFSR_Lbl: process
  begin
    wait until Clk = '1';
    LFSR4_s <= LFSR(LFSR4_s);
  end process LFSR_Lbl;

  LFSR4 <= LFSR4_s;
end TestLFSR_a;
```

Figure 8.4-4 Design of a Four-Bit LFSR Register (ch8_dir\lfsr4.vhd)

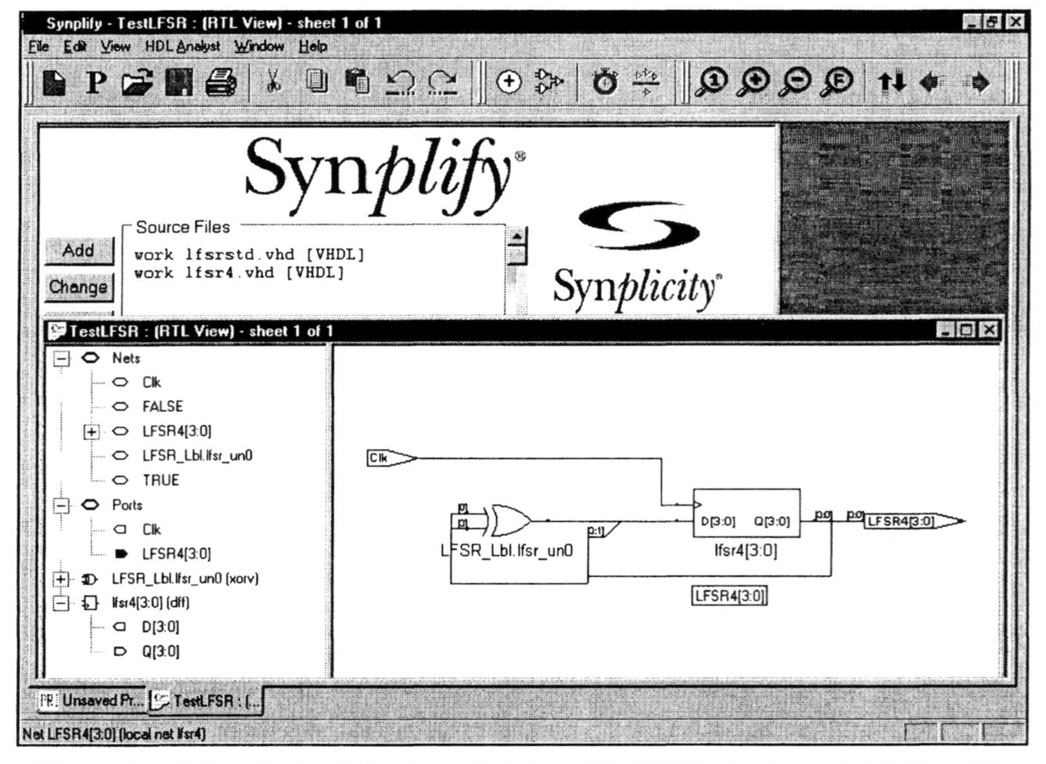

Figure 8.4-5 Synthesized Design of a Four-Bit LFSR Register ***(with Synplify from Synplicity)***

This synthesized four-bit LFSR is slightly different from the design shown in figure 8.4-1. It includes a reset to a value of ONE if the inputs are all zeros.

```
library IEEE;
  use IEEE.Std_Logic_1164.all;
library Work;
  use Work.LfsrStd_Pkg.all;

entity TestLFSR is
end TestLFSR;
architecture TestLFSR_a of TestLFSR is
  signal R32_s : Std_Logic_Vector(31 downto 0) :=
        "00101100010101111110001010100110";
  signal R16_s : Std_Logic_Vector(0 to 15) := "0100110U111Z0100";
  signal R6_s  : Std_Logic_Vector(0 to 5) := (Others => '0');
  signal R8_s  : Std_Logic_Vector(7 downto 0);
  signal R4_s  : Std_Logic_Vector(3 downto 0) := "HL10";
begin
  R32_s <= LFSR(R32_s) after 10 ns;

  R16_s <= LFSR(R16_s) after 10 ns;

  R6_s  <= LFSR(R6_s) after 10 ns;

  R4_s <= LFSR(R4_s) after 10 ns;

  LFSR_Lbl: process
    variable R8_v : Std_Logic_Vector(7 downto 0) := "10101110";
  begin
    R8_v := LFSR(R8_v);
    R8_s <= R8_v;
    wait for 10 ns;
  end process LFSR_Lbl;
end TestLFSR_a;
```

Figure 8.4-6 Testbench for the LFSR function (ch8_dir\lfsr_tb.vhd)

8.4.1 Random Number Generation

The LFSR package can be used to generate random numbers. Conversion functions can be used to convert a *Std_Logic_Vector* to integers. Figure 8.4.1 demonstrates the concept.

```
library IEEE;
  use IEEE.Std_Logic_1164.all;
  use IEEE.STD_LOGIC_UNSIGNED.all;

library Work;
  use Work.LfsrStd_Pkg.all;

entity RandomLFSR is
end RandomLFSR;

architecture RandomLFSR_a of RandomLFSR is
  signal LFSR_s : Integer;
begin
  LFSR_Lbl: process
    variable Random_v : Std_Logic_Vector(10 downto 0)
       := "00001100110";
    variable Rint_v   : Integer;
  begin
    wait for 10 ns;
    Random_v := LFSR(Random_v);
    LFSR_s <= CONV_Integer(Random_v(7 downto 0));
  end process LFSR_Lbl;
end RandomLFSR_a;
```

Figure 8.4.1 Random Number Generator Using LFSR Package (Ch8_dir\random.vhd))

8.5 COMPILATION ORDER

A compilation order must be followed when there are compilation dependencies. If packages are inferred in the entity and or architectures of models, those package declarations must first be compiled because they define the visible contents of the packages. The package bodies need not even exit unless simulation is required. Thus, package bodies can be compiled in any sequence after the compilation of the package declarations. If a package declaration makes use of another package, then that package declaration must be compiled first.

An architecture that instantiates components can be compiled after entities of those components are compiled. The architecture of those components need not be compiled unless simulation is required. In the component declaration section of an architecture, if the naming of the component is the same as the name of the entity it represents, then the default binding for component is assumed. Otherwise, a binding indication is required in the architecture (see chapter 9). Thus, the compilation order must be at least as follows:

1. **Packages declarations** taking account inter-packages dependencies (e.g., if package *A_Pkg* needs declarations from package *B_Pkg*, then package *B_Pkg* must be compiled prior to package *A_Pkg*).
2. **Entities** that make use of the above packages.
3. **Architectures**. Architectures that instantiate components (e.g., testbenches or structural descriptions) may be compiled (but not simulated) without the architectures of the components it uses.
4. **Configurations** (see Chapter 9).

5. **Package Bodies**. Note: Package bodies can be compiled at any time, and do not have to be last.

Typically, package bodies are compiled after package declarations if they are known. In addition, architectures are compiled following the entity declarations if they are defined.

8.5.1 Compilation Rules on Changes

1. If a package declaration is changed, then ALL other packages, entities and architectures that make use of that package must be recompiled. In addition, the package bodies of the dependent package must also be recompiled. In addition, the package body of the modified package must be recompiled. Configuration declarations must also be recompiled.
2. If a package body is changed, then only the package body needs to be recompiled.
3. If a change is made to an entity, then ALL the architectures that make use of that entity (including component declarations and instantiations) must be recompiled. Configuration declarations must also be recompiled.
4. If an architecture is changed, then only that architecture needs to be recompiled.
5. If a configuration is changed, then only that configuration needs to be recompiled.

8.5.2 Automatic Analysis of Dependencies

There exist tools to automatically analyze the dependencies in a VHDL model and create *makefiles*, which greatly eases the task of maintaining a large VHDL model. One such set of tools that is truly vendor independent is "*vmk*" and "*lmk*" from Qualis Design Corporation[15]. These tools allow a user to quickly generate and retarget the *makefiles* from one vendor to another, and maintain concurrent *makefiles* for parallel use of different VHDL toolsets, directly from a single set of VHDL source files.

Some vendors provide mechanisms in their VHDL system to automatically detect a time stamp change in the VHDL source files, and to automatically recompile the necessary files before simulation. However, these systems require that an initial compilation be performed first on all the source files.

[15] Qualis Design Corporation can be reached at info@qualis.com

EXERCISES

1. Write a package declaration and package body that includes the following:
   - State_Typ (Fetch, Decode, FastExecute, SlowExecute)
   - Speed_Typ (Slow, Fast)
   - Function called NextState that accepts two parameters, one of State_Typ, and a one of Speed_Typ and that returns State_Typ. The function operates as follows:

Current state	Speed	Next State
Fetch	-	Decode
Decode	Slow	SlowExecute
Decode	Fast	FastExecute
FastExecute	-	Fetch
SlowExecute	-	Fetch

   - Write an overloaded function "not" that operates on Speed_Typ. Thus,

```
-- if Var is of Speed_Typ, then
Var := not Var; -- if Slow then Var will be Fast
```

2. Write an empty entity and an architecture with a signal of State_Typ initialized to Fetch, and another signal of Speed_Typ initialized to Slow. Write a process that makes use of the package and that performs the following:
   - State signal cycles every 100 ns from fetch to decode to execute.
   - After the execute cycle the Speed signal is changed to the other speed (the "not" of the original speed).

```
STATE: 100 ns Fetch --> 100ns Decode --> 100 ns FastExecute -->
       100ns Fetch --> 100ns Decode --> 100 ns SlowExecute -->

       100 ns Fetch --> 100ns Decode --> 100 ns FastExecute -->
       100ns Fetch --> 100ns Decode --> 100 ns SlowExecute --> -- etc.
```

Use the functions defined in the package.

4. Write two entities and architectures where one component reads text data from a file and transfers this data to other component through a port. The other component displays this data to the transcript window. The 2 entities and architecture must use a package called *Char_Pkg*.

   Q1. Can the type of signal Data_s be Std.TextIO.line? Explain your reasoning.

   Q2. How can the transfer of data between the source and destination be synchronized? Is the diagram below correct? If not, make the necessary additions.

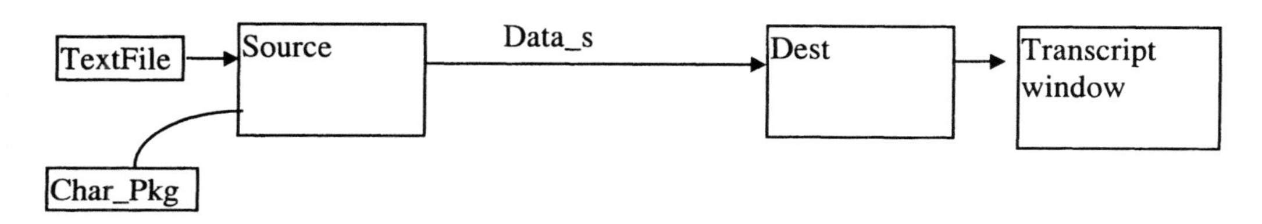

4. Given the entities *A, B,* and the architectures *A_a* and *B_a*, and the packages *A_Pkg* and *B_Pkg* define the compilation order for these VHDL descriptions. In your compilation ordering, identify any descriptions that can be compiled LAST.

```
------------------------------------------------------------------------
--    Project       :  ATEP
--    File name     :  a_e.vhd
--    Title         :  Entity description
--    Description :  Demo of compilation order
------------------------------------------------------------------------
library ATEP_Lib;

entity A is
  use ATEP_Lib.B_Pkg.all;
  use ATEP_Lib.B_Pkg;
end A;
```

```
------------------------------------------------------------------------
--    File name    :  a_a.vhd
------------------------------------------------------------------------
architecture A_a of A is
  use ATEP_Lib.A_Pkg.all;
  use ATEP_Lib.A_Pkg;
begin  --  A_a
end A_a;
```

```
------------------------------------------------------------------------
--    File name    :  b_e.vhd
------------------------------------------------------------------------
library ATEP_Lib;

entity B is
end B;
```

```
------------------------------------------------------------------------
--   File name   :  b_a.vhd
------------------------------------------------------------------------
architecture B_a of B is
  component A
  end component;
begin  --  B_a
  U1: A;
end B_a;
```

```
------------------------------------------------------------------------
--   File name   :  a_p.vhd
------------------------------------------------------------------------
library ATEP_Lib;
  use ATEP_Lib.B_Pkg.all;
  use ATEP_Lib.B_Pkg;

package A_Pkg is
end A_Pkg;
```

```
------------------------------------------------------------------------
--   File name   :  b_p.vhd
------------------------------------------------------------------------
library ATEP_Lib;

package B_Pkg is
end B_Pkg;
```

5. It is desired to define a record type in a manner similar to the following:

```
type CompAttr_Typ is
  record
    Name      : string; -- type string is array (positive range <>) of character;
    Delay     : integer;
    Latency   : integer;
  end record;
```

However, string is unconstrained, and thus illegal. Find a solution so that a variable of type *CompAttr_Typ* can accommodate names of variable lengths. Write an entity/architecture with a process that updates a variable of *CompAttr_Typ* type.

Hint: Use TextIO.Line type, and procedure TextIO.Write to define the contents of the string.

6. If type line is used as a type of a field of a record, can a signal be declared of that record type? Explain your rationale.

9. USER DEFINED ATTRIBUTES, SPECIFICATIONS, AND CONFIGURATIONS

User defined attributes are useful in qualifying or specifying design characteristics that are often used by other tools, such as logic synthesizers. This chapter expands the definition and applications of user defined attributes including attribute declarations and attribute specifications.

A configuration binds instances of components with one of many possible architectures of the components. This chapter discusses configuration specifications with the default and explicit binding. It also addresses a preferred configuration method called the configuration declaration.

9.1 ATTRIBUTE DECLARATIONS

[1] (LRM 4.4) An attribute is a value, function, type, range, signal, or constant that may be associated with one or more named entities in a description. There are two categories of attributes:

1. ***Predefined attributes** provide information about named objects in a description* (e.g., 'delayed, 'last value, 'range, etc.) (see appendix E for a summary of the definitions)
2. ***User-defined attributes** are constants of arbitrary types,* and are used in attribute specifications to **qualify additional information** about an item (such as the capacitance load on ports).

The syntax for user defined attributes is as follows:
attribute_declaration ::=
attribute identifier : type_mark; -- the identifier is the designator of the attribute.

9.2 USER-DEFINED ATTRIBUTES

[1] Attributes may be associated with an entity declaration, an architecture, a configuration, a procedure, a function, a package, a type, a subtype, a constant, a signal, a variable, a component, a label, a literal, a unit, a group, or a file. The association is defined in the specification of the item (see section 9.3).

Object declarations (e.g., variables and signals) types must be defined before declaring objects of those types. In a similar manner, **attributes must first be declared** (so that the type of the attribute designator is known) **before specifying attributes for the objects** (or other named items). The attribute declarations can be located wherever declarations (such as type and subprogram declarations) are made. Thus, attribute declarations can be placed in **packages,** in **entities,** and in **architectures.**

M👍👍 when several attributes related to a common design are declared, use packages for the attribute declarations. In packages, define constants for the values of those attributes if they pertain to the design.

Rationale: *When common attributes and constant values for those attributes are declared in packages, readability and code maintenance is enhanced. When the named entity (e.g., subprogram, architecture, entity, type, ..) uses the attribute package , then the package becomes "bound" to the named entity, and any change to the package declaration requires a recompilation of the files using those attributes.*

Figure 9.2-2 is an example of a package for the attribute declarations of parameters used in the specification of a typical component. Note that the electrical and environmental characteristics, and constants for these characteristics, are declared in this package and will be bound to an entity in the entity declaration. This package makes use of the package *Units_Pkg* shown in figure 9.2-1. Section 9.3 explains how attribute declarations are used for the attribute specifications of named entities.

```
package Units_Pkg is
  type Current_Typ is range -2147483647 to 2147483647
    units
      nA;               -- base unit, nano Amp
      uA  = 1000 nA;    -- micro Amp
      mA  = 1000 uA;    -- milli Amp
      Amp = 1000 mA;    -- Ampere
    end units;
```

```
 type Volt_Typ is range -2147483647 to 2147483647
     units
       uV;                  -- base unit, micro Volt
       mV  = 1000 uV;       -- milli Volt
       V   = 1000 mV;       -- Volt
     end units;

  type DegreeC_Typ is range -1E3 to 1E3
     units
       C;                   -- base unit, Celcius
     end units;

  type Radiation_Typ is range 0 to 1E9
     units
       Rad;                     -- base unit,
       KRad  = 1000  Rad;       -- 1000 rads
       MRad  = 1000 KRad;       -- 10E6  rad
     end units;

  type Frequency_Typ is range 0 to 2147483647
     units
       Hz;                  -- base unit, Hertz
       KHz   = 1000  Hz;
       MHz   = 1000 KHz;
     end units;
end Units_Pkg;
```

Figure 9.2-1 Units Package (ch9_dir\units_p.vhd)

```
library ATEP_Lib;
  use ATEP_Lib.Units_Pkg.all;
  use ATEP_Lib.Units_Pkg;

package FPGA_Pkg is
  attribute Supply_Voltage_Min     : Units_Pkg.Volt_Typ;
  attribute Supply_Voltage_Max     : Units_Pkg.Volt_Typ;
  attribute Operating_Temp_Min     : Units_Pkg.DegreeC_Typ;
  attribute Operating_Temp_Max     : Units_Pkg.DegreeC_Typ;
  attribute High_Level_Vin_Min     : Units_Pkg.Volt_Typ;
  attribute High_Level_Vin_Max     : Units_Pkg.Volt_Typ;
  attribute Low_Level_Vin_Min      : Units_Pkg.Volt_Typ;
  attribute Low_Level_Vin_Max      : Units_Pkg.Volt_Typ;
  attribute High_Level_Vout_Min    : Units_Pkg.Volt_Typ;
  attribute Low_Level_Vout_Max     : Units_Pkg.Volt_Typ;
  attribute Input_Leakage_IMin     : Units_Pkg.Current_Typ;
  attribute Input_Leakage_IMax     : Units_Pkg.Current_Typ;
  attribute Power_Supply_IMin      : Units_Pkg.Current_Typ;
  attribute Power_Supply_IMax      : Units_Pkg.Current_Typ;
  attribute Radiation_Hardness     : Units_Pkg.Radiation_Typ;
  attribute Maximum_Frequency      : Units_Pkg.Frequency_Typ;
  attribute Minimum_Frequency      : Units_Pkg.Frequency_Typ;
  attribute Input_Rise_Time_Max    : time;
  attribute Input_Fall_Time_Max    : time;

  constant  Supply_Voltage_Min_c   : Units_Pkg.Volt_Typ       := 4_750 mV;
  constant  Supply_Voltage_Max_c   : Units_Pkg.Volt_Typ       := 5_250 mV;
  constant  Operating_Temp_Min_c   : Units_Pkg.DegreeC_Typ    := 0      C;
  constant  Operating_Temp_Max_c   : Units_Pkg.DegreeC_Typ    := +70    C;
  constant  High_Level_Vin_Min_c   : Units_Pkg.Volt_Typ       := 2      V;
  constant  High_Level_Vin_Max_c   : Units_Pkg.Volt_Typ       := 5_550 mV;
  constant  Low_Level_Vin_Min_c    : Units_Pkg.Volt_Typ       := -300  mV;
  constant  Low_Level_Vin_Max_c    : Units_Pkg.Volt_Typ       := 800   mV;
  constant  High_Level_Vout_Min_c  : Units_Pkg.Volt_Typ       := 2_400 mV;
  constant  Low_Level_Vout_Max_c   : Units_Pkg.Volt_Typ       := 500   mV;
  constant  Input_Leakage_IMin_c   : Units_Pkg.Current_Typ    := -10   uA;
```

Attribute declarations (IdentifierName + type)

These constants will be used in the attribute specifications

```
   constant  Input_Leakage_IMax_c  : Units_Pkg.Current_Typ   := 10    uA;
   constant  Power_Supply_IMin_c   : Units_Pkg.Current_Typ   := 45    mA;
   constant  Power_Supply_IMax_c   : Units_Pkg.Current_Typ   := 90    mA;
   constant  Radiation_Hardness_c  : Units_Pkg.Radiation_Typ := 10E3 rad;
   constant  Maximum_Frequency_c   : Units_Pkg.Frequency_Typ := 10   MHz;
   constant  Minimum_Frequency_c   : Units_Pkg.Frequency_Typ := 0     Hz;
   constant  Input_Rise_Time_Max_c : time                    := 500   ns;
   constant  Input_Fall_Time_Max_c : time                    := 500   ns;
end FPGA_Pkg;
```

Figure 9.2-2 Package for Attribute Declaration of a Typical FPGA Device (ch9_dir\fpga_p.vhd)

9.3 SPECIFICATIONS

*[1] A specification associates additional information with a previously declared (*named*) entity.* The kinds of specifications include:

1. **Attribute specifications** that associate user-defined attributes with one or more named entities, and define the values of those attributes for those named entities (e.g., label, entity, architecture, signal, etc.). The attribute specification is said to *decorate* the named entity.
2. **Configuration specifications** that *[1] associates binding information with component labels representing instances of a given component declaration.*
3. **Disconnect specifications** is now considered OBSOLETE because of the introduction of Std_Logic_1164 package that utilizes the "Z" state for a disconnect method. The disconnect specification is not discussed in this book.

*[1] A specification always relates to named entities that already exists. Thus a given specification **MUST** either **FOLLOW** or (in certain cases) be contained within the declaration of the named entity to which it relates. Furthermore, a specification **MUST** always appear either immediately within the same declarative part as that in which the declaration of the named entity appears, or (in the case of specifications that relate to design units or the interface objects of design units, subprograms, or block statements) immediately within the declarative part associated with the declaration of the design unit, subprogram body, or block statement.*

9.3.1 Attribute Specifications

Attribute specifications typically add information used by other tools. For example, the attribute *ARRIVAL* for a signal could be used by the synthesizer to optimize the design given the arrival time of a signal. The attributes *PackType* and *PinNumber* could be used by a board placement and route program tool that supports the VHDL attributes to place and route the package of the device onto a circuit board. The syntax for attributes is as follows:

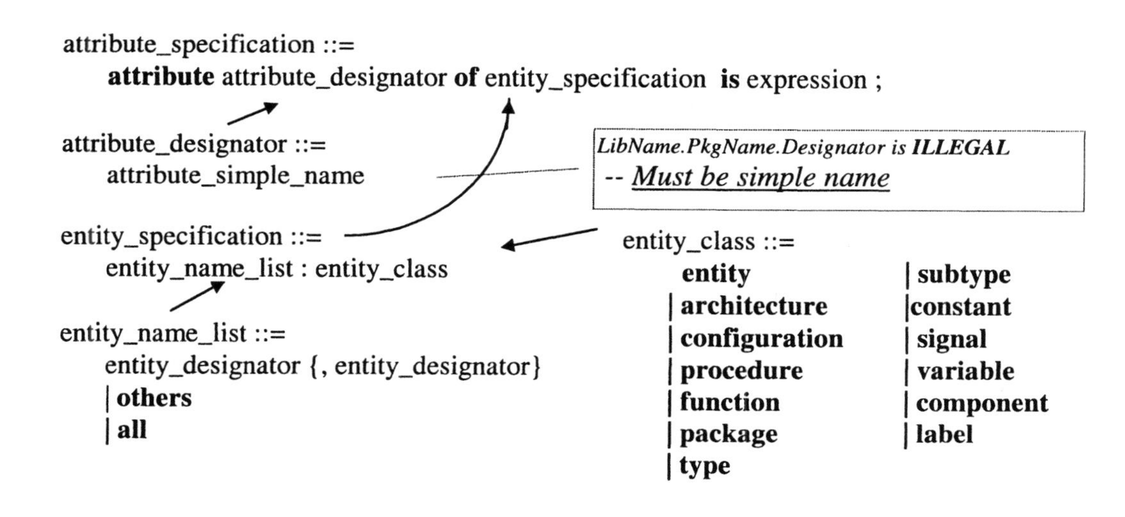

Unlike objects declared in packages (e.g., constants, signals), attribute designator MUST be a simple name (LRM), and the package path cannot be provided. Thus **identify source of package either as comment lines**, or as **single comment line** for a common grouping of attribute specifications.

Rationale: *It is essential to provide good documentation when producing VHDL code. Unlike objects, types, and subprograms declared in packages, VHDL does not allow a complete path when accessing those named entities. The name for an attribute must be simple name. Thus, providing the path in comments is deemed essential for good documentation.*

An architecture can test a user attribute in the same manner as the predefined attributes by naming the identifier with attribute followed by the "tick" symbol (i.e., '), followed by the attribute. For example:

```
-- With this attribute declaration:
attribute InputRiseTime_Max      : time;

-- With the entity_designator Data defined as a port of an entity or a signal
-- of an architecture, the following attribute specification can be defined:
attribute InputRiseTime_Max of Data : signal is 10 ns;

-- The attribute value can be read in the architecture as follows:
if Data'InputRiseTime_Max < 15 ns then ... -- Reading of attribute
```

Table 9.3.1 identifies potential applications and examples for attribute specifications. Figure 9.3 -1 represents the attribute specifications for an FPGA entity, making use of the attribute declaration package and the unit package previously described. Figure 9.3-2 represents examples of attribute declarations, attribute specifications and use of user defined attributes for an entity, variables and procedures.

Table 9.3.1 Potential Applications of Attribute Specifications

Named Entity	Typical Attribute Applications	Attribute Specification Example	Comments
Entity	Characteristics of component (Supply voltages, maximum frequency, temperature range, packaging, size, etc.)	**attribute** Leakage _Min **of** FPGA : **entity is** 10 nA; **attribute** MAX_AREA **of** UART : **entity is** 25.0;	Attributes useful to document operating characteristics of the component. Architecture can make use of the entity attributes to modify timing delays.
Archi-tecture	Identification of technology, or other items related to the architecture (e.g., area)	**attribute** Technology **of** FPGA_a: **architecture is** CMOS;	Attributes could be used by a synthesizer tool to select a library. However, generally a target library is supplied to the synthesizer.
Signal	Characteristics of pins on a port such as capacitance and drive loading, signal arrival time, driver type, pin numbering etc.	**attribute** Pin_Numb **of** Address3 : **signal is** "A1"; **attribute** Arrival **of** Addr3 : **signal is** 55 ns;	Attributes can be used by synthesizer to optimize the design. They could also be used by a circuit board placement and route program.
Label	Characteristics of instances of components	**attribute** DONT_TOUCH **of** U1_Mpy : **label is** true;	Could be used by synthesizer to characterize the handling of instances of components (e.g., avoid optimization, optimize for speed, area, etc.)
proce-dure function	Characteristics of subprograms	**attribute** MaxSpeed **of** ParityTree : **procedure is** true;	Not often used. Could be used by synthesizer to select a particular implementation or optimization.
Confi-guration package type subtype constant com-ponent variable	Attributes used to characterize information about objects or types or configurations.	**attribute** Optimistic **of** MaxDelay_c : **constant is** true; **attribute** Size **of** Line_v : Std.TextIO.line **is** 80;	Not often used. Architecture could test for attributes. Examples: **If** MaxDelay_c'Optimistic **then** ... **if** Line_v'length > Line_v'Size **then** ...

```
library Atep_Lib;
library IEEE;
  use IEEE.Std_Logic_1164.all;
entity FPGA is
  generic (FFdelay_g   : time := 100 ns);
  port (
      Address1   : out    Std_Logic; --  2 bit address
      Address0   : out    Std_Logic; --  2 bit data
      Data1      : inout  Std_Logic; --
      Data0      : inout  Std_Logic; --
      RdWF       : out    Std_Logic; -- Read/ Write active Low
      WeF        : out    Std_Logic; -- Write Enable
      VCC        : in     Std_Logic; -- Supply voltage
      GROUND     : in     Std_Logic);

      attribute Pin_Numb : String;  -- attribute declaration
      attribute Pin_Numb of Address1 : signal is  "A3";  -- attribute specification
      ...
      attribute Pin_Numb of VCC      : signal is  "B4";
      attribute Pin_Numb of GROUND   : signal is  "C4";

   -- NOTE: attribute declarations and constants
   -- are defined in package ATEP_Lib.FPGA_Pkg
   -- Package paths for the constants are not identified because the line
   -- would be very long, and would not add any more information than what is
   -- provided in this comment.
   use Atep_Lib.FPGA_Pkg.all;
   attribute Supply_Voltage_Min   of FPGA : entity is Supply_Voltage_Min_c;
   attribute Supply_Voltage_Max   of FPGA : entity is Supply_Voltage_Max_c;
   attribute Operating_Temp_Min   of FPGA : entity is Operating_Temp_Min_c;
   attribute Operating_Temp_Max   of FPGA : entity is Operating_Temp_Max_c;
   attribute High_Level_Vin_Min   of FPGA : entity is High_Level_Vin_Min_c;
   attribute High_Level_Vin_Max   of FPGA : entity is High_Level_Vin_Max_c;
   attribute Low_Level_Vin_Min    of FPGA : entity is Low_Level_Vin_Min_c;
   ...
   attribute Minimum_frequency    of FPGA : entity is Minimum_frequency_c;
   attribute Input_Rise_Time_Max of FPGA : entity is Input_Rise_Time_Max_c;
   attribute Input_Fall_Time_Max of FPGA : entity is Input_Fall_Time_Max_c;
end FPGA;
```

Figure 9.3 -1 Attribute Specifications for an FPGA Entity, (ch9_dir\fpga_e.vhd)

```
library ATEP_Lib;
-- The Attributes package is supplied on disk in file ch9_dir\attrib_p.vhd
use ATEP_Lib.Attributes.all;   use ATEP_Lib.Attributes;
use ATEP_Lib.Units_Pkg.all;    use ATEP_Lib.Units_Pkg;
use Atep_Lib.FPGA_Pkg.all;     use Atep_Lib.FPGA_Pkg;

entity Attributes is
  port (Address          : out    Bit_Vector(31 downto 0);
        Data             : inout  Bit_Vector(31 downto 0);
        RdWF             : out    Bit);
  attribute Pin_Numb : String ;

  attribute Arrival of Address   : signal is 20.2; -- ns, package ATTRIBUTES
  attribute Arrival of others    : signal is 15.0; -- ns, package ATTRIBUTES
  attribute Load    of all       : signal is 50.0; -- pf, package ATTRIBUTES

  attribute Pin_Numb of RdWf    : signal is "1";
  attribute Pin_Numb of Address : signal is "2 to 33";
  attribute Pin_Numb of Data    : signal is "34 to 66";

  attribute Supply_Voltage_Min  of Attributes : entity
```

Must be simple name. Thus path is not allowed here.

Identify that package the attributes declarations are declared.

```
        is  FPGA_Pkg.Supply_Voltage_Min_c; -- Atep_Lib.FPGA_Pkg
  attribute Input_Rise_Time_Max of Attributes : entity
        is  FPGA_Pkg.Input_Rise_Time_Max_c; -- Atep_Lib.FPGA_Pkg
end Attributes;

architecture Attributes_a of Attributes is
  ----------------------------------------------------------------
  --  Procedure: TestAllOnes
  ----------------------------------------------------------------
  procedure   TestAllOnes (constant Data_c    : in  Bit_Vector;
                           variable AllOnes_v : out boolean) is

    alias    DataA_c    : Bit_Vector(Data_c'length -1 downto 0) is Data_c;
    variable String20_v : string(1 to 20) := "12345678901234567890";
    attribute MaxSize   : natural;   -- attribute specification
    attribute MaxSize of String20_v : variable is 10;  -- attribute declaration

  begin  --  procedure TestAllOnes
    AllOnes_v := true;
    Lp1_Lbl : for Idx_i in DataA_c'range loop
      if DataA_c(Idx_I) = '0' then
        AllOnes_v := false;
        exit Lp1_Lbl;
      end if;
    end loop Lp1_Lbl;
    -- Test of variable attribute
    assert  String20_v'MaxSize >= String20_v'length
      report "String20_v'MaxSize < String20_v'length "
      severity warning;
  end TestAllOnes;

  type Technology_Typ is (CMOS, TTL, ECL);
  attribute Technology : Technology_Typ;
  attribute Technology of TestAllOnes : procedure is CMOS;

begin  --  Attributes_a
  Test_Lbl : process
    variable Arrival_v : real := Address'Arrival;  -- use of atributes
    variable Load_v    : real := Data'Load;
    variable RdWF_v    : real := RdWF'Arrival;
    variable AllOnes_v : boolean;
  begin  --  process Test_Lbl
    if Attributes'Supply_Voltage_Min > 4_500 mV then
      assert false
        report " Minimum supply voltage is too low"
        severity warning;
    end if;
    Address <= X"ABCD_1234" after integer(Address'Load * 0.7) * 1 ns;
    Data    <= X"0123_4567" after integer(Data'Load * 0.7) * 1 ns;
    RdWF    <= '0'          after integer(Data'Arrival * 1.2) * 1 ns;
    if TestAllOnes'Technology = CMOS then
        TestAllOnes (Data_c    => Data,
                     AllOnes_v => AllOnes_v);
    end if;
    wait;
  end process Test_Lbl;
end Attributes_a;
```

Figure 9.3-2 Applications of User Defined Attributes (ch9_dir\attrib2.vhd)

9.4 CONFIGURATION SPECIFICATION

[1] A configuration specification associates binding information with component labels representing instances of a given component declaration. In other words, if an architecture instantiates components, each of which may have several architectural representations (e.g., behavior, gate level), then that architecture can specify a binding for each instance of that component to a particular architecture. *[1] A binding indication associates component instances with a particular design entity.* If an explicit binding indication is not provided, then the default binding indication will apply (see next section). So far, in all the previous examples, the default binding indication was used by the simulator because no explicit binding indication was provided. Figure 9.4 demonstrates various architectures for a component.

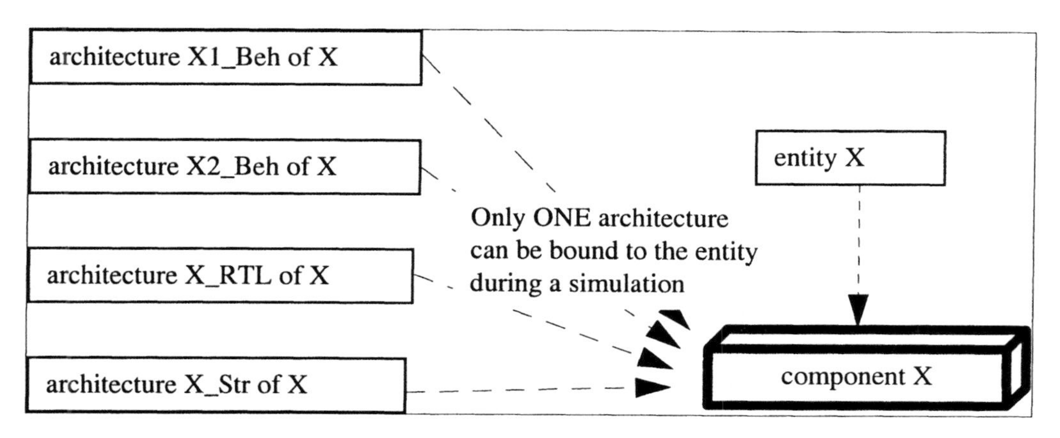

Figure 9.4 Architectures of a Component

9.4.1 Default Binding Indication

In architecture, if no binding indication is provided, and if the component **IDENTIFIER** in the component declaration is the **SAME** as the **ENTITY** name, then the default binding for that component is the ENTITY associated with that component. In addition, its **architecture is the most recently analyzed architecture body** associated with the entity declaration. The architecture identifier is determined during elaboration. The default binding indication includes a default generic map defined in the entity declaration.

M 👍👍 Whenever possible, when declaring a component, use as the component name the entity associated with that component.

***Rationale:** This approach enhances code readability because the name of the entity associated with the component is the identifier for the component. In addition, it allows the use of default binding for simple design cases. In addition, most synthesizers require that the default binding be used.*

Figure 9.4.1 represents a component with two potential architectures *design1_a* and *design2_a*. The component is instantiated in architecture *DesignTop_a*. This component is bound to entity *Design* by default because there is no configuration specification statements. In addition, the architecture for that entity is the most recently compiled version of the architecture. Thus, if file "desgn1_a.vhd" is compiled after "desgn2_a.vhd", then the architecture used during simulation is Design1_a.

```
--   File name   :  desgn_e.vhd
entity Design is
  generic(Limit_g : natural := 20);
end Design;

--   File name   :  desgn1_a.vhd
architecture Design1_a of Design is
begin  --  Design_a
  assert false
    report "This is Design1_a"
    severity note;
end Design1_a;

--   File name   :  desgn2_a.vhd
architecture Design2_a of Design is
begin  --  Design_a
  assert false
    report "This is Design2_a"
    severity note;
end Design2_a;
```

```
--   File name   :  dsgtop_e.vhd
entity DesignTop is
end DesignTop;

--   File name   :  dsgtop_a.vhd
architecture DesignTop_a of DesignTop
is
  component Design
  end component;

begin  --  Design_a
  assert false
    report "This is DesignTop_a"
    severity note;

  U1_Design:  Design;
                            -- default binding.
end DesignTop_a;
```

Figure 9.4.1 Implicit or Default Binding of a Component, (ch9_dir\desgn_e.vhd; ch9_dir\ dsgtop_e.vhd)

9.4.2 Explicit Binding Indication in Configuration Specifications

The explicit binding of components is often used in VHDL to examine the modeling of a different representation of the design under test. These representations could include in a top-down design methodology the modeling of a unit under test (UUT) at the behavioral level, or at the Register Transfer Level (RTL) prior to synthesis, or at the structural level with representations and connection of every gate and flip-flop in the design. Note that synthesizers can automatically write synthesized designs into structural VHDL representation, but it is only machine readable (i.e., it is not for humans). There are two ways to perform binding:

1. **Configuration specification**: This approach is used to specify the bindings of components in the architecture that instantiates the components
2. **Configuration declaration** (see section 9.5): This approach uses a separate design unit to specify the binding of components to entities and architecture.

MÞÞ Use the **configuration declaration** method to bind components to entities/architectures if more than one architecture exists for the entity.

***Rationale:** Configuration declarations represent separate design units and allow for late binding of components.*

The syntax for configuration specification and explicit binding is as follows:

configuration_specification ::=
*   **for** component_specification **use** binding_indication;*

component_specification ::=
  instantiation_list :
      component_name

binding_indication ::=
    entity_aspect [generic_map_aspect]
        [port_map_aspect]

port_map_aspect ::=
    port map (port_association_list)

generic_map_aspect ::=
  generic map (generic_association_list)

instantiation_list ::=
    instantiation_label
  {,instantiation_label }
    | **others**
    | **all**

entity_aspect ::=
    entity entity_name
        [(architecture_identifier)]
    | **configuration** configuration_name
    | **open**

Figure 9.4.2 represents a configuration specification example with components of same and different names as their corresponding entity names. Example of the default configuration specification is also demonstrated.

```
--   File name    :  cfgspec.vhd
--   Title        :  Configuration Specification Example
entity AndGate is
  generic (DlyIO_g:              time := 4 ns);
  port (I1:                      in      Bit;
        I2:                      in      Bit;
        O:                       out     Bit);
end AndGate;
-------------------------------------------------------------------------

entity XorGate is
  generic (DlyIO_g:              time := 4 ns);
  port (I1:                      in      Bit;
        I2:                      in      Bit;
        O:                       out     Bit);
end XorGate;
-------------------------------------------------------------------------

architecture AndGate_a of AndGate is
begin  --  AndGate_a
```

```
  O <= I1 and I2 after DlyIO_g;
end AndGate_a;
-------------------------------------------------------------------------------

architecture XorGate_a of XorGate is
begin  --  XorGate_a
  O <= I1 xor I2 after DlyIO_g;
end XorGate_a;
-------------------------------------------------------------------------------


package Gates_Pkg is
  component WishGate
    generic (Dly_g: time := 4 ns);
    port (A :   in      Bit;
          B :   in      Bit;
          C :   out     Bit);
  end component;
  component AndGate
    generic (DlyIO_g: time := 4 ns);
    port (I1 :  in      Bit;
          I2 :  in      Bit;
          O  :  out     Bit);
  end component;

  component XorGate
    generic (DlyIO_g: time := 4 ns);
    port (I1 :  in      Bit;
          I2 :  in      Bit;
          O  :  out     Bit);
  end component;
end Gates_Pkg;
-------------------------------------------------------------------------------

library ATEP_Lib;
  use ATEP_Lib.Gates_Pkg.all;
  use ATEP_Lib.Gates_Pkg;

entity Structure is
end Structure;

architecture Structure_a of Structure is
-- This architecture implements the equation:
--    Out_s <= ((In1_s and In2_s) and In3_s) xor In4_s;
  signal          And2a_s            : Bit := '0';
  signal          And2b_s            : Bit := '0';
  signal          In1_s              : Bit := '1';
  signal          In2_s              : Bit := '1';
  signal          In3_s              : Bit := '1';
  signal          In4_s              : Bit := '0';
  signal          Out_s              : Bit := '0';
  -- This example demonstrates the maping of generics and ports
  -- that have different names between the component WishGate and
  -- its corresponding mapping to AndGate
  for U1_WishGate: WishGate use
    entity ATEP_Lib.AndGate(AndGate_a)
      generic map(DlyIO_g => Dly_g)
      port map
        (I1 => A,
         I2 => B,
         O  => C);
  -- Since the name of the component is the same as its
  -- corresponding entity, no port map is required.
  for U2_AndGate: AndGate use
    entity ATEP_Lib.AndGate(AndGate_a);

  -- Since NO binding is defined for XorGate, the default bind is
```

This component has a different name and different interfaces than the entity AndGate

Components declared in this package are visible in this architecture. Thus, there is no need to redeclare those components.

Actual component name is "WishGate" mapped into "AndGate"

Formal parameters of **AndGate** mapped (=>) to **Actual** parameters of **WishGate** component

```
  -- used (i.e., XorGate entity and latest compiled XOR architecture
  -- for this entity
begin  -- Structure_a
  U1_WishGate: WishGate                          -- component instantiations
      generic map(Dly_g => 5 ns)
      port map
        (A   => In1_s,
         B   => In2_s,
         C   => And2a_s); -- (In1_s and In2_s )

  U2_AndGate: AndGate
      -- generic map(DlyIO_g => 5 ns)  <----  The generic can be omitted if default
      port map                                is intended, or if configuration
        (I1  => And2a_s,                      declaration redefines it.
         I2  => In3_s,
         O   => And2b_s); -- ((In1_s and In2_s) and In3_s)

  U3_XorGate: XorGate
      generic map(DlyIO_g => 6 ns)
      port map
        (I1  => And2b_s,
         I2  => In4_s,
         O   => Out_s); -- ((In1_s and In2_s) and In3_s) xor In4_s;

end Structure_a;
```

Figure 9.4.2 Configuration Specification Example (ch9_dir\cfgspec.vhd)

9.5 CONFIGURATION DECLARATION

The configuration declaration uses a separate design unit to specify the binding of components to entities and architectures. The syntax is as follows:

configuration_declaration ::=
  configuration identifier **of** entity_name **is**
    configuration_declarative_part
    block_configuration
  end [**configuration**]
    [configuration_simple_name];

configuration_declarative_part ::=
    { configuration_declarative_item }

configuration_declarative_item ::=
    use_clause
    | attribute_specification

block_specification ::=
    architecture_name
    | block_statement_label
    | generate_statement_label
      [(index_specification)]

block_configuration ::=
    for block_specification
    { use_clause }
    {configuration_item }
  end for;

configuration_item ::=
    block_configuration
    | component_configuration

component_configuration ::=
    for component_specification
    [**use** binding_indication;]
    [block_configuration]
  end for;

component_specification ::=
    instantiation_list : component_name

Given the same packages and components used in the previous example, file *cfgdecl.vhd* was modified to reflect the deferred binding of components in architecture *structure_a.* In addition, a configuration design unit was added to show the late binding. Figure 9.5-1 demonstrates the *structure_a* architecture with the deferred binding, and the configuration design unit.

```
library ATEP_Lib;
  use ATEP_Lib.Gates_Pkg.all;
  use ATEP_Lib.Gates_Pkg;

entity Structure is
end Structure;

architecture Structure_a of Structure is
  signal        And2a_s           : Bit := '0';
  signal        And2b_s           : Bit := '0';
  signal        In1_s             : Bit := '1';
  signal        In2_s             : Bit := '1';
  signal        In3_s             : Bit := '1';
  signal        In4_s             : Bit := '0';
  signal        Out_s             : Bit := '0';
begin  --  Structure_a
  U1_WishGate: WishGate
       generic map(Dly_g => 5 ns)
       port map
         (A    => In1_s,
          B    => In2_s,
          C    => And2a_s); -- (In1_s and In2_s)

  U2_AndGate: AndGate
       -- generic map(DlyIO_g => 5 ns )
       port map
         (I1   => And2a_s,
          I2   => In3_s,
          O    => And2b_s); -- ((In1_s and In2_s) and In3_s)

  U3_XorGate: XorGate
       generic map(DlyIO_g => 6 ns)
       port map
         (I1   => And2b_s,
          I2   => In4_s,
          O    => Out_s); -- ((In1_s and In2_s) and In3_s) xor In4_s;
end Structure_a;
```

-- This architecture implements the equation:
-- Out_s <= ((In1_s and In2_s) and In3_s) xor In4_s;

The generic can be omitted if default is intended, or if configuration declaration redefines it.

```
library ATEP_Lib;
configuration Structure_Cfg of Structure is
  use ATEP_Lib.Gates_Pkg.all;
  for Structure_a -- Top level architecture of Structure
    for U1_WishGate: WishGate use
      entity ATEP_Lib.AndGate(AndGate_a )
        generic map(DlyIO_g => Dly_g)
        port map
          (I1 => A,
           I2 => B,
           O  => C);
    end for;
    -- Since the name of the component is the same as its
    -- corresponding entity, this statement binds the entity
    -- to the desired architecture
    for U2_AndGate: AndGate use
      entity ATEP_Lib.AndGate(AndGate_a);
    end for;
    -- Since NO binding is defined for XorGate, the default bind is
    -- used (i.e., XorGate entity and latest compiled XOR architecture
    -- for this entity
  end for;
end Structure_Cfg;
```

Actual component name is *WishGate* mapped into *AndGate*

Formal parameters of **AndGate** mapped (=>) to **Actual** parameters of **WishGate** component

Figure 9.5-1 "Structure_a" Architecture and Configuration Design Unit (ch9_dir\cfgdecl.vhd)

An example of a hierarchical design is shown graphically in Figure 9.5-2. An example that binds hierarchical components in a configuration declaration is shown in figure 9.5-3.

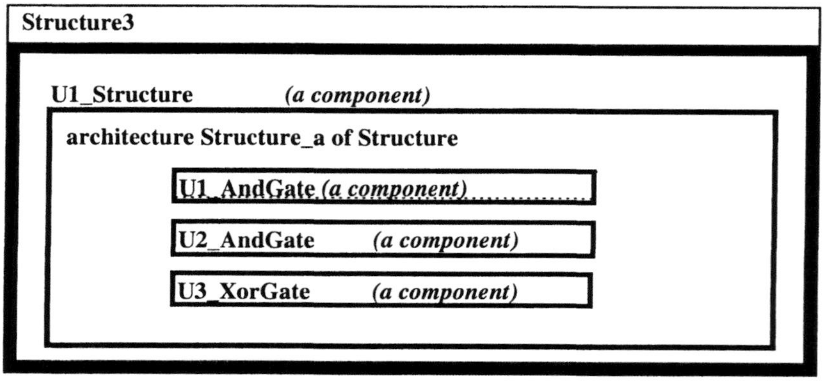

Structure3

U1_Structure (a component)

architecture Structure_a of Structure

U1_AndGate (a component)

U2_AndGate (a component)

U3_XorGate (a component)

Figure 9.5-2 Hierarchical Design

```
... -- File includes all the sublevels
entity Structure3 is
end Structure3;
architecture Structure3_a of Structure3 is
  component Structure
  end component;
begin
  U1_Structure: Structure; -- component instantiation
end Structure3_a;
-----------------------------------------------------------------------------
library ATEP_Lib;
  use ATEP_Lib.Gates_Pkg.all;
configuration Structure3_Cfg of Structure3 is
  for Structure3_a                                   -- struct3_a
    for U1_Structure: Structure use
      entity ATEP_Lib.Structure(Structure_a);
      for Structure_a                                -- Struct_a
        for U1_WishGate: WishGate use                 -- U1_And
          entity ATEP_Lib.AndGate(AndGate_a)
            generic map(DlyIO_g => Dly_g)
            port map
              (I1 => A,
               I2 => B,
               O  => C);
        end for;                                     -- U1_And

        for U2_AndGate: AndGate use         -- U2_And
          entity ATEP_Lib.AndGate(AndGate_a);
        end for;                                     -- U2_And

        for U3_XorGate: XorGate  use        -- U3_Xor
          entity ATEP_Lib.XorGate(XorGate_a);
        end for;                                     -- U3_Xor
      end for;                                       -- Struct_a
    end for;                                         -- U1_Struct
  end for;                                           -- struct3_a
end Structure3_Cfg;
```

Figure 9.5-3 Binding of Components of a hierarchical design with a Configuration Declaration (Ch9_dir\hierarch.vhd)

9.5.1 Binding with configured components

If an architecture instantiates a component that already has a configuration defined, then that configuration can be used for an instance of that component. Figure 9.5.1 represents an example that instantiates the component Structure.

```
library ATEP_Lib;
configuration Structure3_Cfg of Structure3 is
  for Structure3_a
    for U1_Structure: Structure
      use configuration ATEP_Lib.Structure_Cfg;
    end for;
  end for;
end Structure3_Cfg;
```

Figure 9.5.1 Binding with Configured Component, ch9_dir\struct3.vhd

9.5.2 CONFIGURATION OF GENERATE STATEMENTS

Figure 9.5.2 demonstrates the configuration declaration of components instantiated inside a generate statement. A parity tree is used as an example, with two architectures for the exclusive *OR* component.

```
library IEEE;
  use IEEE.Std_Logic_1164.all;

entity XOR_Nty is
  port(A         :  in      Std_Logic;
       B         :  in      Std_Logic;
       Z         :  out     Std_Logic);
end XOR_Nty;

architecture XOR_a of XOR_Nty is
begin  --  XOR_a
  Z <= A xor B;
end XOR_a;

architecture XOR2_a of XOR_Nty is
begin  --  XOR_a
  Z <= ((not A) and B) or (A and (not B));
end XOR2_a;

library IEEE;
 use IEEE.Std_Logic_1164.all;

entity Generate_Nty is
  port(Ain1     :       in      Std_Logic_Vector(31 downto 0);
       Aout     :       out     Std_Logic_Vector(31 downto 0));
end Generate_Nty;
```

Two architectures for same entity.

```
architecture Generate_a of Generate_Nty is
  component XOR_Nty
    port(A         :  in      Std_Logic;
         B         :  in      Std_Logic;
         Z         :  out     Std_Logic);
  end component;

  signal  Temp_s    :  Std_Logic_Vector(31 downto 0);
begin  --  Generate_a
  Aout  <=  Temp_s;
  --  Parity tree, 5 Bit parity
  UK : for K_i in 0 to 3 generate
    UK0 : if K_i = 0 generate
      UXOR : XOR_Nty
               port map(A => Ain1(K_i),
                        B => Ain1(K_i + 1),
                        Z => Temp_s(K_i));
    end generate UK0;

    UK1_3 : if K_i > 0 generate
      UXOR : XOR_Nty
               port map(A => Temp_s(K_i - 1),
                        B => Ain1(K_i + 1),
                        Z => Temp_s(K_i));
    end generate UK1_3;
  end generate UK;
end Generate_a;
library Work;

configuration Parity_Cfg of Generate_Nty is
  for Generate_a
    for UK(0)
      for UK0
        for UXOR : XOR_Nty use
          entity Work.XOR_Nty(XOR_a);
        end for;
      end for;
    end for;

    for UK(1 to 2)
      for UK1_3
        for UXOR : XOR_Nty use
          entity Work.XOR_Nty(XOR2_a);
        end for;
      end for;
    end for;

    for UK(3)
      for UK1_3
        for UXOR : XOR_Nty use
          entity Work.XOR_Nty(XOR_a);
        end for;
      end for;
    end for;
  end for;
end Parity_Cfg;
```


Figure 9.5.2 Configuration of Generate Statement (ch9_dir\genrat_c.vhd)

9.5.3 Deferring the Binding of an Instance of a Component

There can only be one binding for a component in VHDL'87. If a component is bound in an architecture or configuration body then there cannot be another binding in a higher-level configuration body associated with that component. Since it's possible to construct a nested configuration hierarchy, and each level might possibly contain a binding statement associated with a component, the **open** binding statement in a configuration specification (VHDL'93) allows for an explicit binding statement at a higher binding level, thus deferring the binding to a higher level.

Per LRM 12.4.3 *elaboration of component instantiation statement that instantiates a component declaration* ***has no effect*** *unless the component instance is either fully bound to a design entity defined by an entity declaration and architecture body or is bound to a configuration of such design entity.* Thus, the **open** binding can be used to effectively remove a component from an architecture without making modifications to the architecture that instantiates this component.

Figure 9.5.3 represents a configuration that effectively removes a component from an architecture.

```
library ATEP_Lib;
configuration Structure2_Cfg of Structure is
  use ATEP_Lib.Gates_Pkg.all;
  for Structure_a -- Top level architecture of Structure
    for U1_WishGate: WishGate use
      entity ATEP_Lib.AndGate(AndGate_a)
        generic map(DlyIO_g => Dly_g)
        port map
          (I1 => A,
           I2 => B,
           O  => C);
    end for;

    for U2_AndGate: AndGate use
      entity ATEP_Lib.AndGate(AndGate_a);
    end for;

    -- U3_XorGate is effectively removed from the model.
    for U3_XorGate: XorGate  use open;
    end for;
  end for;
end Structure2_Cfg;
```

Figure 9.5.2 Configuration to Remove a Component from Architecture

(ch9_dir\deferd_c.vhd)

EXERCISES

1. In a package declare attributes necessary to **position** a component (X and Y coordinates of type natural) on a circuit board. Define a Counter entity and specify the attribute **position** to be (10, 15). Define a simple counter architecture (0 to 3, output is type natural) in which the delay of the value assigned to the counter output is 1.4 times the generic delay if the counter is located on the COLD side of the board (X> 5). Otherwise, the delay is the value of the generic.

2. Define another counter architecture whose delay is insensitive to location.

3. Embed this counter component in a testbench that instantiates the two flavors of counters (i.e., two instances of the component, one for each architecture). Use a configuration specification to bind each instance of the counter entity to each counter architecture.

4. Define another testbench without the configuration specification. Define a configuration declaration to bind each instance of the counter entity to each counter architecture.

10. DESIGN FOR SYNTHESIS

Synthesis is the process of translating a design from a hardware description (such as VHDL) into a circuit design using components from a specified library (e.g., TTL, FPGA, ASIC library for a specific technology). VHDL code written for synthesis is not necessarily compatible among synthesizers from different vendors. Each vendor imposes its own sets of rules in the VHDL style, VHDL constructs, and pragmas (i.e., comment directives) to direct the compiler in certain directions. Synthesis is a moving technology with guidelines continuously changing. The reader must study the vendor guidelines and restrictions to perform synthesis with the vendor's toolset. This chapter provides some elementary guidelines in using VHDL for circuit synthesis along with a summary of the VHDL constructs that are typically synthesizable.

The reader is strongly encouraged to study *the IEEE P1076.6 Standard for VHDL Register Transfer Level Synthesis* prepared by the VHDL Synthesis Interoperability Working Group of the Design Automation Standards Committee. "This standard describes a standard syntax and semantics for VHDL *RTL* synthesis. It defines the subset of *IEEE 1076* (VHDL) that is suitable for *RTL* synthesis and defines the semantics of that subset for the synthesis domain. This standard is based on the standards *IEEE 1076, 1164*, and *1076.3*. The purpose of this standard is to define a syntax and semantics that can be used in common by all compliant *RTL* synthesis tools to achieve uniformity of results in a similar manner to which simulation tools use the *IEEE 1076* standard. This will allow users of synthesis tools to produce well defined designs whose functional characteristics are independent of a particular synthesis implementation by making their designs compliant with this standard". This standard was approved in 1998, and could be obtained from the Institute of Electrical and Electronics Engineers, Inc., 345 East 47th Street, New York, NY 10017, USA (http://stdsbbs.ieee.org/).

10.1 CONSTRUCTS FOR SYNTHESIS

VHDL is a very rich language. Not all constructs defined in the language are synthesizable. In fact, RTL level of modeling requires a specific set of templates to specify combinational and sequential logic. This sections reviews the various VHDL constructs for synthesis, and defines the synthesizable VHDL templates. It also addresses the guidelines for the design of state machines, and methods to infer registers, latches, and the asynchronous reset or set of registers.

Table 10.1-1 summarizes the generally supported and unsupported synthesis constructs. The user must verify the synthesis rules with the tool of interest.

Table 10.1-1 Supported and Unsupported Synthesis Constructs

CONSTRUCT	COMMENTS AND EXAMPLE
Library	Supported, but generally restricted to *Work* with access to *IEEE*
Packages	
IEEE	*Std_Logic_1164, Std_Logic_Unsigned,* *Std_Logic_Signed, Std_Logic_Arith* *Numeric_Bit, Numeric_Std* (Not yet fully adopted by all synthesis vendors)
Standard	**Supported** (See types and subtypes restrictions)
TextIO	Unsupported
User package	**Supported**, provided the package **does not** include: -- Global signal or shared variable declarations -- Deferred constants -- Unsupported type declarations -- Unsupported constructs in subprograms (see subprograms)
Entity	**Supported**, provided port are of supported types
Architecture	**Supported**, however default binding (last compiled) takes effect.
Signal, Variable, Generic	**Supported** (Signal kind register and bus – Unsupported)
Types and Subtypes	*Integer, enumeration, Bit, Std_Logic, Bit_Vector, Std_Logic_Vector, Boolean,* One-dimensional array of above types Record type is allowed, but record aggregate assignments are not allowed. Physical type – ignored Time, real, and multidimensional arrays are not allowed
Alias declarations	Not Allowed
Signal, Variable, Generic Initialization	Signal and variable initialization is not allowed anywhere Constant initialized to formal parameter or to port is unsupported Generic initialization is allowed
Object Classes	Constants, signals and variables are allowed Files are not allowed
Operators	**Logical:** (e.g., *and, or*), relational (e.g., =, /=, >), concatenation, [*, /, **, *mod, rem*] (restricted to operands of power of 2 for shifts), *abs, not,* operator overloading **Arithmetic** *,+, - (require use of arithmetic packages)

Assignment	Signal, variable, conditional signal, selected signal, and Aggregate [e.g., (A, B, C) <= K(1 **to** 3) **and** M(1 **to** 3);] "AFTER clause" -- IGNORED
Subprogram	Supported if declared in packages or in declaration part of architect. Multiple wait statements are not allowed *now* function -- Unsupported User defined resolution functions – Unsupported
Attributes	Supported for <u>subset</u> of predefined attributes *'base, 'left, 'right, 'high, 'low, 'range, 'reverse_range, 'length,* [*'event, 'stable*] allowed in *if* and *wait* statements *'pos, 'val, 'leftof, 'rightof, 'succ, 'pred, 'image, 'value, 'active, 'transaction, 'delayed, 'quiet, 'last_event, 'last_active, 'last_value* Vendor's defined attributes (typically in packages) User defined attributes are unsupported
Sequential Statements	- *Wait*, signal assignment (No *transport* or *after* clause), variable assignment, procedure call (with some exceptions), *if, case, loop* (restriction on loop index of integer type), *next, exit, return, null* - *Assert, after* -- Ignored or unsupported
Concurrent statements Process	Process with sensitivity list must contain all signals on the right hand side of signal assignments within the process for proper simulation. Sensitivity list is ignored.
Concurrent signal assignment	Supported (guarded and transport unsupported)
Concurrent procedure	Supported (No multiple *wait* statement allowed)
Component Instantiation	Supported (Type conversion on formal port not allowed)
Generate	Supported
BLOCK	Supported except: Block guards and register -- Unsupported Ports and generics in blocks -- Unsupported
Concurrent Assertion	IGNORED or unsupported
Configuration	Not supported

10.2REGISTER INFERENCE

A clock process is defined as a process having **ONE** and **ONLY ONE** of the following statements:

- *wait until Clock = '1'; -- rising edge of the clock*
- *wait until Clock'event and Clock = '1'; -- rising edge of the clock*
- *wait until Rising_Edge(Clock); -- from Std_Logic_1164 package*
- *if Clock'event and Clock = '1' then ; -- rising edge of the clock*
- *elsif Clock'event and Clock = '1' then ; -- rising edge of the clock*

- *wait until Clock = '0'; -- falling edge of the clock*
- *wait until Clock'event and Clock = '0'; -- falling edge of the clock*
- *wait until Falling_Edge(Clock); -- from Std_Logic_1164 package*
- *if Clock'event and Clock = '0' then -- falling edge of the clock*
- *elsif Clock'event and Clock = '0' then -- falling edge of the clock*

10.2.1 Signals Assignments in Clocked Process

A register is always inferred when a signal is assigned a value in a clocked process. Figure 10.2.1 is an example of a register implied by a signal assignment within a clocked process.

```
library IEEE;
  use IEEE.Std_Logic_1164.all;

entity FFSIG is
  port (Clk     : in      std_logic;
        D       : in      std_logic;
        Q       : out     std_logic);
end FFSIG;

architecture FFSIG_a of FFSIG is
begin  --  FFSIG_a
  FFSIG_Proc : process
  begin  --  process FFSIG_Lbl
    wait until Clk'EVENT and Clk = '1';
    Q  <= D;
  end process FFSIG_Proc;
end FFSIG_a;
```

Figure 10.2.1 Register Implied by Signal Assignment in a Clocked Process (Ch10_dir\ffsig.vhd)

10.2.2 Variable assignments in clocked process

The rules for register inference of variables are different than for signals. In clocked processes the following rules apply to variables:

1. **Reading a variable before assigning a value to that variable implies** reading the "old" value of that variable, or a **register implementation** for that variable. In some cases, the synthesizer may optimize out the register implied by the variable. This optimization is tool dependent.
2. If a **variable is assigned or written before it is read**, and there is NO PATH where the variable is first read before the variable assignment, then that variable is used as temporary storage and **NO register is implied for that variable**. Thus if the value of that variable is later used in the equation for a signal assignment, then a register implementation is made for the assigned signal only (but not for the variable). If the variable is used in the control element (e.g., **if** or **case** statement) then the variable is used in the generation of the control logic. Thus, If a register is not intended (e.g., using a variable as a temporary placeholder or computational element) it is important to assign the variable prior to reading that variable.

Figure 10.2.2-1 represents the use of a variable with a variable assignment prior to reading the variable (i.e., no implied register). Figure 10.2.2-2 represents a synthesized implementation of that design. Figure 10.2.2-3 represents the use of a variable with a variable assignment after the reading of the variable, thus inferring a register on inputs *J* and *K*. Figure 10.2.2-4 is a synthesized implementation of that second design (Notice the implied Flip-Flops following the *J* and *K* inputs). Figure 10.2.2-5 is an example of a variable in a clocked process whose register implementation is often optimized out by smart synthesizer. Figure 10.2.2-6 is an RTL view of that counter.

M 👍👍 Avoid using variables in clocked processes.

Rationale: *Variables may produce unexpected additional registers. This may be caused by a user error or by the synthesizer tool.*

```
...
entity FF is
  port (Clk      : in      Std_Logic;
        ReseT    : in      Std_logic;
        J        : in      Std_logic;
        K        : in      Std_logic;
        Q        : out     Std_Logic);
end FF;

architecture FF_a of FF is
  signal Q_s  : Std_Logic;
begin  --  FF_a

  FF_Lbl : process
    variable    JK_v      : Std_Logic_Vector(1 downto 0);
  begin  --  process FF_Lbl
    wait until Clk'event and Clk = '1';
    JK_v := (J & K);
    if ReseT = '1' then
      Q_s  <= '0';
    else
      case JK_v is   -- reading the variable
        when "00"    =>  Q_s  <=  Q_s;
        when "01"    =>  Q_s  <= '0';
        when "10"    =>  Q_s  <= '1';
        when "11"    =>  Q_s  <= not Q_s; -- signal needed for reading Q output
        when others  =>  Q_s  <= 'X';
      end case;                              --  JK
    end if;
  end process FF_Lbl;

  -- Concurrent statement
  Q  <= Q_s;
end FF_a;
```

Variable assignment before read of variable, thus NO registers are Implied

Any signal assignment implies a register in a clocked process.

Figure 10.2.2-1 Variable Assignment Prior To Reading the Variable used as Control Element (ch10_dir\ff_ea.vhd)

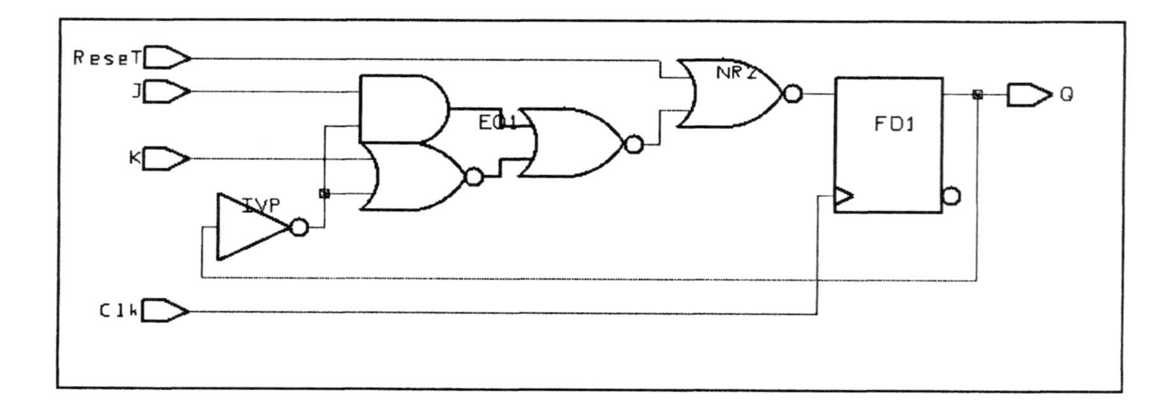

Figure 10.2.2-2 RTL Synthesized Implementation of Design in Figure 10.2.2-1 ***(with Synplify from Synplicity)***

```
...
entity FF is
  port (Clk      : in      Std_Logic;
        ReseT    : in      Std_logic;
        J        : in      Std_logic;
        K        : in      Std_logic;
        Q        : out     Std_Logic);
end FF;

architecture FF_a of FF is
  signal Q_s  : Std_Logic;
begin  --  FF_a
  FF_Lbl : process
    variable    JK_v     : Std_Logic_Vector(1 downto 0);
  begin  --  process FF_Lbl
    wait until Clk'event and Clk = '1';
    if ReseT = '1' then
      Q_s  <= '0';
    else
      case JK_v is
        when "00"     =>  Q_s  <=  Q_s;
        when "01"     =>  Q_s  <= '0';
        when "10"     =>  Q_s  <= '1';
        when "11"     =>  Q_s  <= not Q_s; -- signal needed for reading
        when others   =>  Q_s  <= 'X';
      end case;                            --  JK
    end if;

    JK_v := (J & K); -- Variable assignment after read of variable.
                     -- Thus,  memory implied by variable.  This may not be
                     -- desired in this design
  end process FF_Lbl;
  -- Concurrent statement
  Q  <= Q_s;
end FF_a;
```

In clocked process, reading variable before assigning a value IMPLIES registers on J and K inputs

Figure 10.2.2-3 Variable Assignment After Reading the Variable (ch10_dir\ff2_ea.vhd)

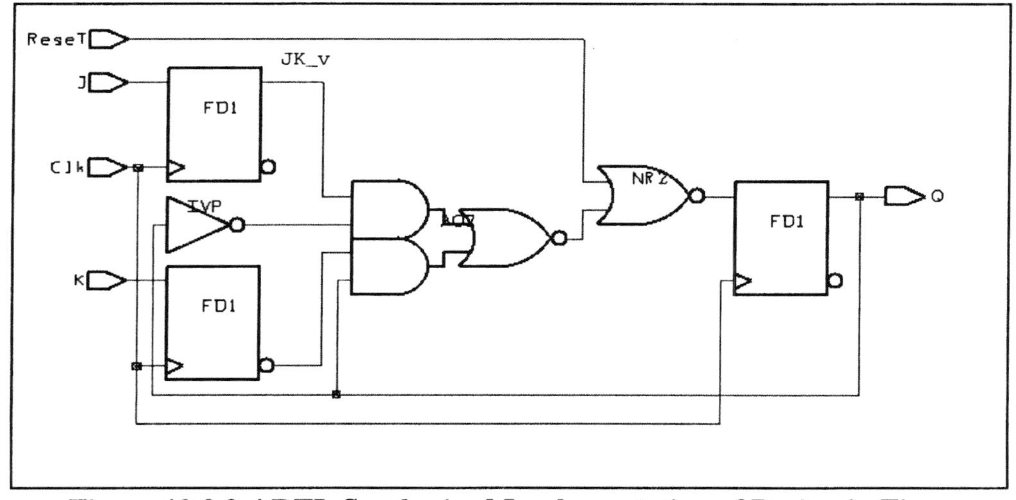

Figure 10.2.2-4 RTL Synthesized Implementation of Design in Figure 10.2.2-3 ***(with Synplify from Synplicity)***

```
library IEEE;
 use IEEE.Std_Logic_1164.all;
 use IEEE.Std_Logic_Unsigned.all;
entity RegVar is
  port (Clk   : in  Std_Logic;
        Reset : in  Std_Logic;
        Count : out Std_Logic_Vector(3 downto 0));
end RegVar;

architecture RegVAr_a of RegVar is
begin  -- RegVAr_a
  RegVAr_Proc : process
    variable Count_v : Std_Logic_Vector(3 downto 0);
  begin  -- process RegVAr_Proc
    wait until Clk = '1';
    if Reset = '1' then
      Count_v := (others => '0');
      Count <= Count_v;
    else
      Count_v := Count_v + 1;
      Count <= Count_v;
    end if;
  end process RegVAr_Proc;
end RegVAr_a;
```

Figure 10.2.2-5 Variable in a Clocked Process whose Register Implementation is Optimized out by Smart Synthesizer (Ch10_dir\regvar.vhd)

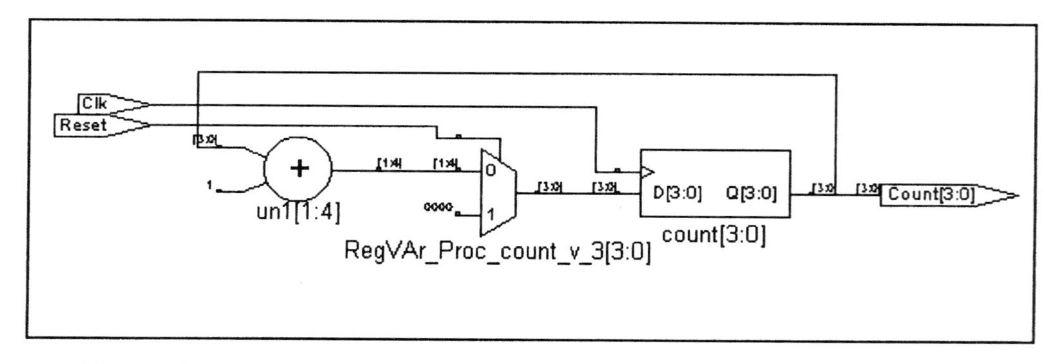

Figure 10.2.2-6 RTL View of Counter ***(with Synplify from Synplicity)***

10.2.3 Asynchronous Reset or Set of Registers

The asynchronous set or reset of registers is performed with an **if** statement prior to the **elsif** clocked statement. The **if** statement separates the asynchronous from the synchronous portions of the design. Figure 10.2.3 provides an example of such a methodology.

```
entity FF_Async is
  port(Clk    : in   bit;
       D      : in   bit;
       Reset  : in   bit;
       Q      : out  bit);
end FF_Async;
```

```
architecture FF_Async_a of FF_Async is
begin
  Async_Lbl: process(Clk, Reset)
  begin
    if Reset = '1' then
      Q <= '0';
    elsif Clk'event and Clk = '1' then
      Q <= D;
    end if;
  end process Async_Lbl;
end FF_Async_a;
```

Figure 10.2.3 Asynchronous Reset or Set of Registers (ch10_dir\ffasync.vhd)

10.2.4 Synchronous Reset or Set of Registers

A synchronous reset or set is performed after the *wait until clock* statement. Figure 10.2.4 demonstrates the concept.

```
entity FF_Sync is
  port(Clk    : in   bit;
       D      : in   Bit_Vector(7 downto 0);
       Reset  : in   bit;
       Q      : out  Bit_Vector(7 downto 0));
end FF_Sync;

architecture FF_Sync_a of FF_Sync is
begin
  Sync_Lbl: process
  begin
    wait until Clk = '1';
    if Reset = '1' then
      Q <= (others => '0');
    else
      Q <= D;
    end if;
  end process Sync_Lbl;
end FF_Sync_a;
```

Figure 10.2.4 Synchronous Reset of Register (ch10_dir\ffSync.vhd)

10.3COMBINATIONAL LOGIC INFERENCE

The easiest method to imply combinational logic is to use concurrent signal assignments. Logic synthesizers minimize the equations defined in the concurrent signal assignments and produce a circuit with all the constraints imposed on the design. A concurrent signal assignment is implicitly sensitive to ALL the signals used in the waveforms. Section 6.2.2 describes the concurrent signal assignments. These include the conditional signal assignment and the selected signal assignment statements. Figure 10.3-1 provides some application examples of concurrent signal assignments. Figure 10.3-2 is the RTL view of the design.

```
library IEEE;
  use IEEE.std_logic_1164.all;
entity Combinational is
  port (A : in  Std_Logic_Vector(7 downto 0);
        B : out Std_Logic_Vector(3 downto 0));
end Combinational;
```

```
architecture Combinational_a of Combinational is
  constant Ones_c   : Std_Logic_Vector(A'length - 1 downto 0)
    := (others => '1');
begin  -- Combinational_a
  -- Conditional signal assignments
  B(0) <= '1' when A = Ones_c else
          '0';

  B(1) <= (A(0) and A(1)) or A(2);

  -- Selected signal assignments
  with A(7 downto 4) select
    B(3 downto 2) <=
    "00" when "0000",
    "01" when "0101",
    "10" when "1010",
    "11" when others;
end Combinational_a;
```

Figure 10.3-1 Examples of Concurrent Signal Assignments (Ch10_dir\ combinational.vhd

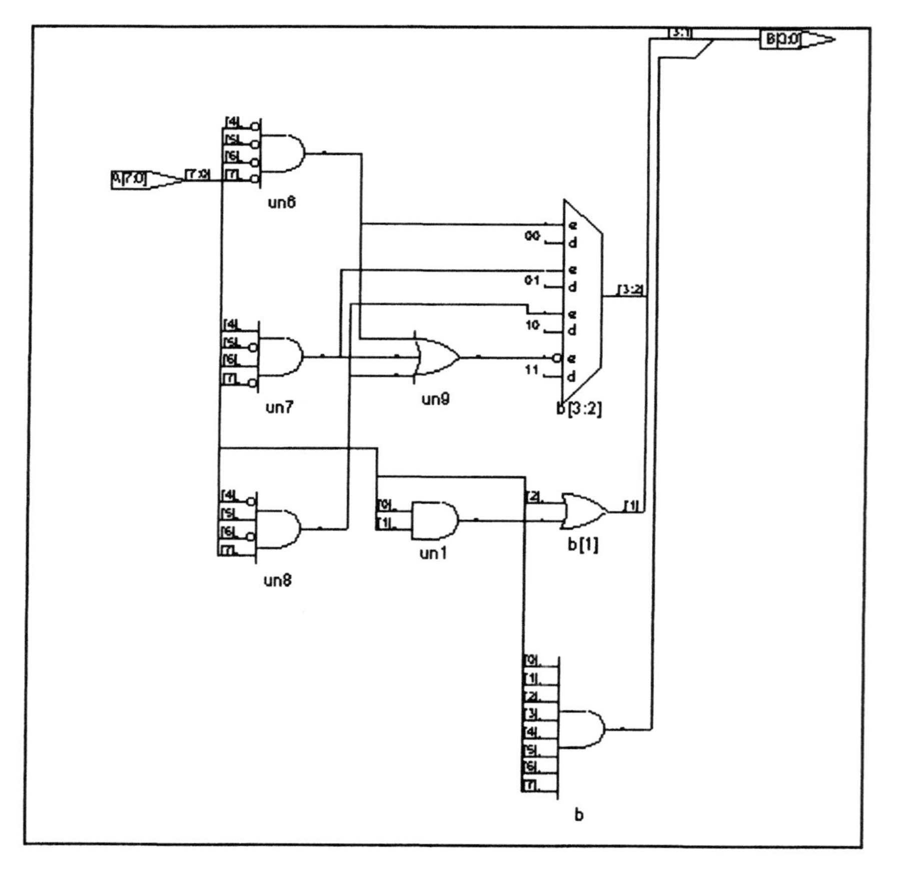

Figure 10.3-2 is the RTL view of the Combinational Design ***(with Synplify from Synplicity)***

Another method to infer combinational logic is through the use of processes. In non-clocked processes, combinational logic is inferred when the signal assignments are complete (i.e., assigned under all conditions). Processes are convenient to describe complex combinational logic (see section 10.5, state machine designs).

Clocked processes can also infer combinational logic that drives the registers. Figure 10.3-2 is an example of a timer with a synchronous load and increment logic. The output of that logic is stored into the register *DataOut_r*. Figure 10.3-3 is the RTL view of the synthesis process.

```
library IEEE;
  use ieee.std_logic_1164.all;
  use IEEE.Std_Logic_Unsigned.all;
entity Timer is
  generic (Size_g : Integer := 5);
  port (DataIn   :  in   Std_Logic_Vector(Size_g -1 downto 0);   -- Load Data
        LdF      :  in   Std_Logic;                     -- Load, active Low
        Clk      :  in   Std_Logic;                      -- CLock, rising edge
        DataOut  :  out  Std_Logic_Vector(Size_g - 1 downto 0) );
end Timer;

architecture Timer_Rtl of Timer is
  signal DataOut_r : Std_Logic_Vector(Size_g - 1 downto 0);
begin  -- Timer_Rtl
  Count_Proc : process
  begin  -- process COunt_Proc
   wait until Clk = '1';
   if LdF = '0' then
     DataOut_r <= DataIn;
   else
     DataOut_r <= DataOut_r + 1;
   end if;
  end process Count_Proc;

  DataOut <= DataOut_r;
end Timer_Rtl;
```

Combinational logic driving a register

10.3-2 Timer with a Synchronous Load and Increment Logic (ch10_dir\timer.vhd)

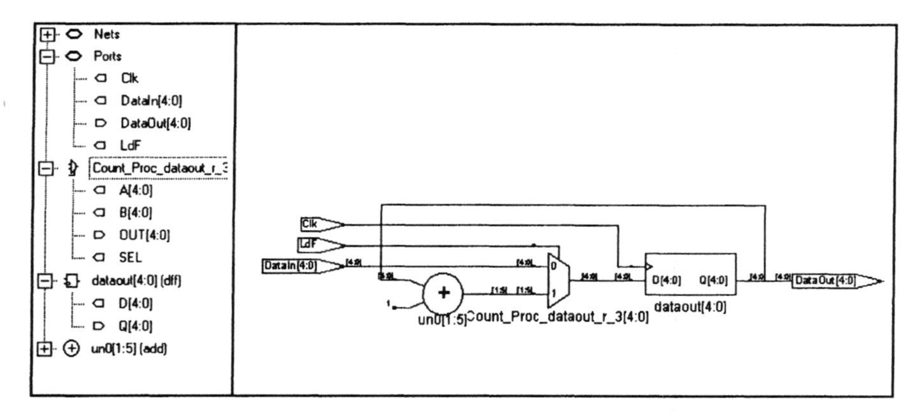

Figure 10.3-3 RTL view of the Synthesis Process *(with Synplify* from *Synplicity)*

10.3.1 Latch Inference and Avoidance

A latch is inferred for signals when they are not assigned under all conditional statements. For example: **if** SomeSignal = '1' **then** -- latch is inferred for SomeSignal
X <= '0';
end if;

𝕸 👍👍 Latches must be avoided in synchronous designs.
Rationale: *Latches infer feedback, cause difficulties in timing analysis (timing is ambiguous), and in test insertion applications.*
Using clocked registers instead of latches helps scan insertion, test coverage, and static timing analysis[16].

𝕸 👍👍 Latches can be avoided by using any of the following coding techniques: [17]

- Assign a default value at the beginning of a process.
- Assign outputs for all input conditions.
- Use *else* (instead of *elsif*) for the final priority branch.

Figure 10.3.1-1 provides examples of latch implications for signals and variables. Figure 10.3.1-2 is the RTL view of the synthesis process detailing the latches. Figure 10.3.1-3 uses the default assignment at the beginning of the process to prevent a latch. and 10.3.1-4 2 is the RTL view of the synthesis demonstrating that the latch was avoided.

```
entity Combinational is
  port(A      : in   Bit;
       B      : in   Bit;
       C      : in   Bit_Vector(3 downto 0);
       T      : out  Bit_Vector(1 downto 0));
end Combinational;

architecture Incomplete_a of Combinational is
begin
  -- Incomplete process implying a latch  for T because
  -- if inputs A, B, C are equal to 001 then T must hold the
  -- previous value of T, thus implying a latch.
  Combntl_Proc: process(A, B, C)
  begin
    if A = '1' then
      T <= "00";
    elsif B  = '1' then -- here if A = '0'
      T <= "01";
    elsif C(0) = '0' then  -- here if A = '0' and B = '0'
      T <= "10";
    end if;
  end process Combntl_Proc;
end Incomplete_a;
```

T is a latch because it is not assigned under all conditions

Figure 10.3.1-1 Example of Incomplete Signal Assignments (ch10_dir\incomplete.vhd)

---

[16] **The Reuse Methodology Manual: A Constructive Critique,** Janick Bergeron, http://www.qualis.com

[17] **Reuse Methodology Manual**, Michael Keating, Pierre Bricaud, Kluwer Academic Publishers, 1998, ISBN 0-7923-8175-0

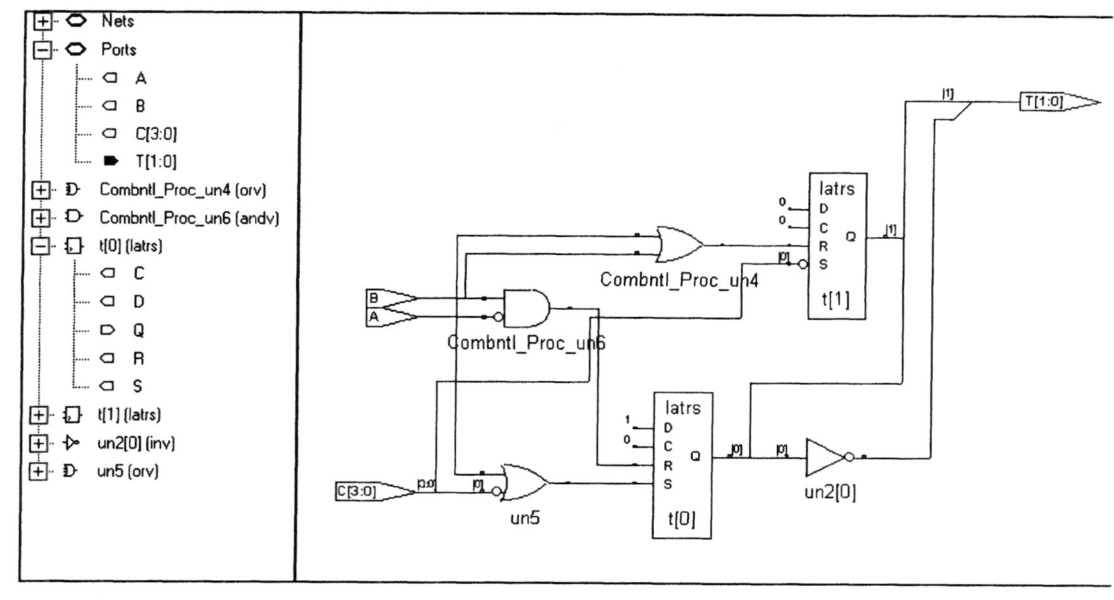

10.3.1-2 RTL View of the Synthesis Process Detailing the Latches for the Combinational Entity ***(with Synplify from Synplicity)***

```
entity Combinational is
  port(A       : in   Bit;
       B       : in   Bit;
       C       : in   Bit_Vector(3 downto 0);
       T       : out  Bit_Vector(1 downto 0));
end Combinational;

architecture Complete_a of Combinational is

begin
  Combntl_Proc: process(A, B, C)
  begin
    T <= "11";           -- default assignment to avert a latch
    if A = '1' then
      T <= "00";
    elsif B  = '1' then -- here if A = '0'
      T <= "01";
    elsif C(0) = '0' then  -- here if A = '0' and B = '0'
      T <= "10";
    end if;
  end process Combntl_Proc;
end Complete_a;
```

Figure 10.3.1-3 Default Assignment at the Beginning of the Process to Prevent a Latch (ch10_dir\complete.vhd)

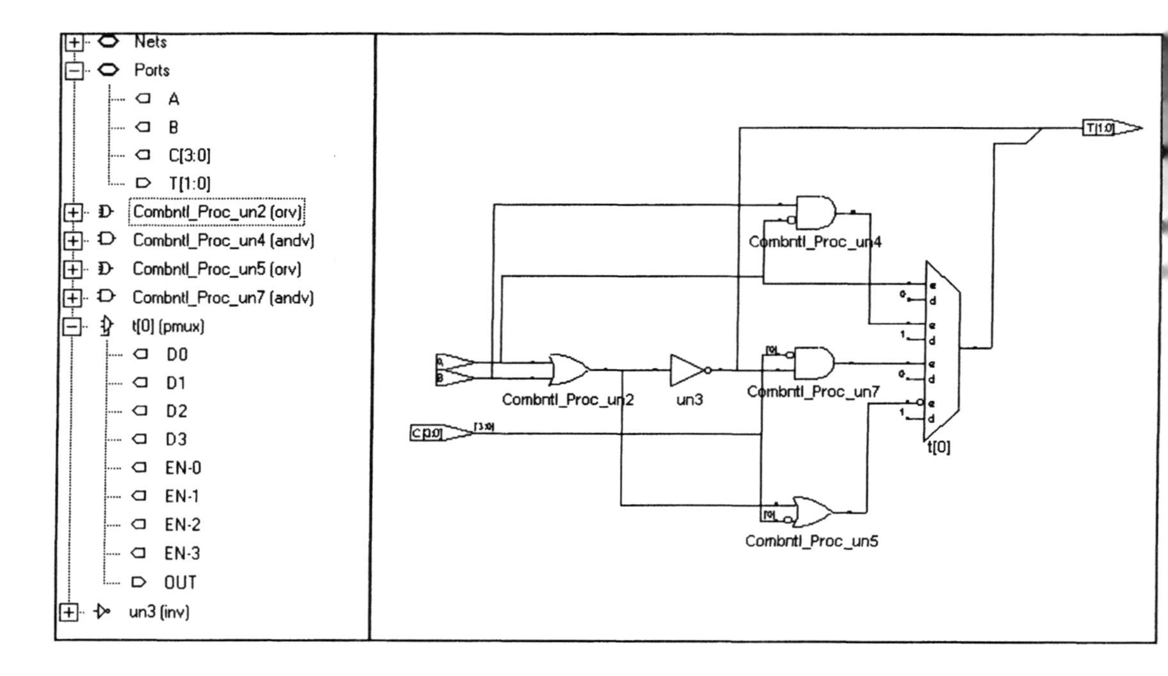

Figure 10.3.1-4 RTL View of the Synthesis Process Detailing NO Latches for the Combinational Entity *(with Synplify from Synplicity)*

10.3.2 Variable

In non-clocked processes, when a variable is written before it is read, that variable implies combinational logic.

In non-clocked processes variables must be updated (or written) before they are read.
***Rationale*: *If a* variable is read before its value is assigned a value, then simulation mismatch will result between the RTL model and the synthesized model.**

Figure 10.3.2-1 demonstrated the potential error condition when a variable is read before it is written. Figure 10.3.2-2 demonstrated the correct coding style, while Figure 10.3.2-3 shows the synthesized RTL view.

```
entity VarError is
  port(A      : in   Bit;
       B      : in   Bit;
       C      : in   Bit_Vector(3 downto 0);
       Q      : out  Bit_Vector(1 downto 0));
end VarError;

architecture VarError_a of VarError is
begin
  Var_Proc : process (A, B, C)
    variable Var : Bit_Vector(1 downto 0);
  begin  -- process Var_Proc
    if Var = "00"  then
      Q <= C(1 downto 0);
    else
      Q <= C(3 downto 2);
    end if;
    var := A & B;
  end process Var_Proc;
end VarError_a;
```


Figure 10.3.2 Error Condition When a Variable is Read Before it is Written (ch10_dir\varerror.vhd)

```
entity VarOK is
  port(A      : in   Bit;
       B      : in   Bit;
       C      : in   Bit_Vector(3 downto 0);
       Q      : out  Bit_Vector(1 downto 0));
end VarOK;

architecture VarOK_a of VarOK is
begin
  Var_Proc : process (A, B, C)
    variable Var : Bit_Vector(1 downto 0);
  begin  -- process Var_Proc
    var := A & B;
    if Var = "00"  then
      Q <= C(1 downto 0);
    else
      Q <= C(3 downto 2);
    end if;
  end process Var_Proc;
end VarOK_a;
```

Figure 10.3.2-2 Correct Coding Style Using Variables (ch10_dir\varok.vhd)

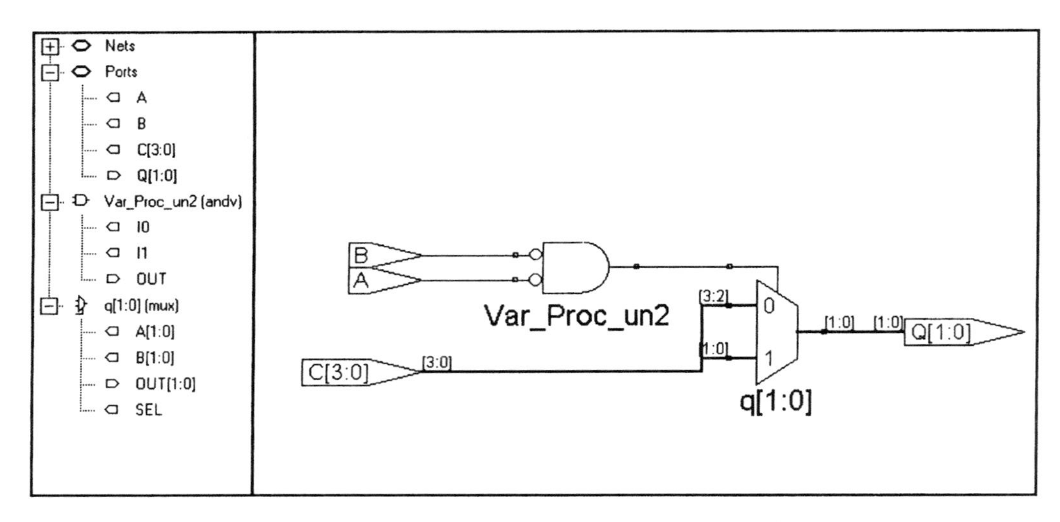

10.3.2-3 RTL View of the Synthesis Process for the VarOK and VarError Entity *(with Synplify from Synplicity)*

10.4 STATE MACHINE

A state machine is a sequential logic where its outputs are a function of not only its inputs but also the present condition or state of the circuit. The state of a circuit is stored in state registers. Two VHDL styles are used to define a sequential logic:

1. **Implicit finite state machine** descriptions that contain no explicit states register. Those registers are implied by the control flow. This type of design is only synthesizable with behavior compiler tools, but not RTL synthesis tools.

2. **Explicit finite state machine** descriptions that use an explicit state register to hold an encoded representation of the current state. For each state machine, a state register is defined in a clocked process. Combinational logic can be expressed either external to this clocked process, or can be implied from within the clocked process. This style of coding is synthesizable with RTL synthesizers.

Figure 10.4 is an example of two identical representations of a state machine, an implicit state machine in architecture *Implicit_a*, and an explicit state machine in architecture *Explicit_a*. Note that the **implicit** state machine model uses a procedural flow of control to determine the sequences of states. This flow is controlled by the introduction of multiple *wait until clock'event* statements. However, the **explicit** state machine restricts the use of a single *wait until clock'event* statement.

```
entity ImplExpl is
  port (Red     : out    bit;
        Yellow  : out    bit;
        Green   : out    bit;
        Reset   : in     bit;
        Clk     : in     bit);
end ImplExpl;
```

```
architecture Implicit_a of ImplExpl is
begin  --  Implicit_a
  ----------------------------------------------------------------------------
  --  Process:    Implicit_Lbl
  --  Purpose:    Red, Yellow, and Green ports are
  --              in turn ON unless there is a Reset.
  --              Once a cycle starts, it must complete
  ----------------------------------------------------------------------------
  Implicit_Lbl : process

  begin  --  process Implicit_Lbl
    wait until Clk'event and Clk = '1';
    if Reset = '1' then    -- RST state
      Red    <= '0';
      Yellow <= '0';
      Green  <= '0';

    else
      Red    <= '1';        -- RedON
      Yellow <= '0';
      Green  <= '0';
      wait until Clk'event and Clk = '1';

      Red    <= '0';        -- YelON
      Yellow <= '1';
      Green  <= '0';
      wait until Clk'event and Clk = '1';

      Red    <= '0';        -- GrnON
      Yellow <= '0';
      Green  <= '1';
    end if;
  end process Implicit_Lbl;
end Implicit_a;
```

Implicit Finite State Machine
Multiple **wait until** clock statements

```
architecture Explicit_a of ImplExpl is
  type State_Typ is (RST, RedOn, YelOn, GrnOn);
  signal State_s  : State_Typ;  -- State register
begin  --  Explicit_a
  ----------------------------------------------------------------------------
  --  Process:    Explicit_Lbl
  --  Purpose:    Red, Yellow, and Green ports are
  --              in turn ON unless there is a Reset.
  --              Once a cycle starts, it must complete
  ----------------------------------------------------------------------------

  Explicit_Lbl : process

  begin  --  process Explicit_Lbl
    wait until Clk'event and Clk = '1';
    if Reset = '1' and State_s = GrnOn then
      Red    <= '0';
      Yellow <= '0';
      Green  <= '0';
      State_s  <= RST;


   elsif Reset = '0' and
        (State_s = RST or State_s = GrnOn) then
      Red    <= '1';
      Yellow <= '0';
      Green  <= '0';
      State_s  <= RedOn;

    elsif State_s = RedOn then
```

Explicit states of state machine

Explicit Finite State Machine
Single **wait until** clock statement

```
        Red    <= '0';
        Yellow <= '1';
        Green  <= '0';
        State_s  <=  YelOn;

      elsif State_s = YelOn then
        Red    <= '0';
        Yellow <= '0';
        Green  <= '1';
        State_s  <= GrnOn;
      end if;
    end process Explicit_Lbl;
end Explicit_a;
```

Figure 10.4 Implicit and Explicit Representations of a State Machine (ch10_dir\ implexpl.vhd)

10.5RTL STATE MACHINE DESIGN STYLES

Since a state machine design includes a state register and combinational logic to determine the next state, several issues often arise as to style.

- Should there be one process that incorporates the state register and the combinational logic?
- Should there be two processes, one for the state register and another one for the combinational logic?
- How should the combinational logic be described?
- How should the outputs be defined?
- How can lockup states be prevented? Lockup states are unused hardware states that can't be transitioned out of.

These issues are address in this section.

10.5.1 State Machine Styles

There are two types of finite state machines:

- **Mealy machine**: Where the outputs are a function of both the inputs and the current machine state (i.e., from state registers and combinational logic). Figure 10.5.1-1 shows a data flow view of the Mealy machine.
- **Moore machine**: Where the outputs are a function of the current machine state. Typically, in a Moore machine the outputs are issued directly from a register to minimize the output delay incurred by additional levels of logic. This is in contrast to the Mealy machine. Figure 10.5.1-2 shows a data flow view of the Moore machine.

There are several coding styles for the design of the state machine and the output logic. These include:

1. **One process only** handles both state transitions and outputs.
2. **Two processes,** one for the clocked process to update the state register and synchronous outputs, and another combinational process for conditionally to derive deriving the next machine state and to update the outputs.
3. **Three (or more) processes (and concurrent signal assignments)** that encompass a clocked process for updating the state register, another clocked process for the

synchronous outputs, and potentially other processes for the combinational logic.

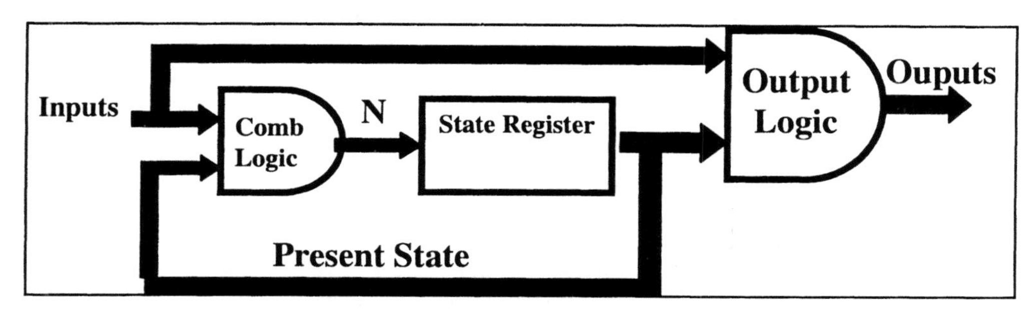

Figure 10.5.1-1 Data Flow view of a Typical Mealy Machine

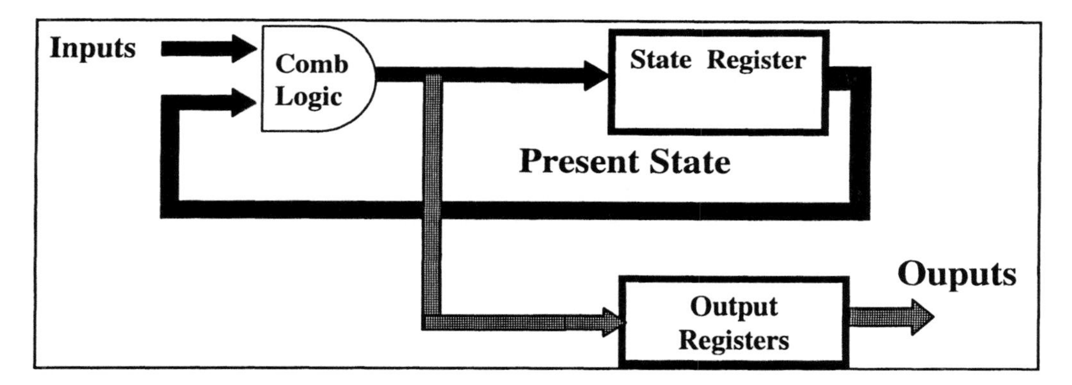

Figure 10.5.1-2 Data Flow view of a Typical Moore Machine

Methods two and three represent the most straightforward approaches because they separate the synchronous registers from the combinational logic used in the computation of the next state and the outputs. In addition, the output of the combinational process that determines the next state can be used as inputs to other processes to determine outputs or other control logic (see Figure 10.5.1-2). Method one makes this step difficult because there is no way to get into the implied combinational logic of a process from another process. This recommendation matches the recommendation proposed in the *Reuse Methodology Manual (ISBN 0-7923-8175)* that recommends keeping the FSM logic and non-FSM logic in separate modules.

M 👍👍 In state machine designs, provide two processes, one for the state register, one for the combinational logic. In addition provide an equivalent process (clocked process or concurrent signal assignments) for the outputs (see Figure 10.5.1-3). The output clocked process may imply combinational logic that supplies the data into the output registers.

Rationale: *This approach clearly separates the combinational aspect of the design from the sequential aspect. It also allows for the easy access of combinational signals from within the combinational process if necessary.*

M 👍👎 When defining state machines, encode the states of states using the enumeration data type. Attempt to organize the enumerations so that the first enumeration ('left) is the default or benign state. Some synthesizers allow the manual encoding of those states into binary values using pragmas. Refer to vendor's manual for implementation and guidelines. If a user is concerned with potential lock of state machine, the number of enumerations should be a power of two (see 10.5.2).

Rationale: *This approach provides commonality in code writing. Some vendors encode the first enumeration as a zero that tends to correspond to the reset or benign state. If the FSM erroneously enters an illegal state, the FSM shall return to a safe state.*

Figure 10.5.1-4 provides a sample of a state machine requirement. Figure 10..5.1-5 is a VHDL model of the sample state machine implemented as a Moore machine with non-registered outputs.

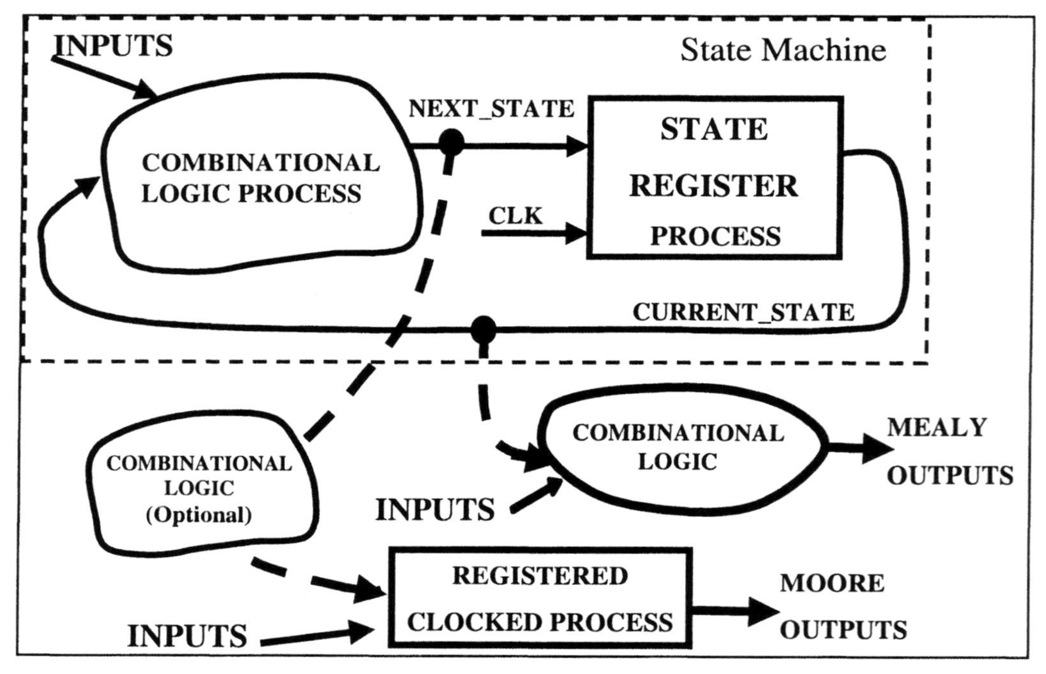

Figure 10.5.1-3 Recommended State Machine Architectural Styles

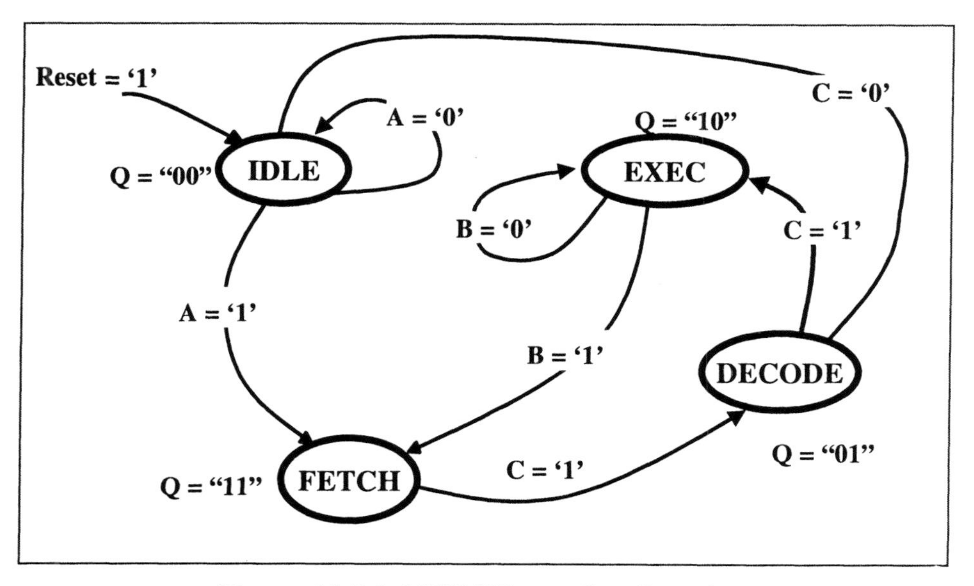

Figure 10.5.1-4 FSM Example – Requirements

```
library IEEE;
  use IEEE.Std_Logic_1164.all;
entity FSM is
  port (A      :  in  Std_Logic;
        B      :  in  Std_Logic;
        C      :  in  Std_Logic;
        Reset  :  in  Std_Logic;
        Clk    :  in  Std_Logic;
        Q      :  out Std_Logic_Vector(1 downto 0));
end FSM;

architecture FSM_a  of FSM is
  type States_Typ is (IDLE, FETCH, DECODE, EXEC);
  signal State_r : States_Typ;
  signal NextState : States_Typ;
begin  -- FSM_a
  States_Proc : process
  begin  -- process States_Proc
    wait until Clk = '1';
    State_r <= NextState;
  end process States_Proc;

  -- Compute next state
  extState_Proc : process (State_r, A, B, C, Reset)
  begin  -- process extState_Proc
    NextState <= State_r; --
    if Reset = '1' then
      NextState <= IDLE;
    else
      case State_r is
        when  IDLE =>
          if A = '1' then
            NextState <= FETCH;
          end if;


        when FETCH =>
          if C = '1' then
            NextState <= DECODE;
```

State register clocked process

Default assignment to prevent a latch and to simplify coding because there is no need to specify the conditions that stay in the current state.

Computation of next state

```
          end if;

        when DECODE =>
          if C = '0' then
            NextState <= IDLE;
          else
            NextState <= EXEC;
          end if;

        when EXEC =>
          if B = '1' then
            NextState <= FETCH;
          end if;
      end case;
    end if;
  end process extState_Proc;

  -- Combinational outputs
  with State_r select
    Q <= "00" when IDLE,
         "11" when FETCH,
         "10" when EXEC,
         "01" when DECODE;
end FSM_a ;
```

Selected signal assignment for the definition of the outputs based on the value of the state register

Figure 10.5.1-5 Moore Machine, Non-Registered Outputs (Ch10_dir\fsmmoore0.vhd)

Figure 10.5.1-6 represents the RTL view of that machine. Figure 10.5.1-7 is the VHDL code for the Moore machine with registered outputs, and Figure 10.5.1-8 is the RTL view of that machine.

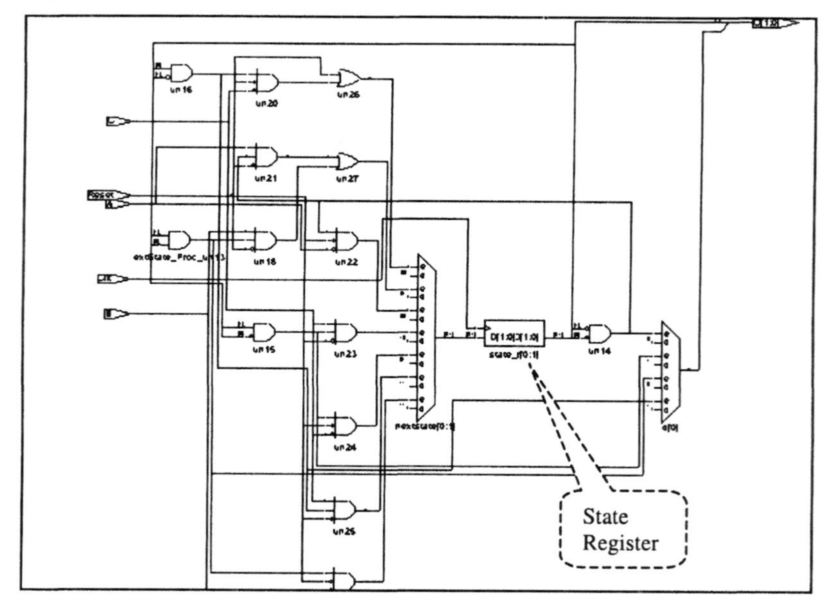

Figure 10.5.1-6 RTL view of the Moore Machine, Non-Registered Outputs, Sequential Encoding Style ***(with Synplify from Synplicity)***

```
architecture FSM_a  of FSM is
  type States_Typ is (IDLE, FETCH, DECODE, EXEC);
  signal State_r : States_Typ;
  signal NextState : States_Typ;
begin  -- FSM_a
  States_Proc : process
  begin  -- process States_Proc
    wait until Clk = '1';
    State_r <= NextState;
  end process States_Proc;

  -- Compute next state
  extState_Proc : process (State_r, A, B, C, Reset)
  begin  -- process extState_Proc
    NextState <= State_r; --
    if Reset = '1' then
      NextState <= IDLE;
    else
      case State_r is
        when  IDLE =>
          if A = '1' then
            NextState <= FETCH;
          end if;

        when FETCH =>
          if C = '1' then
            NextState <= DECODE;
          end if;

        when DECODE =>
          if C = '0' then
            NextState <= IDLE;
          else
            NextState <= EXEC;
          end if;

        when EXEC =>
          if B = '1' then
            NextState <= FETCH;
          end if;
      end case;
    end if;
  end process extState_Proc;

  -- Moore machine for outputs
  Outputs_Proc : process
  begin  -- process Outputs_Proc
    wait until Clk = '1';
    case NextState is
      when IDLE =>
        Q <= "00";
      when FETCH =>
        Q <= "11";
      when EXEC =>
        Q <= "10";
      when DECODE =>
        Q <= "01";
    end case;
  end process Outputs_Proc;
end FSM_a ;
```

Default assignment to prevent a latch and to simplify coding because there is no need to specify the conditions that stay in the current state.

Computation of next state

Clocked process for the definition of the outputs based on the value of the next state

Figure 10.5.1-7 VHDL -- Machine, Registered Outputs (ch10_dir\fsmmoore)

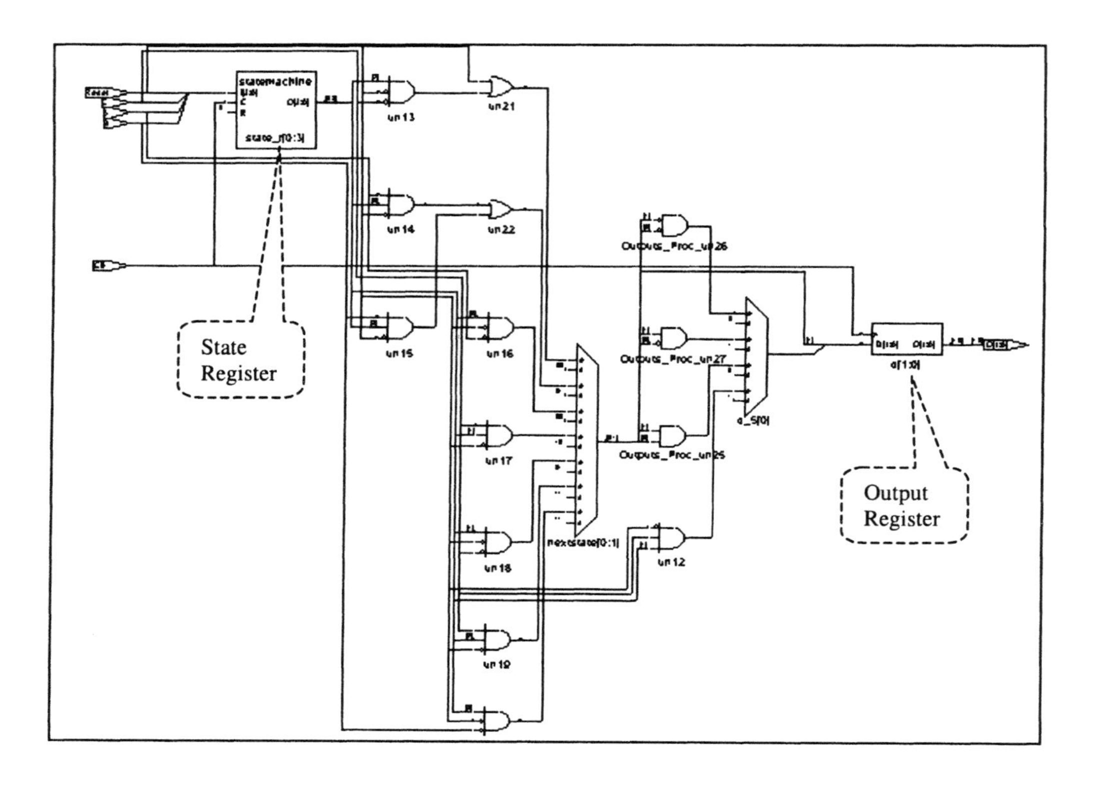

Figure 10.5.1-8 RTL view of the Moore Machine, Registered Outputs, One-Hot Encoding ***(with Synplify from Synplicity)***

10.5.2 Safe FSM with No Lock up

It is possible in VHDL to write a state machine with unused implemented states. For example, if a state machine (FSM) has five states with an enumeration type of five enumerations (e.g., *S0, S1, S2, S3, S4*), then binary encoded machine would be implemented with three bits, most likely encoded as follows:

S0="000"; S1="001"; S2="010"; S3="011"; S4="100".

Three bits define eight states. What happens if the state register enters one of those undefined state (e.g., "101" or a "110" or "111")? The action to be taken in one of those undefined states is tool dependent and is unpredictable. For binary encoded machines, some synthesizers will eliminate "unreachable" states to achieve "quality of results".

𝔐 👍👍 For mission critical designs, is not acceptable to be locked into an unpredictable state if the FSM enters an unused hardware state. To overcome this problem, it is important to follow the following recommendations:

- Define the number of enumeration for states of the FSM to be a power of two. For example, an FSM with five states would have eight enumerations.
- In the *case* statement of the FSM, use the *when others* statement to define a recovery for the unexpected states. For the five states example, after defining the cases for

the five used states, use the *when others* case choice to define the course of action (and next recovery state) for the error condition.

One-hot encoded FSMs are more difficult to control than sequential encoding in the event of a hardware error in one of the bits of the state register. In one-hot encoding, only one of the bits of the state register is at a logical ONE, all other bits are at logical ZERO. A sixteen-state machine would use sixteen registers. To reliably detect an error in the state register, it is necessary to perform a sum of all the bits of the state register, and compare that sum against the value of ONE. If the sum is unequal to ONE, then an error condition exists. The cost of the adder in terms of delays may not overcome the speed advantage of the one-hot encoding scheme. Thus, for non-mission critical designs, one-hot encoding scheme is acceptable. For mission critical designs use a binary encoding scheme.

10.6 ARITHMETIC OPERATIONS

The arithmetic operations are specified in several packages. Most synthesizer vendors have adopted the Synopsys packages including:

- Std_Logic_Arith
- Std_Logic_Signed
- Std_Logic_Unsigned

Many vendors are also adopting the IEEE standard packages including:

- Numeric_Bit.
- Numeric_Std

These packages define the *Signed* and *Unsigned* types, and a set of overloaded operators that operate on *Signed* , *Unsigned, Std_Logic_Vector,* and *Integer* types. Figure 10.6-1 is demonstrates the use of the *Std_Logic_Unsigned* and *Std_Logic_Arith* packages in a model that performs addition, subtraction, and multiplication. Figure 10.6-2 is an RTL view of the model.

```
library IEEE;
  use IEEE.Std_Logic_1164.all;
  use IEEE.Std_Logic_Arith.all;
  use IEEE.Std_Logic_Unsigned.all;
entity TestArith is
  port (Asigned :  in  Signed(15 downto 0);
        Bsigned :  in  Signed(15 downto 0);
        Cstd    :  in  Std_Logic_Vector(15 downto 0);
        Dstd    :  in  Std_Logic_Vector(15 downto 0);
        Qab     :  out Signed(15 downto 0);
        Mab     :  out Signed(31 downto 0);
        Qcd     :  out Std_Logic_Vector(15 downto 0);
        Mcd     :  out Std_Logic_Vector(31 downto 0)
        );
end TestArith;

architecture TEstArith_a  of TestArith is
begin  -- TEstArith_a
    Qab <= Asigned + Bsigned;

    Mab <= Asigned * Bsigned;

    Qcd <= Cstd - Dstd;

    Mcd <= Cstd * Dstd;
end TEstArith_a ;
```

Figure 10.6-1 Arithmetic Model (ch10_dir\testarith.vhd)

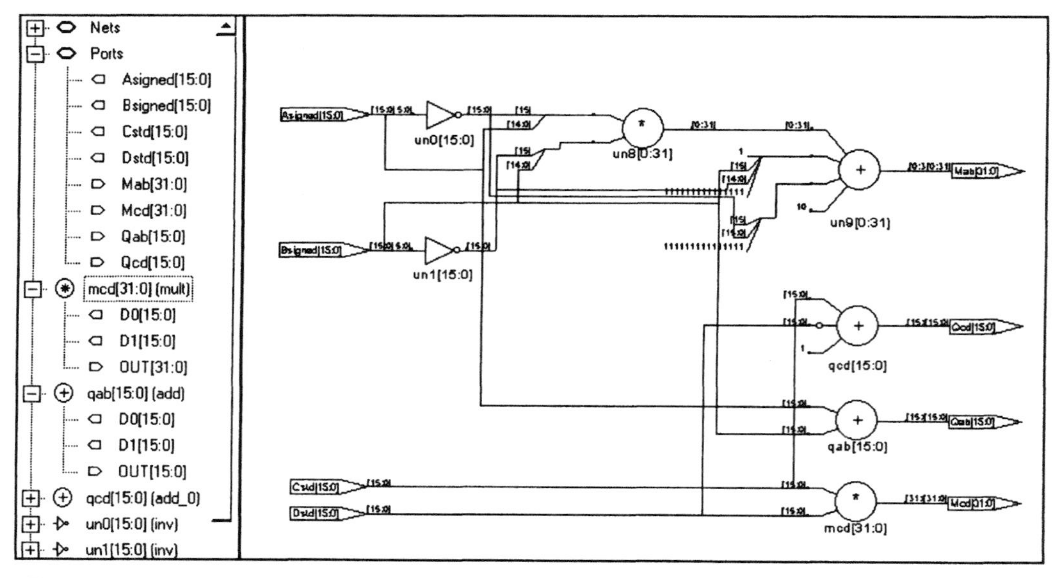

Figure 10.6-2 RTL View of Arithmetic Model ***(with Synplify from Synplicity)***

EXERCISES

1. Implement in VHDL the following state machine.

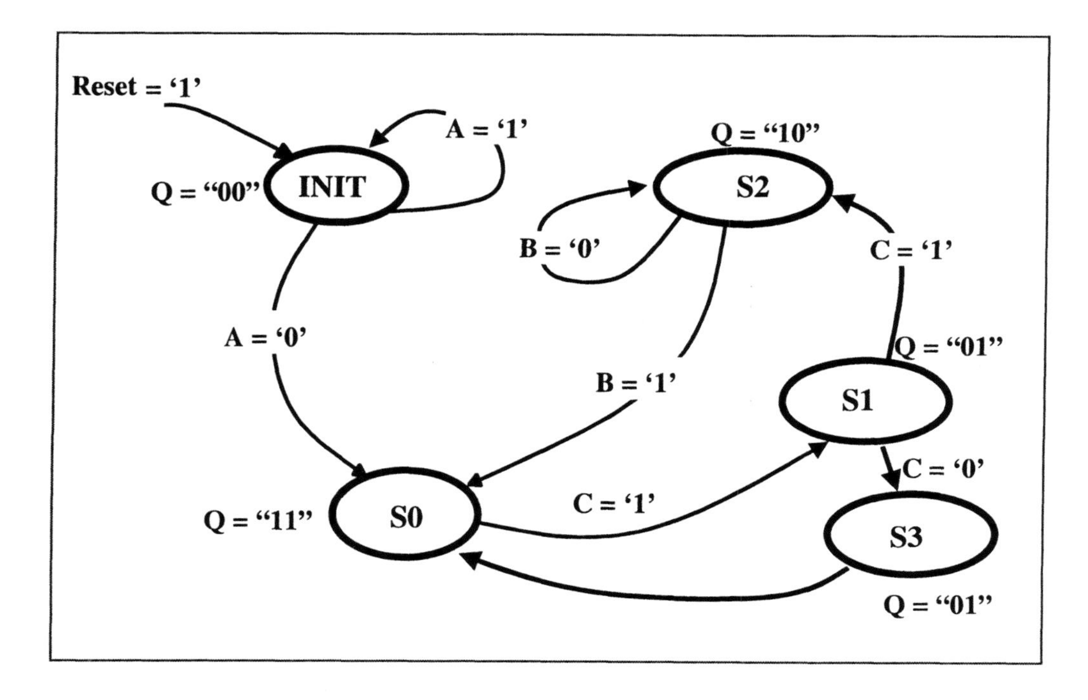

2. Explain why the following code is NOT synthesizable. Make the necessary changes to make it synthesizable.

```
---------------------------------------------------------------------------
--
--    Project     :  ATEP
--    File name   :  waveform.vhd
--    Title       :  Why is this VHDL code Not Synthesizable?
--    Description :  Change code to synthesizable logic
--
---------------------------------------------------------------------------
entity WaveForm is
  generic (Delay_g:               time := 100 ns);
  port (Clk:                      in      bit := '1';
        Wave:                     out     integer := 0);
end WaveForm;

architecture WaveForm_a of WaveForm is

begin  --  WaveForm_a
  -------------------------------------------------------------------------
  --  Process:   GenWave_Lbl
  --  Purpose:   Generate a set of waveforms synchronous to the clock
  -------------------------------------------------------------------------
  GenWave_Lbl : process

  begin  --  process GenWave_Lbl
    wait until Clk'event and Clk = '1';  -- rising edge of clock
    Wave <= 1 after Delay_g;
    wait until Clk'event and Clk = '1';  -- rising edge of clock
    Wave <= 2 after 2 * Delay_g;
    wait until Clk'event and Clk = '1';  -- rising edge of clock
    Wave <= 5 after Delay_g;
  end process GenWave_Lbl;
end WaveForm_a;
```

11. FUNCTIONAL MODELS AND TESTBENCHES

A *Functional Model* (FM) is a model of a component that represents both the interfaces of the component and the internal operation or structure of the component. This operation can be described at various levels including *Instruction Set Architecture* (ISA) level, *RTL*, and gate level. A *Bus Functional Model* (BFM) is a subset of the *Functional Model* in that it only models the bus interfaces and bus transactions of the component rather than its operation.

A VHDL testbench is an environment to simulate and verify the operation of the *Unit Under Test* (UUT) that represents a design in consideration. A testbench may make use of FM and BFM models.

This chapter explains discusses various approaches and methodologies in constructing FMs, BFMs, and testbenches.

11.1 TESTBENCH MODELING

11.1.1 Testbench Overview

The testbench is a VHDL component (i.e., entity/architecture) that instantiates the UUT. A traditional approach to verify the functionality of a design is to use a vendor specific *Simulation Control Language* (SCL)The simulation outputs are presented either in text form, or in waveforms displayed with an interface built around the vendor's software application. These stimulus test vectors are typically written in scripts written in the vendor's SCL. Verification is performed either visually by analyzing the timing waveforms, or semi-automatically by analyzing processed simulation output listings using a user program written in languages such as *'C'* or *'awk'*. The advantages may be faster simulations because the SCL is tailored to the vendor's environment. The disadvantages include lack of interaction between the *UUT* and the *UUT*'s environment, the lack of automatic verification, and the lack of portability. However, in some situations, comparing strobed files might still be necessary because stimulus test vectors and stimulus response vectors might be available.

The recommended testbench design approach is as follows:

M👍👍 In the design of testbenches use a **complete VHDL simulation environment** where the **stimuli and verification processes are described in VHDL**. Do NOT use any SCL.

Rationale: *VHDL is superior to vendor specific SCLs for the creation of testbenches in the following respects:*

1. ***POWERFUL CONSTRUCTS*** *- VHDL's powerful instruction set and ability to model the concurrent nature of hardware provides the capability to create test models that* ***surround*** *and* ***interact*** *with the UUT in a manner similar to the environment of the UUT.* ***Interaction*** *also means that modification can more easily be made to adapt to design changes. In addition,* ***interaction*** *allows for a comprehensive functional verification of the design and significant cost savings for repeated regression tests.*
2. ***PORTABILITY -*** *Testbenches can be used in any simulator supporting the VHDL standard.*
3. ***COMMON LANGUAGE -*** *Common language for the design of the UUT and the testbench enhances productivity because designers need to learn only one language.*

Thus, the testbench is a VHDL environment that instantiates the UUT as a component and uses any of the VHDL constructs (e.g., components instantiation, processes, concurrent signal assignments) to generate stimuli to the UUT in accordance to the interface protocol. The purpose of the testbench includes, in priority order, the following:

1. **Stimulus generator**(s) to drive the UUT under a variety of test conditions (e.g., normal transactions, error transactions) and configurations (e.g., minimum/maximum delays, fault condition, scenarios).

2. **Verifier** to automatically verify that the *UUT* meets the specifications, and to log all errors (e.g., protocol, signal drive, setup and hold timing).

3. **Report generator** to log the simulation transactions of the UUT in a concise human readable format (e.g., report of meaningful transactions versus individual log of signals).

Some guidelines provided by Peter Sinander from the European Space Agency (ESA) on the design of testbenches include the following:

M A testbench shall be a distinct design unit, separated from the model or package to be verified, and placed in a design library separate from the model itself.

Rationale: *The distinct design unit and library isolate the UUT from any architecture where the UUT may be used or simulated.*

M If the testbench incorporates models of components surrounding the model to be tested, they only need to incorporate functions and interfaces required to properly operate the model under test. It is not necessary to develop complete VHDL models of them (i.e., BFMs are sufficient).

Rationale: *Given the same computing resources, BFMs provide for faster simulation and may be easier to control than a complete modeling of the interface device. However, if the model exists in some acceptable form, then there is no reason to develop one. For example a hardware model (HML) or gate level model of the interface device may be available, and is more accurate; however, it may require additional computing resources.*

M If external stimuli (e.g., memory data) are required, the source of the file shall be implemented in ASCII format to ensure portability. The ASCII file may be converted to binary format using a VHDL model.

Rationale: *External data is usually generated by non-VHDL sources such as the output of a software program or actual scene data. The model must be designed to easily adapt to changes in the data contents. A file is the most natural method as a means to transport this information into the model and avoids SCLs. Binary files execute faster than text files because the information is defined in the proper structure (e.g., Std_Logic_Vector, Boolean, or records). Binary files can be derived from the text file through a VHDL model.*

M👍👍 Testbenches should be self-checking, reporting success or failure for each sub-test.

Rationale: *In-line automated verification allows for fast and accurate verification that the UUT performs as intended, and that the UUT interface protocols meet the specifications. This is very useful for regression tests where iterative tests of a design is performed with some minor changes in the timing phase and/or in the data value of stimulus vectors. Manual verification through visual inspection of waveforms is subject to human fatigue and is inefficient. In addition, automation allows for a reduction of future maintenance effort because it enables fast and reliable verification of a model when modifications are introduced.*

Figure 11.1.1 represents a generic testbench architecture that incorporates the UUT, the BFMs, and other VHDL constructs.

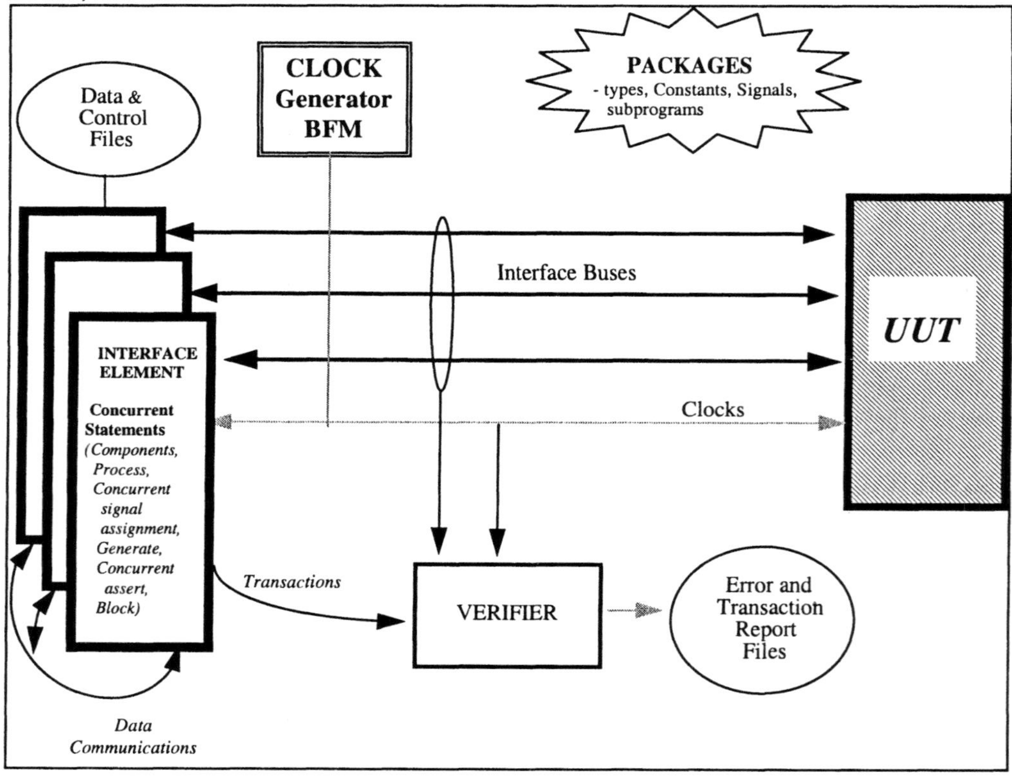

Figure 11.1.1 Generic Testbench Architecture

11.1.2 Testbench Design Methodology

M 👍👍 Use an orderly approach to arrive at an architecture of a testbench. Four basic steps are identified and are shown in Figure 11.2.2. These include:

1. **Validation plan.**
2. **List of errors to be detected.**
3. **Architecture block diagram.**
4. **Actual testbench design.**

RATIONALE: *Good planning tends to yield more efficient and thorough verification of a design.*

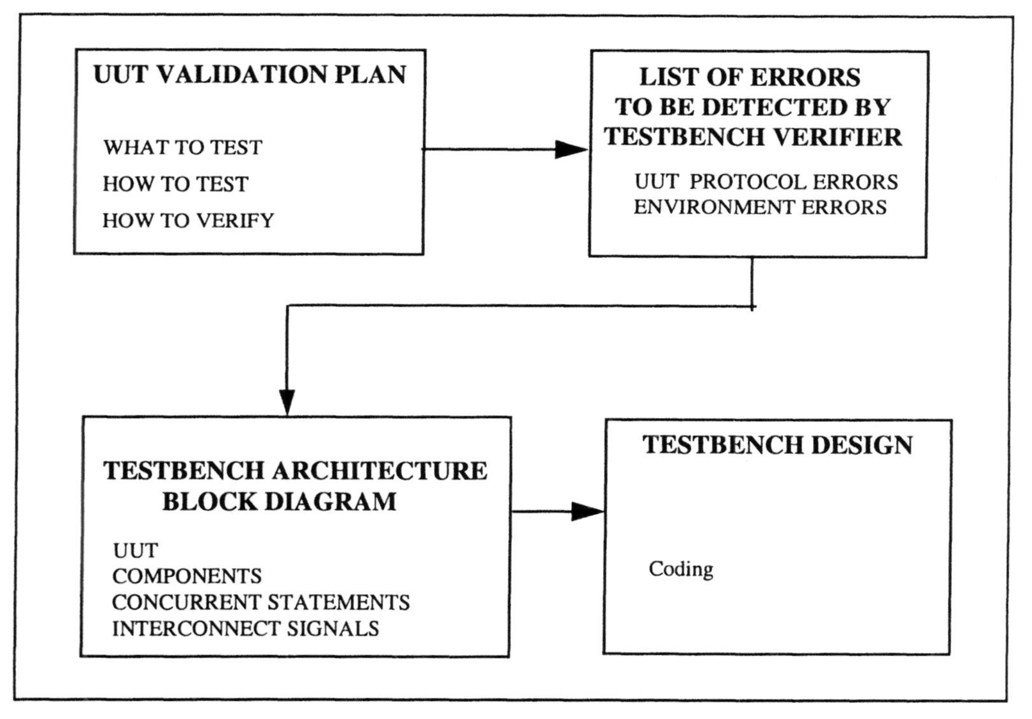

Figure 11.2.2 Testbench Design Methodology

11.1.2.1 Validation Plan

Because of the complexity of the tests, a test plan insures that the testbench includes all the necessary functions and validation tests. While this formalism might seem cumbersome to implement, it is generally the most efficient approach to verify a design "completely". A validation plan defines:

1. **WHAT** should be tested in the unit under test,
2. **METHODS** used for the tests,
3. **VALIDATION** approach to ensure that the UUT meets the requirements.

Thus, the validation plan should define **the classes of tests, the test mechanisms, and the validation approaches**. The test plan does specify the actual test vectors. The test scenarios define the lists of tests to be performed to verify that the UUT meets the requirements. Note that in addition to testing normal operation, it is very important to test incorrect operations, such as illegal input combinations and sequences. These are also known as "corner testing" because they "involve the most interaction between blocks"[18]. The clearest method to define such tests is to build a table that defines the requirements and the tests required verifying such requirements. Figure 11.1.2.1 represents a typical testbench for a UART interfaced to a processor. Table 11.1.2.1 represents an example of a test scenario definition table for a UART. The table is derived from a UUT interface specification, not included in this document (but available from various UART manufacturers).

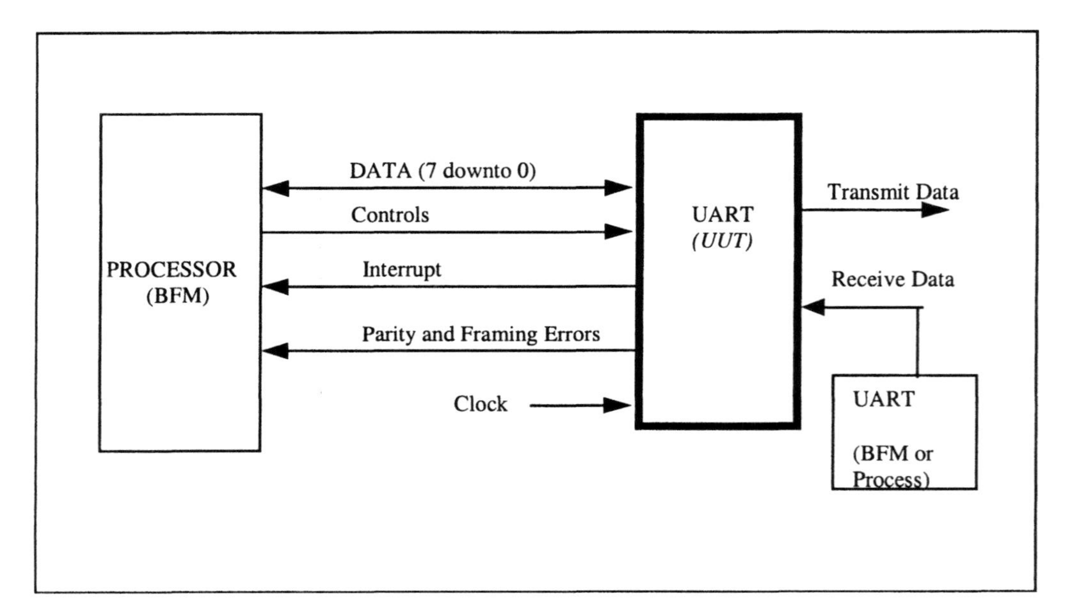

Figure 11.1.2.1 Typical Testbench for UART Interfaced to a Processor

[18] *Reuse Methodology Manual,* Michael Keating, Pierre Bricaud, Kluwer Academic Pblishers, ISBN 0-7923-8175-0, 1998

Table 11.1.2.1 Example of a Test Scenario Definition Table.

REQ #	REQUIREMENT DESCRIPTION	TEST METHOD	VALIDATION METHOD
X.A	Interface Requirements	Processor BFM to Read and Write UUT internal registers. Processor programs UUT into its various modes. A UART BFM sends serial data to UUT	Processor BFM verifies contents of UART registers. Protocol verification of transmit data performed by a protocol bus verifier in testbench. - The UART BFM is programmed to send data with good protocol, and data with errors in the protocol (e.g., framing error, parity error, clock rate errors, etc.). Bus protocol bus verifier detects errors.
Y.Z	AC TIMING	Signals on interfaces shall be issued with MINIMUM delays, MAXIMUM delays, TYPICAL delays, and ERROR delays under the control of generic parameters defined in the configuration.	- Automatic timing checks and error reporting of timing violations in a report file. - Automatic protocol verification to verify impact of timing delays.

11.1.2.2 List of errors to be detected

A list of errors to be detected by the testbench provides the following:

1. Defines testbench validation results.
2. Guides UUT designers for errors to detect.
3. Enhances testbench design review process to ensure completeness of tests.

A sample set of errors to be automatically detected by the UART testbench is shown in table 11.1.2.2

Table 11.1.2.2 Summary of Bus Errors to be Detected by the UART Testbench.

#	ERROR	REQ #
# 1	"UART Failed to Detect a Framing Error"	X.--
# 2	"UART Failed to Detect a Parity Error"	X.--
# 3	"UART Failed to send a Serial Word commanded by Processor"	X.--
# 4	"UART Failed to Alert Processor of a Received Word "	X.--
# 5	"UART Failed to Report an Overrun Error "	X.--
# 6	"UART Transmitted data at an Incorrect Baud Rate"	X.--
# 7	"UART Reports Parity Error when Programmed to Ignore Parity"	X.--
# 8	"UART Set-up Timing Error on Input ____" *(all UART inputs)*	Y.--
#9	"UART Failed to meet the required HOLD time on output ____" *(all UART Outputs)*	Y.--

11.1.2.3 Architecture block diagram

An accurate testbench block diagram that shows the relationship between different VHDL modules provides for good documentation during the review and analysis stages of the testbench.

11.1.2.4 Testbench design

This step of the testbench design process provides for the VHDL coding of the testbench, and the coding of the test scenarios.

11.1.3 Testbench Architectures

A testbench is a self-contained environment that includes all the essential elements to stimulate and verify a UUT design. The following guidelines are provided:

M👍👍 Include the following elements in a testbench:

1. A ***Unit Under Test.*** This *UUT* can be represented at any level including RTL, Gate Level (GL), combination of both *RTL* and GL, or even Hardware Modeler (HML).
2. A set of models that emulate the **bus interfaces** and **bus transactions** to the UUT. These models could be full models if they exist and if simulation speed and resources allow the use of such models. Otherwise, BFMs can be used.
3. A **clock generator** to supply clocks to the system.
4. A **bus verifier** to perform timing and protocol checks on all the results and data. The bus verifier may also generate bus transaction reports (if necessary).

Rationale:

1. *The UUT is the whole purpose of the testbench.*
2. *BFMs represent the environment that interfaces to the UUT*
3. *Clocks represent another environment to the UUT.*

4. *Bus verifier performs automatic verification function.*

11.1.3.1 Typical Testbench Architecture

Figure 11.1.3.1 represents an example of a testbench for the verification of a generic UUT. This testbench includes the following models:

1. **The Unit Under Test.** In this example, the UUT is represented at any desired level (Behavioral, RTL, Structural, or gate level).

2. **Processor BFM.** This BFM interfaces with the Processor BUS, and represents the local bus interfaces to the UUT. This bus includes the processor data and control signals (Read/write, chip select, etc.). Bus signals are typically of type *IEEE 1164 Standard Logic* or *Standard_Logic_Vector*. The BFM may interface with a simulation control file to control the environment, the tests, and the data items to be used as test vectors.

3. **Clock Generator Process.** This process supplies clock signals to the components in the testbench.

4. **Other BFMs.** This design represents another interface to the UUT.

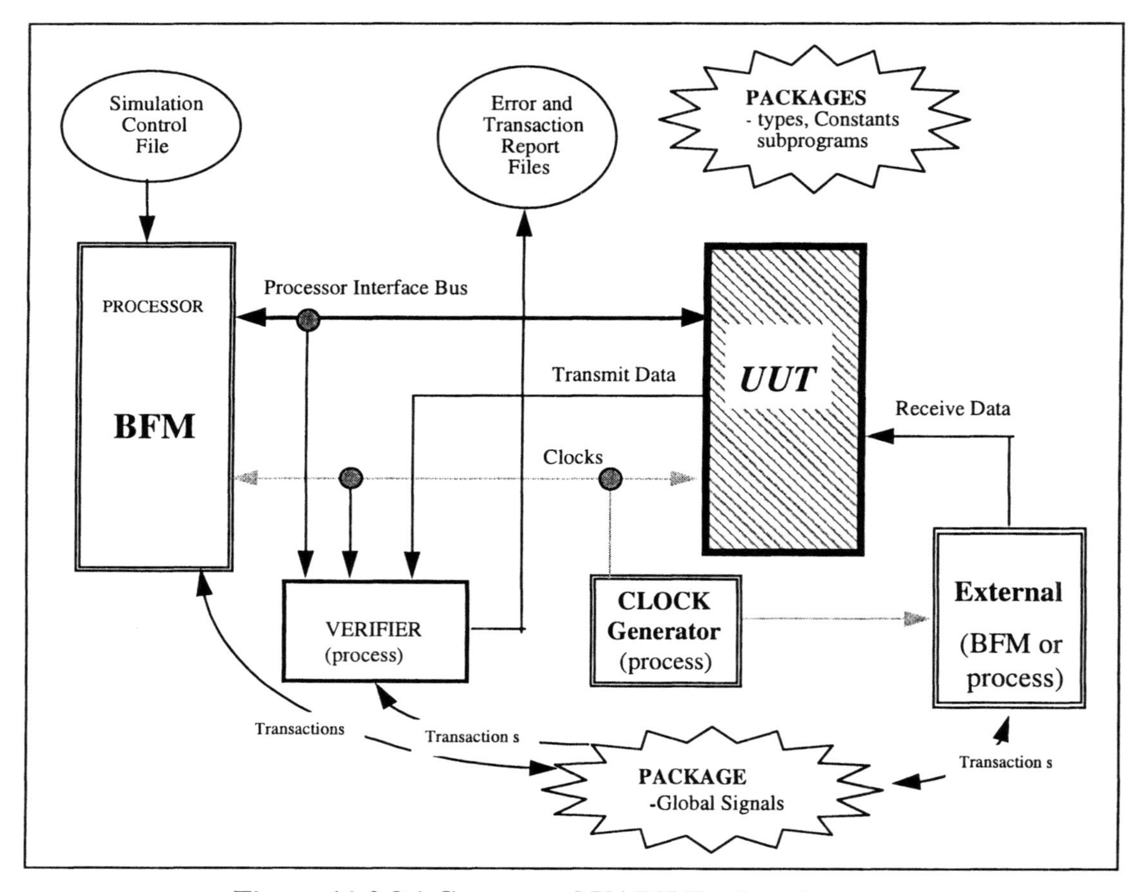

Figure 11.2.3.1 Conceptual UART Testbench

5. **VERIFIER.** The verifier automatically verifies that the UUT is operating correctly. The verifier also logs the actual characters sent and received by the UUT.

6. **Packages.** Packages define data types, constants, and subprograms.

7. **Global Signals package.** This package declares global signals. These global signals are used as inter-BFM communication links. Examples of such links include:

   a) **Control.** The Processor BFM instructs the External BFM to start issuing a transaction with certain characteristics, such as data content, parity state, number of stop bits, error condition, etc.

   b) **Monitoring.** The External BFM alerts the Verifier that it sent data to the UUT. The verifier can use this information, the UUT Transmit Data, and the processor bus transactions to automatically determine if the UUT is operating correctly.

c) **Transactions Synchronization between BFMs.** In this example, the processor BFM controls the start and type of transactions to be issued by the External BFM.

d) **Data flow between BFMs**. Because of the complexity of the designs, multiple BFMs are used to represent the various interfaces to the UUT. In some circumstances, there is a need to pass information from one BFM onto another, thus emulating a data flow between BFMs. In this example, the processor passes data to the external BFM to direct it to issue a transaction over its bus interface.

M 👍 👎 Use **global signals** instead of ports of a component when communication channels represent **abstract transactions** that are **NOT** part of a UUT bus protocol. **ALWAYS** include the **package path** in naming the global signal.
Rationale:

1. *Using **ports** as communication channel between BFMs **clutters** the testbench with ports and signals NOT directly related to the UUT.*
2. *When making design updates to the BFMs, the **addition of ports** to enhance BFM functionality (such as inter-BFM communications for control or data transfer) is a **major effort** because it is a testbench redesign. Specifically, **ports** must be added to the component **declaration and instantiations,** and **signals** must be added to the testbench. However, **the addition of global signals does not involve any changes to the testbench.** The changes are in the package declarations and BFM architectures only.*
3. *The package path for the global signals provides for good documentation and readability.*

The complete development of the UART model and its testbench is defined in chapter 12.

11.1.4 FM/BFM Modeling Requirements

FMs and *BFMs* are very useful because they enable the modeling of the interfaces of complex components, thus providing a "simulatable specification" of the interfaces. Unlike *FMs*, *BFMs* do not necessarily emulate the operations of the component, and therefore can be built relatively quickly and can be quite efficient from a simulation viewpoint. *FMs* and *BFMs* are very useful in testbenches where they can be used to verify the operation of the units under test (*UUT*). BFMs are generally easier to construct because the internal complexity of highly integrated devices are generally known by the vendor only, however the signal interfaces and timing parameters are defined in vendor's specification sheets (even though they may not necessarily be sufficient enough to create a BFM).

The functional model (BFM, and sometimes the FM) operational requirements include the following:

1. It must have a means to issue transactions (i.e., the scenarios) on the interface signals (e.g., READ, WRITE, *IDLE, DMA*).
2. It must reflect the correct cycle timing (i.e., phase and clock cycle delays) on the interfaces.
3. If required, it should reflect the correct propagation delays on the interface signals. These delays should be selectable for minimum, typical, maximum, or user defined delays.
4. It must react to the environment just like a real device. For example, a user wishes to issues 1000 *READ* and 1000 *WRITES* transactions, and a *Direct Memory Access* (DMA) request signal is asserted by the environment after the first transaction. In that situation the BFM must acknowledge the DMA, tri-state the appropriate signals, and stop issuing bus transactions until bus ownership is returned to the microprocessor.
5. If data-flow from one model to another model is a requirement, then a means must be provided to transfer the data across the BFMs.
6. It should be able to force bus errors to verify the operations of the *UUTs* when exposed to errors (e.g., parity errors).
7. It should detect and report bus transaction errors. Note that some of the verification and error reporting functions may be performed by concurrent VHDL statements (e.g., components, procedures), and can be used collectively with the FM/BFM bus error monitors to implement the full error detection requirements of the model.
8. It should be able to log the bus transactions into a file in a human readable format for further analysis of the tests run on the UUT.
9. It may emulate the execution of instructions if mandated by the project.

11.2 SCENARIO GENERATION SCHEMES

The purpose of the BFM is to emulate interactive transactions to the UUT. Various schemes are examined in this section. The vehicle used to demonstrate the various scenario generation schemes is the testing of a memory device. The memory is a good UUT model because:

1. It provides a *WRITE* interface, where data (one or more words) is transferred from the testbench to the UUT. The source of the data is defined within the testbench from either files, or code, or random functions.
2. It provides a *READ* interface, where data (one or more words) is transferred from the UUT to the testbench.
3. It allows for *IDLE* cycles between memory transactions. These *IDLE* cycles can be specified from either files, or code, or random functions.

In this model, the scenario generator models shall provide those three transactions*: Write data, Read data, Idle cycles.* The scheduling of transactions shall be pseudo-random. Figure 11.2-1 represents a functional model of a RAM memory initialized from a file. This model uses deferred constants to allow for memories of varying widths and depths. The model uses a variable to represent the storage elements because variable are significantly more storage efficient than signals. The model also uses a generic for the definition of file containing the initialization data. This scheme can accommodate multiple instances of the memory component, each with its own initialization data.

```
--    Description :  Model a memory initialized by a file.
--    Memory Operation:    Memory is initialized at startup from file.
--    Compiled Library:    ATEP_Lib
--
--                         READ          <- write at this edge
--              RdWrF    --------+       +------
--                               +_______+
--
--              CeF     ---+  +---+         +------
--                         +--+   +--------+
library IEEE;
  use IEEE.Std_Logic_1164.all;

package Mem_Pkg is
   constant DataWidth_c  : Natural; -- Data width
   constant AddrWidth_c  : Natural; -- Address width
   constant MaxDepth_c   : Natural;
end Mem_Pkg;

package body Mem_Pkg is
   constant DataWidth_c  : Natural  := 8;  -- deferred constant
   constant AddrWidth_c  : Natural  := 16; -- Address width
   constant MaxDepth_c   : Natural  := 2 ** AddrWidth_c; -- 64K
end Mem_Pkg;
```

```
library IEEE;
  use IEEE.Std_Logic_1164.all;
  use IEEE.Std_Logic_TextIO.all;
library ATEP_Lib;
  use ATEP_Lib.Mem_Pkg.all;
  use ATEP_Lib.Mem_Pkg;

entity Memory is  -- VHDL'87 file definition format
  generic (FileName_g : String:= "memdata_.txt");
  port(
    MemAddr        : in    Natural;
    MemData        : inout Std_Logic_Vector(Mem_Pkg.DataWidth_c - 1 downto 0);
    RdWrF          : in    Std_Logic;
    CeF            : in    Std_Logic);  -- Chip Select
end Memory;

architecture Memory_a of Memory is
begin  --  Memory_a
  Memory_Proc : process(MemAddr, MemData, RdWrF, CeF)
    use Std.TextIO.all; -- localize TextIO to this process
    use Std.TextIO;

    subtype Data_Typ is
        Std_Logic_Vector(Mem_Pkg.DataWidth_c - 1 downto 0);
    subtype MemSize_Typ is Integer range 0 to Mem_Pkg.MaxDepth_c - 1;
    type Mem_Typ is array(MemSize_Typ) of Data_Typ;

    variable Memory_v  : Mem_Typ;
    variable Initialization_v : Boolean := true;
    -- file     MemData_f : TextIO.text is in FileName_g; -- VHDL'87
    file     MemData_f : TextIO.text open Read_Mode is FileName_g; --VHDL'93

    procedure MemoryData -- Preload memory from a file,
        -- (variable FileName_f  : in     TextIO.Text; VHDL'87
        (file FileName_f  :  TextIO.text;
         variable Memory_v    : inout Mem_Typ) is

      variable InLine_v      : TextIO.line;
      variable MemAddress_v : MemSize_Typ := 0;  -- 0 to 64k
      variable Word_v        : Data_Typ;
      variable WordIndex_v   : natural;
    begin
      File_Lp : while not TextIO.Endfile(FileName_f) loop
        -- Read 1 line from the input file
        TextIO.ReadLine(FileName_f, InLine_v);

        -- Test if InLine_v is an empty line,
        next File_Lp when InLine_v'length = 0;          -- null line
        if InLine_v'length < DataWidth_c then -- error
          assert False
            report "Data width in file is less than defined width"
            severity Warning;
          next File_Lp;
        -- detect line with space as last character
        elsif InLine_v(DataWidth_c) = ' ' then -- error
          assert false
            report "Blanks in Data"
            severity warning;
          next File_Lp;
        end if;

        -- Read a data word
        IEEE.STD_Logic_TextIO.Read(InLine_v, Word_v);
        -- IEEE.STD_Logic_TextIO.HexRead(InLine_v, Word_v);
        -- Store word in memory and increment address index
        Memory_v(MemAddress_v) := Word_v;
        MemAddress_v := (MemAddress_v + 1) mod Mem_Pkg.MaxDepth_c;
```

Memory declaration

Procedure to initialize memory from data stored in a file

Empty line test

Check that line has data to avoid a fatal error.

Use HexRead for Hex data

```
      end loop File_Lp;
    end MemoryData;
  begin  -- process Memory_Proc
    -- Load memory if in initialization
    if Initialization_v then
      MemoryData(FileName_f  => MemData_f,
                 Memory_v    => Memory_v);
      Initialization_v := false;
    end if;

    -- Normal operation
    if CeF = '0' then -- memory transaction
      if RdWrF'event and
         RdWrF'last_value = '0'and   -- positive transition
         RdWrF = '1' then            -- of RdWrF
        Memory_v(MemAddr) := MemData;  -- WRITE OPERATION
      elsif RdWrF = '1' then   -- memory read
        MemData <= Memory_v(MemAddr);
      else
        MemData <= (others => 'Z');     -- 8/15/98
      end if;
    else
      MemData <= (others => 'Z');
    end if;
  end process Memory_Proc;
end Memory_a;
```

Functional algorithm for the memory

Figure 11.2-1 RAM Functional Model (ch11_dir\memory.vhd)

The RAM model requires a text file to initialize the memory. The creation of this text file can be tedious if done manually because of the size of the memory. However, this process can be facilitated with a VHDL model that automatically generates this text file with a user-defined algorithm. Figure 11.2-2 provides a sample of such code where the data for each address of the RAM is the same as the address. It makes use of the image package to write the text string into the desired format, binary or hexadecimal. The file name, depth, initial data value, and increment are passed as generic parameters. This model can be modified as necessary to automate the data generation process.

```
library IEEE;
  use IEEE.Std_Logic_1164.all;
  use IEEE.STD_LOGIC_UNSIGNED.all;

library Std;
  use Std.TextIO.all;

library Work;
  use Work.Image_Pkg.all;

--  use Work.rsg_util.all;
entity GenData is
    generic (FileName_g    : String;
             InitValue_g   : Std_Logic_Vector;
             Depth_g       : Natural;
             Increment_g   : Integer);
end GenData;
```

```
architecture GenData_a of GenData is
  constant Width_c : Natural := InitValue_g'length;
begin  -- GenData_a
  Memory_Proc: process
 --   file     RamData_f    : Std.TextIO.Text is out FileName_g;  -- '87
     file     RamData_f    : Std.TextIO.Text open
                                Write_Mode is FileName_g;  --'93
     variable Addr_v : Natural := 0;     -- ram address
     variable L   : Std.TextIO.Line;
     variable Data_v : std_logic_vector(Width_c downto 1) := InitValue_g;
  begin
       Addr_v := 0;
       RamWriteUp_Lp : for I in 0 to Depth_g - 1 loop
         Write(L, Image(Data_v)); -- use HexImage function for hex format
         WriteLine(RamData_f, L);
         Data_v := Data_v + Increment_g;
       end loop RamWriteUp_Lp;
     assert False
       report "Done writing data to file " & FileName_g
       severity note;
     wait;
    end process Memory_Proc;
end GenData_a;
------------------------------------------------------------------------
library IEEE;
  use IEEE.std_logic_1164.all;
entity GenDataTop is
end GenDataTop;

architecture GenDataTop_a of GenDataTop is
  component GenData
    generic (FileName_g    : String;
             InitValue_g   : Std_Logic_Vector;
             Depth_g       : Natural;
             Increment_g   : Integer);
  end component;
begin   -- DataGenTop_a
  U_GenData : GenData
        generic map
            (FileName_g    => "c:\gendata\test.dat",
             InitValue_g   => "00000",
             Depth_g       => 16,
             Increment_g   => 1);
end GenDataTop_a;

configuration Test_cfg of GenDataTop is
  for GenDataTop_a
    for U_GenData : GenData use entity Work.GenData(GenData_a)
      generic map
            (FileName_g    => "c:\gendata\test1.dat",
             InitValue_g   => "000000000",
             Depth_g       => 255,
             Increment_g   => 1);
    end for;
  end for;
end Test_cfg;
```

Figure 11.2-2 Text Data Generation (ch11_dir\gendata.vhd)

11.2.1 Scenario Generation Model : VHDL Code

This technique makes use of VHDL as the interactive waveform generation source. Figure 11.2.1 is a graphical view for this approach.

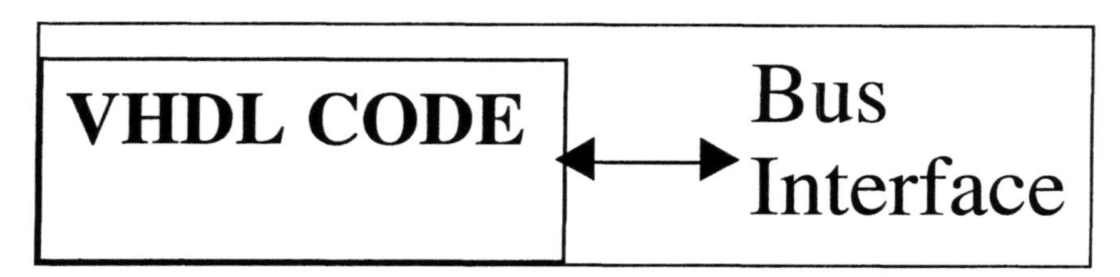

Figure 11.2.1 VHDL Model as Source of Interactive Waveform Generation

There are two generally accepted modeling approaches for the coding of the VHDL code as source of scenario generator:
1. *Waveform generator,*
2. *Client/server.*

11.2.1.1 Waveform Generator

This approach is the simplest in that the scenario definitions and implementation is combined into simple statements to generate waveforms. In other words, this method combines the transactions (e.g., *READ, WRITE)* and protocol (actual fluctuations of signals on the interfaces). The simplest implementation of the waveform generator is the application of a process with signal assignments placed between *wait* statements. This method is fairly straightforward, but is generally not recommended because it is difficult to read and maintain. It is shown here for completion of potential methods. Figure 11.2.1.1-1 provides a graphical view of the testbench architecture, while 11.2.1.1-2 demonstrates actual code.

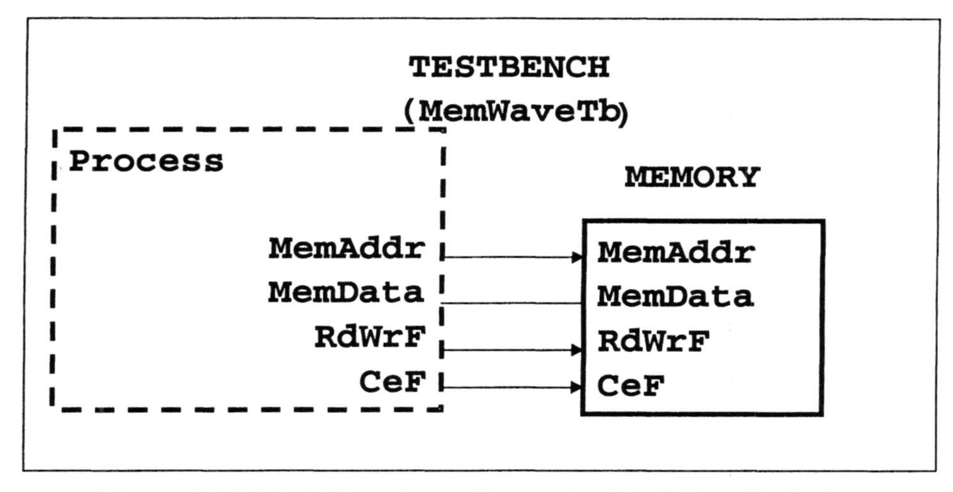

Figure 11.2.1.1-1 Waveform Generator Modeling Technique

```
library IEEE;                                  --   used because Std_Logic
  use IEEE.Std_Logic_1164.all;                   --   signals are resolved
  use IEEE.std_logic_arith.all;   -- for access to conversion function
library ATEP_Lib;
  use ATEP_Lib.Mem_Pkg.all;
  use ATEP_Lib.LfsrStd_Pkg.all;

entity MemWaveTB is
end MemWaveTB;

architecture MemWaveTB_a of MemWaveTB is
  use Std.TextIO.all; use Std.TextIO;

  constant MaxWordSize_c   : Natural := Atep_Lib.Mem_Pkg.DataWidth_c;
  signal MemAddr          : Natural;
  signal MemData          : Std_Logic_Vector(MaxWordSize_c - 1 downto 0)
                                  := (others => 'Z' );
  signal RdWrF            : Std_Logic := '1';
  signal CeF              : Std_Logic := '1';  -- Chip Select
  signal Clk              : Std_Logic := '1';  -- for delay computations

  component Memory
    generic (FileName_g     : String := "memdata1.txt");
    port (MemAddr         : in    Natural;
          MemData         : inout Std_Logic_Vector
                            (ATEP_Lib.Mem_Pkg.DataWidth_c - 1 downto 0);
          RdWrF           : in    Std_Logic := '1';
          CeF             : in    Std_Logic := '1');  -- Chip Select
    end component;

begin  -- MemWaveTB_a
  --  Clock generation, used for delays
  Clk <= not Clk after 10 ns;

  U1_Memory: Memory
    generic map
      (FileName_g     => "memdata1.txt")
    port map
      (MemAddr  => MemAddr,
       MemData  => MemData,
       RdWrF    => RdWrF,
       CeF      => CeF);

  WaveGen_Proc: process
   variable Addr_v : Natural;
   variable LfsrAddr_v : Std_Logic_Vector
              (ATEP_Lib.Mem_Pkg.AddrWidth_c - 1 downto 0) :=
                              (0      => '1',
                               others => '0');
   variable LfsrData_v : Std_Logic_Vector
              (ATEP_Lib.Mem_Pkg.DataWidth_c - 1 downto 0) :=
                               (0      => '0',
                                others => '1');
   variable LfsrDelay_v : Std_Logic_Vector(7 downto 0) := "10101010";
   variable RandDelay_v  : Natural;
```

UUT Instantiation

```
  begin
    -- compute random address
    LfsrAddr_v := LFSR(LfsrAddr_v);
    Addr_v := CONV_INTEGER(UNSIGNED(LfsrAddr_v));
    MemAddr <= Addr_v;
    -- compute random data
    LfsrData_v := LFSR(LfsrData_v);
    MemData <= LfsrData_v;

-- generate the WRITE operation
    (RdWrF, CeF)   <= Std_Logic_Vector'("10");
    wait until Clk = '1';
    (RdWrF, CeF)   <= Std_Logic_Vector'("00");
    wait until Clk = '1';
    (RdWrF, CeF)   <= Std_Logic_Vector'("10");
    wait until Clk = '1';
    (RdWrF, CeF)   <= Std_Logic_Vector'("11");
    MemData <= (others => 'Z');
    wait until Clk = '1';

    -- compute random delay
    LfsrDelay_v := LFSR(LfsrDelay_v);
    RandDelay_v :=  CONV_INTEGER(UNSIGNED(LfsrDelay_v(3 downto 0)));
    DelayWrite_Lp : for I in 1 to RandDelay_v loop
      wait until Clk = '1';
    end loop DelayWrite_Lp;

    -- Read Data from written location
    (RdWrF, CeF)   <= Std_Logic_Vector'("10");
    wait until Clk = '1';
    (RdWrF, CeF)   <= Std_Logic_Vector'("11");
    MemData <= (others => 'Z');
    wait until Clk = '1';
    -- compute random delay
    LfsrDelay_v := LFSR(LfsrDelay_v);
    RandDelay_v :=  CONV_INTEGER(UNSIGNED(LfsrDelay_v(3 downto 0)));
    DelayRead_Lp : for I in 1 to RandDelay_v loop
      wait until Clk = '1';
    end loop DelayRead_Lp;
  end process WaveGen_Proc;

end MemWaveTB_a;
```


Figure 11.2.1.1-2 Waveform Generation Method (ch11_dir\memwavetb.vhd)

11.2.1.2 Client/Server

This method is an object-oriented approach similar in concept to what is implemented in object-oriented languages. For example, in *JAVA,* objects belong to classes and object instances communicate among each other through *methods.* One object (the *client*) can make requests (e.g., request for data transactions) from the other object (the *server*) through those *methods.* It is possible to apply those techniques in VHDL for testbench designs, where a *client* object (*process* or *component*) communicates with a *server* object (*process* or *component*) through signals of record type to emulate the call method. The *client*, or executive, makes high level transaction requests (e.g., *Read, Write*) to the *server*, or support unit. The *server* detects the arrival of new messages and honors the requests by providing the actual bus interfaces (the *handshakes* and *protocol*) to the unit under test. Thus, the *server* emulates the implementation of the *methods.* The *server* also collects any interface data (using the *protocol*) and transfers that information to the *client* through a signal. In summary, the *client* (or master) initiates *TASKS* (or high level scenarios) to the *server* (or slave) that provides the stimulus to the UUT interfaces. The object oriented approach for the design of scenario generators enhances the concepts of model reusability and maintainability. Figure 11.2.1.2 provides a graphical view of the *client/server* modeling technique. The *clients* and *servers* can be modeled in VHDL as processes or components.

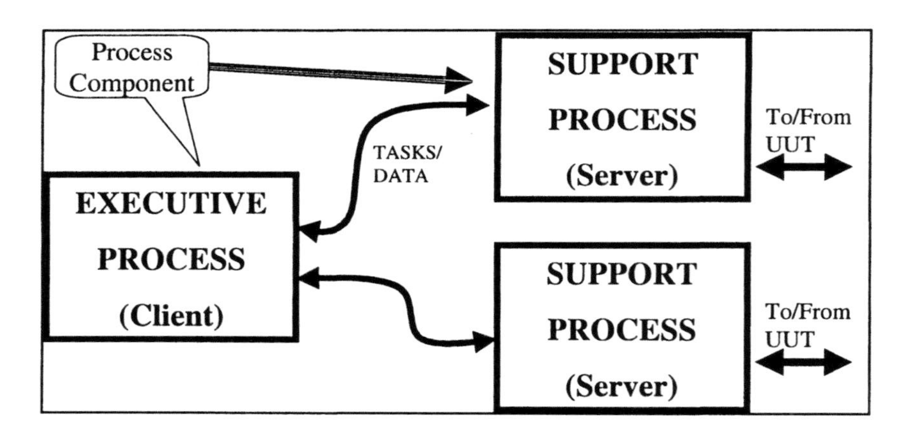

Figure 11.2.1.2-1 Client/Server Modeling Technique.

A component implementation for the *client* and server provides several advantages:

1. **Functional Separation.** The component clearly separates the **bus protocol** function from the **instruction or data** generation function. Unlike concurrent statements that have visibility over all the ports of the BFM entity and all the signals of the BFM architecture, the architecture of the component has limited visibility on only the ports of the components.

2. **Flexibility and Adaptability to changes.** A change to the bus protocol with no change to the interface ports can easily be handled with a configuration declaration that maps the architecture of the server component to a newly developed architecture.

3. **Information Hiding.** A component represents a level of hierarchy and represents a single instantiation into the BFM architecture. A process is represented with in-line code that can clutter the readability of the model.

Figure 11.2.1.2-2 represents a detailed graphical view of the memory testbench using the *client/server* method. Figure 11.2.1.2-3 represents the *client* package and the *client* model. Figure 11.2.1.2-4 represents the server model, and Figure 11.2.1.2-5 represents the VHDL testbench. The testbenches described in this chapter do not demonstrate the use of a verifier component because the emphasis is on the methodologies for bus functional models. An example of the verifier model is supplied in chapter 11.

Key design points in using the *Client* modeling technique include the following:

1. The *client* is responsible for the creation of the *tasks* that will be executed by the *server.*
2. A *task* typically consists of a record where the elements of that record provide the necessary information consumed by the *server.* For example, these elements could contain an address, a data item, an array of data, a command, and other parameters. The task used for the memory test consists of four fields: **Instruction** (read, write, idle), the **memory address, memory data** to be written, and the **number of delay cycles** for an idle instruction.
3. *Task* preparation is usually performed in a process using a variable of the *task* type. Information for the creation of the *task* can be from external files, external Programming Interface Language (PLI), or from VHDL code.
4. Once a *task* is prepared, the variable representing that *task* is assigned onto the *task* signal.
5. After assignment of the *task*, the process must wait until completion of that *task* detected with an *End Of Task* (EOT) signal generated by the server.
6. To facilitate the debugging of the model, and to avoid the creation on an implicit EOT signal, the EOT signal is a short one-nanosecond pulse of type *Boolean.* Thus, the process can use the statement *"wait until EOT;".*
7. Once the *EOT* is received, a new *task* can be generated.

Key points about the *server* modeling technique include the following:

1. The *server* waits for the arrival of a *task.* Since the *tasks* may have identical parameters (e.g., read of data from the same address twice), the *server* must wait on the *task transaction (TaskSignal'transaction).*
2. Once a *task transaction* is detected, the *server* must first decode the instruction passed into the *task*, and then execute that instruction as required. The *server* may also extract other information from the *task* as required. For example, if the *task* instruction is a *WRITE* to memory, then the memory address and data must be extracted from the *task.*
3. Once the *server* completes the execution of the *task* (e.g., wrote data to the memory), it then notifies the *client* of the completion of the task. This is accomplished through the assignment of a short pulse on the *EOT* signal. In VHDL the code is: *EOT <= True, FALSE after 1 ns;*
4. If data is collected by the *server* (e.g., data read from memory as a result of a *READ* instruction), then that data is collected into variables and assigned for delivery to the *client.* The data structure used for the signal carrying the information from the *client* depends upon the complexity of that information. It could also of a record type. In the memory example, a *task type* is used because the *server* returns to the *client* not only the read data, but also the address and instruction used in that transaction.

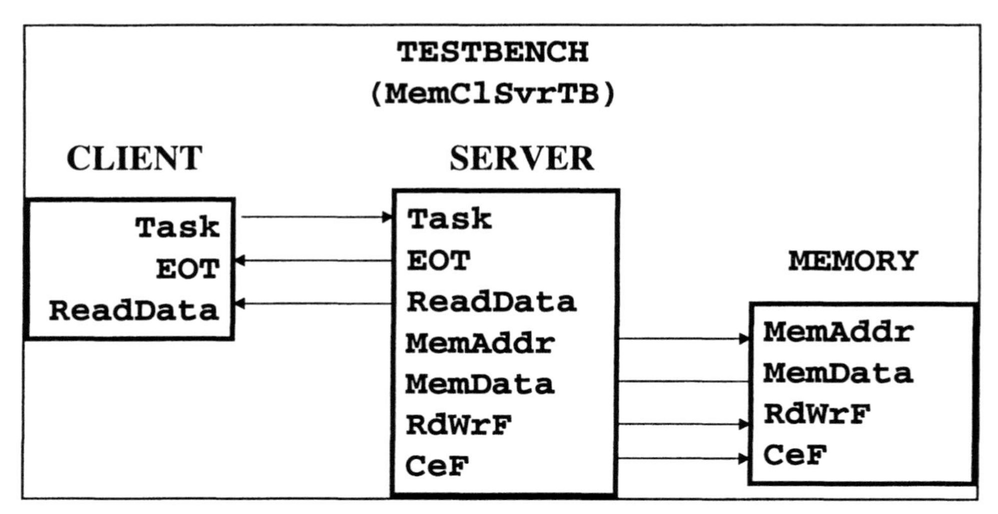

Figure 11.2.1.2-2 TestBench with Client/Server Architecture

```
-- in file client_p.vhd
library IEEE;
  use IEEE.Std_Logic_1164.all;
library ATEP_Lib;
  use ATEP_Lib.Mem_Pkg.all;
package Client_Pkg is
  -- Instruction read, write, idle for n cycles, return data from memory
  type Instr_Typ  is (READ, WRITE, IDLE);
  type Task_Typ is record
      Instr : Instr_Typ;
      Addr  : Natural;
      Data  : Std_Logic_Vector(Atep_Lib.Mem_Pkg.DataWidth_c - 1 downto 0);
      Delay : Natural;                        -- in clock cycles
    end record;
end Client_Pkg;
-----------------------------------------------------------------------
-- in file client.vhd
library IEEE;
  use IEEE.Std_Logic_1164.all;
  use IEEE.std_logic_arith.all;    -- for access to conversion function
library ATEP_Lib;
  use ATEP_Lib.Mem_Pkg.all;
  use ATEP_Lib.LfsrStd_Pkg.all;
  use Atep_Lib.Client_Pkg.all;
  use ATEP_Lib.Image_Pkg.all;

  use Std.TextIO.all;
entity Client is
  port (Task      : out ATEP_Lib.Client_Pkg.Task_Typ;  -- Task to server
        EOT       : in  Boolean;
        ReadData  : in ATEP_Lib.Client_Pkg.Task_Typ);  -- Data to client
end Client;
```

TASK Type

```
architecture Client_a of Client is
begin  -- Client_a
  TaskGen_Proc : process
   variable LfsrAddr_v : Std_Logic_Vector
                                    (ATEP_Lib.Mem_Pkg.AddrWidth_c - 1 downto 0) :=
                                       (0       => '1',
                                        others => '0');
   variable LfsrData_v : Std_Logic_Vector
                                    (ATEP_Lib.Mem_Pkg.DataWidth_c - 1 downto 0) :=
                                       (0       => '0',
                                        others => '1');
   variable LfsrDelay_v : Std_Logic_Vector(7 downto 0) := "10101010";
   variable Task_v : Task_Typ;
   variable L  : Std.TextIO.Line;
  begin  -- process TaskGen_Proc
    -- WRITE  Instruction
    Task_v.Instr := WRITE;

    -- compute random address
    LfsrAddr_v := LFSR(LfsrAddr_v);
    Task_v.Addr := CONV_INTEGER(UNSIGNED(LfsrAddr_v));

    -- compute random data
    LfsrData_v := LFSR(LfsrData_v);
    Task_v.Data := LfsrData_v;

    -- Send the task and wait until completed
    Task <= Task_v;
    wait until EOT;
    -- ==============================
    -- compute random delay
    Task_v.Instr := IDLE;
    LfsrDelay_v := LFSR(LfsrDelay_v);
    Task_v.Delay :=  CONV_INTEGER(UNSIGNED(LfsrDelay_v(3 downto 0)));
    Task <= Task_v;
    wait until EOT;
    -- ==============================
    -- Read Data from written location
    -- Task variable holds the address already
    Task_v.Instr := READ;
    Task <= Task_v;
    wait until EOT;

    -- ==============================
    -- Test if correct data was written
    if ReadData.Instr = READ then
      if ReadData.Data /= Task_v.Data then
        Std.TextIO.Write(L, "Time = " & ATEP_Lib.Image_Pkg.Image(Now) &
      "  Data not written correctly in memory" &
        "Written data = " &  ATEP_Lib.Image_Pkg.HexImage(Task_v.Data) &
        "read data = " & ATEP_Lib.Image_Pkg.HexImage(ReadData.Data));
        Std.TextIO.WriteLine(Output, L);
     else
            Std.TextIO.Write(L, "Time = " & ATEP_Lib.Image_Pkg.Image(Now) &
               "  Data written at " &
                 ATEP_Lib.Image_Pkg.Image(Task_v.Addr) & " = " &
                 ATEP_Lib.Image_Pkg.HexImage(ReadData.Data));
        Std.TextIO.WriteLine(Output, L);
      end if;
    end if;
```

Task Preparation using variable

Task Initiation and waiting for task completion.

```
    -- ==============================
    -- compute random delay
    Task_v.Instr := IDLE;
    LfsrDelay_v := LFSR(LfsrDelay_v);
    Task_v.Delay :=  CONV_INTEGER(UNSIGNED(LfsrDelay_v(3 downto 0)));
    Task <= Task_v;
    wait until EOT;
  end process TaskGen_Proc;
end Client_a;
```

Figure 11.2.1.3 Client Package and Client Model (ch11_dir\client_p.vhd, ch11_dir\client.vhd)

```
library IEEE;                                  --   used because Std_Logic
  use IEEE.Std_Logic_1164.all;                  --   signals are resolved
  use IEEE.std_logic_arith.all;   -- for access to conversion function
library ATEP_Lib;
  use ATEP_Lib.Mem_Pkg.all;
  use ATEP_Lib.LfsrStd_Pkg.all;
  use Atep_Lib.Client_Pkg.all;

library ATEP_Lib;
  use ATEP_Lib.Image_Pkg.all;

entity Server is
  port (MemAddr  : out   Natural;
        MemData  : inout Std_Logic_Vector
                          (ATEP_Lib.Mem_Pkg.DataWidth_c - 1 downto 0);
        RdWrF    : out    Std_Logic := '1';
        CeF      : out    Std_Logic := '1';  -- Chip Select
        Task     : in   ATEP_Lib.Client_Pkg.Task_Typ;  -- Task to server
         EOT      : out  Boolean;
         ReadData : out  ATEP_Lib.Client_Pkg.Task_Typ);  -- Data to client
end Server;

architecture Server_a of Server is
  signal Clk            : Std_Logic := '1';  -- for delay computations
begin  -- Server_a
  --  Clock generation, used for delays
  Clk <= not Clk after 10 ns;

  ServeMem_Proc : process
    variable ReadData_v : Task_Typ; -- collects read data
  begin  -- process ServeMem_Proc
    wait on Task'transaction;
    case Task.Instr is
      when READ             =>
         MemAddr <= Task.Addr;
         MemData <= (others =>      'Z');
         -- Read Data from written location
         (RdWrF, CeF) <= Std_Logic_Vector'("10");
         wait until Clk = '1';
         ReadData_v := Task;
         ReadData_v.Instr := READ;      -- redundant
         ReadData_v.Data  := MemData;
         ReadData <= ReadData_v;
         (RdWrF, CeF) <= Std_Logic_Vector'("11");
         wait until Clk = '1';
         EOT <= TRUE,
                FALSE after 1 ns;
```

Wait until arrival of a new task. Task'transaction is needed to detect new tasks assignments is same data.

EOT is a short pulse to create transactions for easy detection in Client. It also enhances debugging.

```
      when WRITE =>
         MemAddr <= Task.Addr;
         MemData <= Task.Data;
         -- generate the WRITE operation
         (RdWrF, CeF)   <= Std_Logic_Vector'("10");
         wait until Clk = '1';
         (RdWrF, CeF)   <= Std_Logic_Vector'("00");
         wait until Clk = '1';
         (RdWrF, CeF)   <= Std_Logic_Vector'("10");
         wait until Clk = '1';
         (RdWrF, CeF)   <= Std_Logic_Vector'("11");
         MemData <= (others => 'Z');
         wait until Clk = '1';
         EOT <= TRUE,
                FALSE after 1 ns;

      when IDLE  =>
        -- compute delay
        DelayWrite_Lp : for I in 1 to Task.Delay loop
          wait until Clk = '1';
        end loop DelayWrite_Lp;
        wait until Clk = '1';
        EOT <= TRUE,
               FALSE after 1 ns;

    end case;
  end process ServeMem_Proc;
end Server_a;
```

Figure 11.2.1.2-4 Server Model (Ch11_dir\ server.vhd)

```
library IEEE;                                   --  used because Std_Logic
  use IEEE.Std_Logic_1164.all;                   --   signals are resolved
library ATEP_Lib;
  use ATEP_Lib.Mem_Pkg.all;
  use ATEP_Lib.LfsrStd_Pkg.all;
  use ATEP_Lib.Mem_Pkg.all;

entity MemClSvrTB is
end MemClSvrTB;

architecture MemClSvrTB_a of MemClSvrTB is
  use Std.TextIO.all; use Std.TextIO;

  constant MaxWordSize_c  : natural := Atep_Lib.Mem_Pkg.DataWidth_c;
  signal MemAddr        : natural;
  signal MemData        : Std_Logic_Vector(MaxWordSize_c  - 1 downto 0);
  signal RdWrF          : Std_Logic := '1';
  signal CeF            : Std_Logic := '1';  -- Chip Select
  signal EOT            :  Boolean;
  signal ReadData       :  ATEP_Lib.Client_Pkg.Task_Typ;
  signal Task           :  ATEP_Lib.Client_Pkg.Task_Typ;

  component Memory
    generic (FileName_g    : String := "memdata1.txt");
    port (MemAddr         : in    natural;
          MemData         : inout Std_Logic_Vector
                                 (ATEP_Lib.Mem_Pkg.DataWidth_c - 1 downto 0);
          RdWrF           : in    Std_Logic := '1';
          CeF             : in    Std_Logic := '1');  -- Chip Select
    end component;

  component Client
    port(
        Task     : out ATEP_Lib.Client_Pkg.Task_Typ;  -- Task to server
         EOT      : in  Boolean;
         ReadData : in ATEP_Lib.Client_Pkg.Task_Typ
```

```
    );
  end component;

  component Server
    port(
        MemAddr   : out   Natural;
        MemData   : inout Std_Logic_Vector
                           (ATEP_Lib.Mem_Pkg.DataWidth_c - 1 downto 0);
        RdWrF     : out    Std_Logic := '1';
        CeF       : out    Std_Logic := '1';  -- Chip Select
        Task      : in   ATEP_Lib.Client_Pkg.Task_Typ;  -- Task to server
         EOT      : out  Boolean;
         ReadData : out  ATEP_Lib.Client_Pkg.Task_Typ
    );
  end component;

begin  --  MemClSvrTB_a

  U1_Memory: Memory
    generic map
      (FileName_g     => "memdata1.txt")
    port map
      (MemAddr   => MemAddr,
       MemData   => MemData,
       RdWrF     => RdWrF,
       CeF       => CeF);

  Client_I: Client
    port map (
        Task              => Task,
        EOT               => EOT,
        ReadData          => ReadData
    );

  Server_I: Server
    port map (
        MemAddr           => MemAddr,
        MemData           => MemData,
        RdWrF                 => RdWrF,
        CeF               => CeF,
        Task              => Task,
        EOT               => EOT,
        ReadData          => ReadData
    );
end MemClSvrTB_a;
```

Figure 11.2.1.2-5 VHDL testbench with Client/Server (Ch11_dir\ memclsvrtb.vhd)

Table 11.2.1.2 shows the compilation script for Model Technology's *ModelSim*. Figure 11.2.1.2-6 is a sample of the simulation results.

Table 11.2.1.2 Compilation Script for Client/Server Model (ch11\clsvr.do)

```
vcom -work atep_lib C:/SRC_DIR/CH8_DIR/image_pb.vhd
vcom -work atep_lib C:/SRC_DIR/CH8_DIR/lfsrstd.vhd
vcom -work atep_lib C:/SRC_DIR/CH11_DIR/memory.vhd
vcom -work atep_lib C:/SRC_DIR/CH11_DIR/client_p.vhd
vcom -work atep_lib C:/SRC_DIR/CH11_DIR/client.vhd
vcom -work atep_lib C:/SRC_DIR/CH11_DIR/server.vhd
vcom -work atep_lib C:/SRC_DIR/CH11_DIR/memclsvrtb.vhd
```

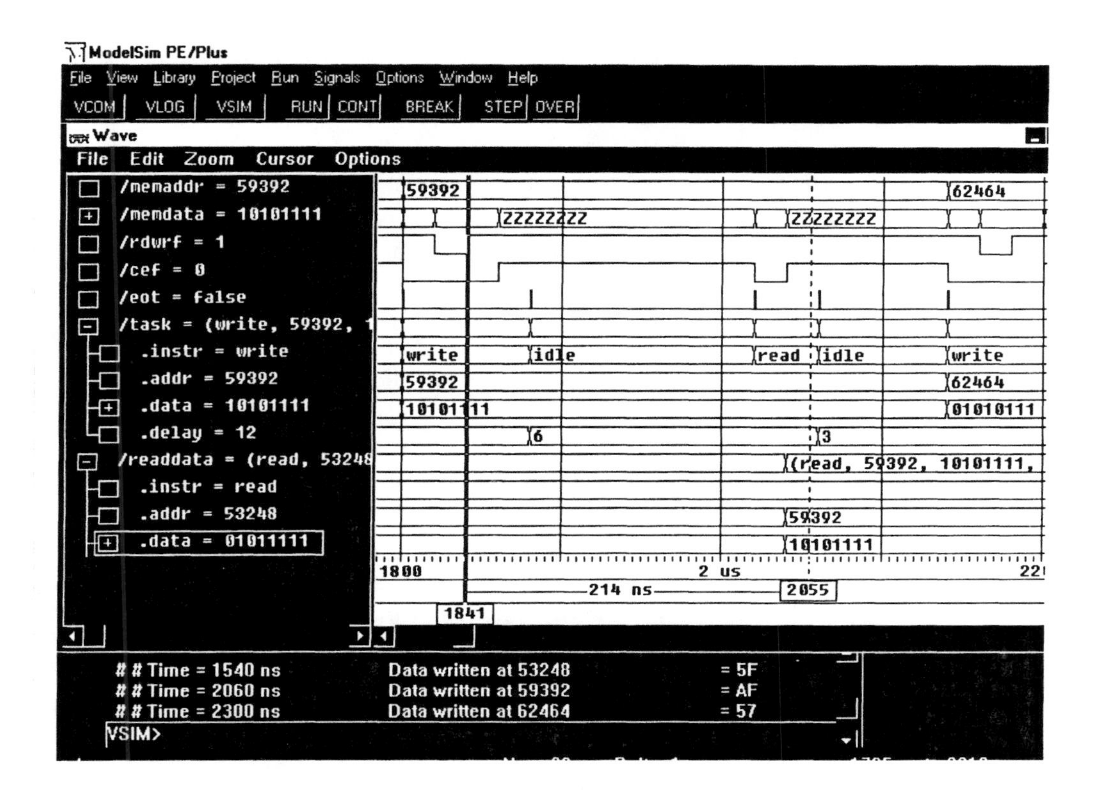

Figure 11.2.1.2-6 Simulation Results with Client/Server Method ***(with ModelSim from Model Technology).***

11.2.2 Scenario Generation Model : Text Command File

This method makes use of text files to define the set of transactions to be performed on the bus interface that connects to the UUT. With this method, the user need not know VHDL to define the sequence of transactions. Text files are in human readable format, and can easily be understood by the design team. In addition, if the VHDL model that reads the text files uses generics for the identity of those files, then no recompilation is necessary when different files are selected. Configuration declarations can be used to select the desired files. Text files can also be modified without the need for recompilation of the model.

However, text command files create other problems. First, the model becomes more complex because a parser needs to be defined to read the lines from the file, parse the contents of each line, convert that information from strings to the correct data types, and transfer that information to the VHDL code. Another issue is the need to learn another stimulus language created by the user. In the memory example, the instructions include *READ, WRITE,* and *IDLE.* For complex designs, the instruction set may not be wide enough to support all the requirements. For example, how can looping be achieved with text files? Does the instruction set support new features such as the data with "walking ones" or "walking zeros"? A third issue in the use of text file is the slow down caused by TEXTIO and the data processing performed by the instruction parser.

Figure 11.2.2-1 provides an outline of the Text Command File processing.

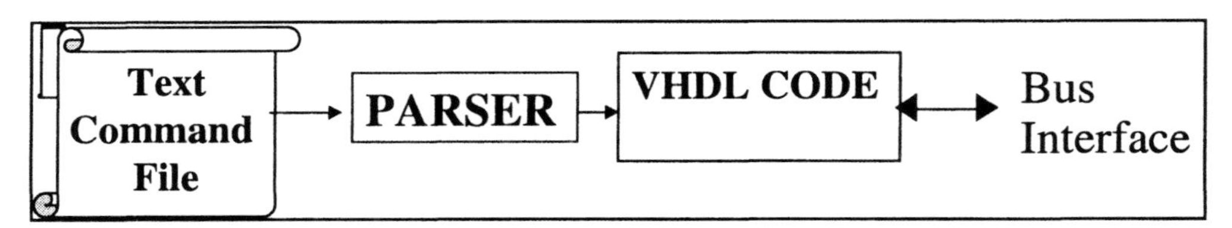

Figure 11.2.2-1 Text Command File Processing

Key design points in using the *Text Command* modeling technique include the following:

1. *Parser* should be designed as a component to enhance reusability. For example, a different architecture could be used to read binary files instead of text files (this is explained in the next section).
2. *Parser* requires an instruction dictionary to facilitate the decoding of the instructions. In this example, the definition of the instruction dictionary necessary to interface to a memory device is defined in packages. This dictionary includes:
   - The instruction set defined as an enumerated data type such as READ, *WRITE, IDLE.*
   - A string of six-character subtype to help in the search of a legal instruction string such as *subtype Strng6_Typ is String(1 to 6).*
   - An array type where the index is the enumeration instruction type, and the elements of the array represent the six character string, such as *type Strng2Instr_Typ is array (Client_Pkg.Instr_Typ) of Strng6_Typ;*
   - A constant table lookup to help in the search of a legal instruction string, and to translate a successful search to the enumerated instruction, such as

```
constant Instr2Strng_c : Strng2Instr_Typ
        := (READ  => "READ  ",
            WRITE => "WRITE ",
            IDLE  => "IDLE  ");
```

- A constant table Type to determine the number of parameters for each instruction. This table is optional, but may be needed to generalize the parse algorithm. For example,

  type Param_Typ is array (Client_Pkg.Instr_Typ) of Natural;
  constant Instr2Param_c : Param_Typ
  := (READ => 1, -- param = address
  WRITE => 2, -- param = address + data
  IDLE => 1); -- param = # cycles

  The search for a legal instruction and the number of parameters for the found instruction can be performed in a loop as shown below:

```
IsLegal_v := False;
FindInstr_Lp : for I in Instr_Typ loop
  if InstrStrng_v = Instr2Strng_c(I) then
    IsLegal_v := True;
    Instr_v := I;
    NumbParam_v := Instr2Param_c(I);
    exit FindInstr_Lp;
  end if;
end loop FindInstr_Lp;
```

- The *parser* component interfaces with the model that requests the data through a set of handshakes. Data is provided through a port of a record type.

Figure 11.2.2-2 provides the source code for the parser. The *parser* interfaces with the *Client* as shown in the testbench in Figure 11.3.2-3. The client makes a data request through the *ReqData* signal. The *parser* reads data from the file, parses it into the correct task type format, and assigns it onto the *FileTask* signal. The *parser* then asserts the *TaskRdy* to inform the *client* that the parsed file task is ready. The *parser* also provides and *EOD* signal to inform the *client* that an end of file was reached.

```
-- file parser_p.vhd
library IEEE;
  use IEEE.Std_Logic_1164.all;
library ATEP_Lib;
  use ATEP_Lib.Client_Pkg.all;  use ATEP_Lib.Client_Pkg;

package Parser_Pkg is
  -- Instruction
  subtype Strng6_Typ is String(1 to 6);
  type Strng2Instr_Typ is array (Client_Pkg.Instr_Typ) of Strng6_Typ;
  -- Type to determine # parameters
  type Param_Typ is array (Client_Pkg.Instr_Typ) of Natural;
  -- Instruction dictionary
  constant Instr2Strng_c : Strng2Instr_Typ
      := (READ  => "READ  ",
          WRITE => "WRITE ",
          IDLE  => "IDLE  ");

  constant Instr2Param_c : Param_Typ
      := (READ  => 1,                    -- param = address
          WRITE => 2,                    -- param = address + data
          IDLE  => 1);                   -- param = # cycles
end Parser_Pkg;
```

Figure 11.2.2-2a Parser Package for Text Command File processing (ch11_dir\parser_p.vhd)

```
-- file parser.vhd
library IEEE;
  use IEEE.Std_Logic_1164.all;
  use IEEE.STD_LOGIC_TEXTIO.all; use IEEE.STD_LOGIC_TEXTIO;

library ATEP_Lib;
  use ATEP_Lib.Parser_Pkg.all; use ATEP_Lib.Parser_Pkg;
  use ATEP_Lib.Client_Pkg.all; use ATEP_Lib.Client_Pkg;
  use ATEP_Lib.Mem_Pkg.all;    use ATEP_Lib.Mem_Pkg;
entity Parser is
  generic (FileName_g : String := "instr.txt");
  port(
    ReqData       : in   Boolean;
    FileTask      : out  Client_Pkg.Task_Typ;
    TaskRdy       : out  Boolean;        -- task ready
     EOD          : out  Boolean);       -- end of data
end Parser;
```

```
architecture Parser_TextFile of Parser is
begin  --  Parserory_a
  Parser_Proc : process
    use Std.TextIO.all; -- localize TextIO to this process
    use Std.TextIO;

 -- file ParserData_f : TextIO.text is in FileName_g; -- VHDL'87
    file ParserData_f : TextIO.text open Read_Mode is FileName_g;
      --VHDL'93
    variable InLine_v   :   TextIO.line;  -- TextIO line
    variable Task_v     :   Client_Pkg.Task_Typ;
    variable IsLegal_v  :   Boolean;
    variable EOD_v      :   Boolean;
    variable IsValidLine_v  : Boolean := True;

    procedure ParseData
      (variable InLine_v   : inout TextIO.line;  -- TextIO line
       variable Task_v     : inout Client_Pkg.Task_Typ;
       variable IsLegal_v  : inout Boolean) is
```

```
      variable InstrStrng_v : Parser_Pkg.Strng6_Typ;   -- String of 6 char
      variable NumbParam_v  : Natural;   -- # of parameters in instruction
      variable Instr_v   : Instr_Typ;

    begin
        IsLegal_v := False;
        -- Read first string of 6 characters
       Std.TextIO.Read(InLine_v, InstrStrng_v);

        -- Test if read word is legal
        IsLegal_v := False;
        FindInstr_Lp : for I in Client_Pkg.Instr_Typ loop
          if InstrStrng_v = Instr2Strng_c(I) then
              -- string  is legal?
            IsLegal_v := True;
            Instr_v := I;
            NumbParam_v := Instr2Param_c(I);
            exit FindInstr_Lp;
          end if;
        end loop FindInstr_Lp;

        -- Test for legal instruction and report error
        if not IsLegal_v then
          assert False report "Error in instruction" &
                              "Instruction = " & InstrStrng_v
            severity Warning;
          return;
        else
          -- Read other parameters
          -- Assume that all parameters do exist.  Otherwise,
          -- must do a test for legality.
          case Instr_v is
            when WRITE =>
              Task_v.Instr := WRITE;
              Std.TextIO.Read(InLine_v, Task_v.Addr);
          -- Can also use STD_LOGIC_TEXTIO.HREAD for HEX formatted data
              STD_LOGIC_TEXTIO.Read(InLine_v, Task_v.Data);

            when READ  =>
              Task_v.Instr := READ;
              Std.TextIO.Read(InLine_v, Task_v.Addr);

            when IDLE  =>
              Task_v.Instr := IDLE;
              Std.TextIO.Read(InLine_v, Task_v.Delay);
          end case;
          return;
        end if;
    end ParseData;


begin  --  process Parser_Proc
   File_Lp : while not TextIO.Endfile(ParserData_f) loop
   if IsValidLine_v then
     wait until ReqData;                   -- Wait for a request
   end if;
   -- Read 1 line from the input file
   TextIO.ReadLine(ParserData_f, InLine_v);
   -- Test if InLine_v is an empty line,
   if InLine_v'length = 0 then  -- null line then
     IsValidLine_v := False;
     next File_Lp;
   end if;

   -- Test for comment line
   if (InLine_v(1) = '-' and InLine_v(2) = '-') then
     IsValidLine_v := False;
```

```
      next File_Lp;
    end if;
    -- OK line
    IsValidLine_v := True;
    ParseData
      (InLine_v    => InLine_v,
       Task_v      => Task_v,
       IsLegal_v   => IsLegal_v);

     if IsLegal_v then
       FileTask <= Task_v;
       TaskRdy <= True, False after 1 ns;
     end if;
    end loop File_Lp;

    EOD <= True;
    wait;
  end process Parser_Proc;
end Parser_TextFile;
```

Call to *ParseData* Procedure

Figure 11.2.2-2b Parser Model for Text Command File processing (ch11_dir\ parser.vhd)

The architectural view of the memory testbench that makes use of the text command file with a parser component is shown in Figure 11.2.2-3. The *Client* component interfaces are expanded to communicate with the *Parser*, and obtain the next instruction and data in the proper format. Figure 11.2.2-4 provides the new client model, and Figure 11.2.2-5 is the VHDL testbench model. Figure 11.2.2-6 is a configuration declaration for the testbench. Table 11.2.2 represents the compilation order. Figure 11.2.2-7 shows an example of an instruction file.

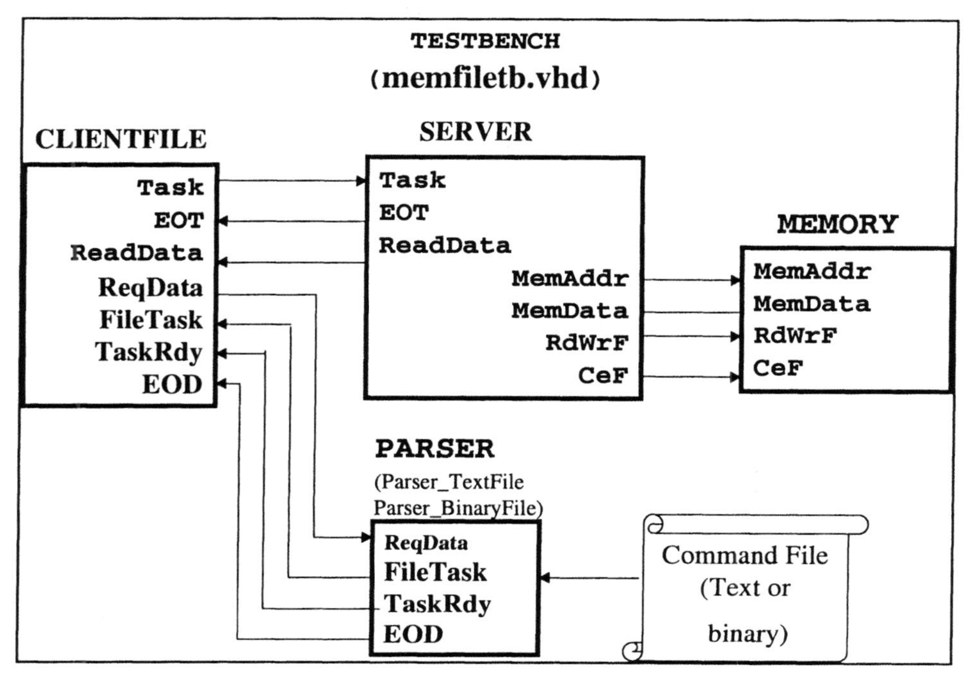

Figure 11.2.2-3 Architectural view of the Memory Testbench Using the Text or Binary Command File with a Parser

```
-- file clientfile.vhd
library IEEE;                  --
  use IEEE.Std_Logic_1164.all;             --
  use IEEE.std_logic_arith.all;   -- for access to conversion function
library ATEP_Lib;
  use ATEP_Lib.Mem_Pkg.all;
  use Atep_Lib.Client_Pkg.all;   use Atep_Lib.Client_Pkg;
entity ClientFile is
  port (Task      : out Client_Pkg.Task_Typ;  -- Task to server
        EOT       : in  Boolean;
        ReadData  : in  Client_Pkg.Task_Typ;  -- Data to client
        ReqData   : out Boolean;              -- request for data
        FileTask  : in  Client_Pkg.Task_Typ;  -- file data
        TaskRdy   : in  Boolean;              -- task ready
        EOD       : in  Boolean);    -- end of data
end ClientFile;
```

```
architecture ClientFile_a of ClientFile is
begin  -- ClientFile_a
  TaskGen_Proc : process
  begin  -- process TaskGen_Proc
    -- Request an instruction from the parser
    ReqData <= True, False after 1 ns;
    wait until TaskRdy or EOD;           -- new task or end of data file
    if EOD then
      assert False report "End of Instruction stream in file"
        severity Note;
      wait;
    end if;
    Task <= FileTask;
    wait until EOT;
  end process TaskGen_Proc;
end ClientFile_a;
```

Figure 11.2.2-4 Client Model for File Data Processing (Ch11_dir\clientfile.vhd)

```
library IEEE;
  use IEEE.Std_Logic_1164.all;
library ATEP_Lib;
  use ATEP_Lib.Mem_Pkg.all;
  use Atep_Lib.Client_Pkg.all;   use Atep_Lib.Client_Pkg;
entity MemFileTB is
end MemFileTB;

architecture MemFileTB_a of MemFileTB is
  use Std.TextIO.all; use Std.TextIO;

  constant MaxWordSize_c  : natural := Atep_Lib.Mem_Pkg.DataWidth_c;
  signal MemAddr        : natural;
  signal MemData        : Std_Logic_Vector(MaxWordSize_c  - 1 downto 0);
  signal RdWrF          : Std_Logic := '1';
  signal CeF            : Std_Logic := '1';  -- Chip Select
  signal EOT          :  Boolean;
  signal ReadData     :  ATEP_Lib.Client_Pkg.Task_Typ;
  signal Task         :  ATEP_Lib.Client_Pkg.Task_Typ;
  signal EOD          :  Boolean;
  signal FileTask     :  Client_Pkg.Task_Typ;
  signal ReqData      :  Boolean;
  signal TaskRdy      :  Boolean;

  component Memory
    generic (FileName_g    : String := "memdata1.txt");
    port (MemAddr        : in    natural;
          MemData        : inout Std_Logic_Vector
                           (ATEP_Lib.Mem_Pkg.DataWidth_c - 1 downto 0);
          RdWrF          : in    Std_Logic := '1';
          CeF            : in    Std_Logic := '1');  -- Chip Select
    end component;
```

```
component ClientFile
      port(
        Task     : out Client_Pkg.Task_Typ;  -- Task to server
         EOT      : in  Boolean;
         ReadData : in  Client_Pkg.Task_Typ;  -- Data to client
        ReqData  : out Boolean;            -- request for data
        FileTask : in  Client_Pkg.Task_Typ;  -- file data
        TaskRdy  : in  Boolean;            -- task ready
         EOD      : in  Boolean
       );
  end component;

component Server
    port(
        MemAddr   : out   Natural;
        MemData   : inout Std_Logic_Vector
                          (ATEP_Lib.Mem_Pkg.DataWidth_c - 1 downto 0);
        RdWrF     : out    Std_Logic := '1';
        CeF       : out    Std_Logic := '1';  -- Chip Select
        Task      : in    ATEP_Lib.Client_Pkg.Task_Typ;  -- Task to server
         EOT       : out  Boolean;
         ReadData : out  ATEP_Lib.Client_Pkg.Task_Typ
    );
  end component;

  component Parser
    generic (FileName_g : String := "instr.txt");
    port(
      ReqData        : in   Boolean;           -- request for data
      FileTask       : out  Client_Pkg.Task_Typ;  -- file data
      TaskRdy        : out  Boolean;           -- task ready
       EOD            : out  Boolean
      );
  end component;

begin  --  MemFileTB_a

  U1_Memory: Memory
    generic map
      (FileName_g     => "memdata1.txt")
    port map
      (MemAddr  => MemAddr,
       MemData  => MemData,
       RdWrF    => RdWrF,
       CeF      => CeF);

  ClientFile_I: ClientFile
    port map (
        Task              => Task,
        EOT               => EOT,
        ReadData          => ReadData,
        ReqData           => ReqData,
        FileTask          => FileTask,
        TaskRdy           => TaskRdy,
        EOD               => EOD
      );
```

```
Server_I: Server
    port map (
        MemAddr         => MemAddr,
        MemData         => MemData,
        RdWrF               => RdWrF,
        CeF             => CeF,
        Task            => Task,
        EOT             => EOT,
        ReadData        => ReadData
    );

  Parser_I: Parser
  generic map (FileName_g =>  "instr.txt")
  port map (
        ReqData         => ReqData,
        FileTask        => FileTask,
        TaskRdy         => TaskRdy,
        EOD             => EOD
  );
end MemFileTB_a;
```

Figure 11.2.2-5 VHDL Testbench Model with Client Handling File Commands through a Text File Parser (Ch11_dir\memfiletb.vhd)

```
library IEEE;
  use IEEE.Std_Logic_1164.all;
configuration MemFile_Cfg of MemFileTB is
  for MemFileTB_a
     for U1_Memory: Memory use
       entity ATEP_Lib.Memory(Memory_a)
         generic map(FileName_g     => "memdata1.txt");
    end for;

     for  ClientFile_I: ClientFile use
       entity ATEP_Lib.ClientFile(ClientFile_a);
    end for;

    for Server_I: Server use
      entity ATEP_Lib.Server(Server_a);
    end for;

    for Parser_I: Parser use
      entity ATEP_Lib.Parser(Parser_TextFile)
        generic map (FileName_g =>  "instr.txt");
    end for;
  end for;
end MemFile_Cfg;
```

Figure 11.2.2-6 Configuration Declaration for the Testbench (ch11_dir\memfile_c.vhd)

```
-- Sample instruction for memory access
WRITE 10 10101010
READ  15
IDLE  5
READ  10
READ  11
WRITE 9  10111000
IDLE  1
READ  9
```

Figure 11.2.2-7 Contents of an Instruction File

Table 11.2.2 Compilation Script for Text file Model (ch11_dir\memfile.do)

```
vcom -work atep_lib CH8_DIR/image_pb.vhd
vcom -work atep_lib CH8_DIR/lfsrstd.vhd
vcom -work atep_lib memory.vhd
vcom -work atep_lib client_p.vhd
vcom -work atep_lib clientfile.vhd
vcom -work atep_lib server.vhd
vcom -work atep_lib parser_p.vhd
vcom -work atep_lib parser.vhd
vcom -work atep_lib memfiletb.vhd
vcom -work atep_lib memfile_c.vhd
```

Figure 11.2.2-8 provides a sample of the simulation of the configuration file. Notice the handshaking between the *ClientFile* and the *Parser* components.

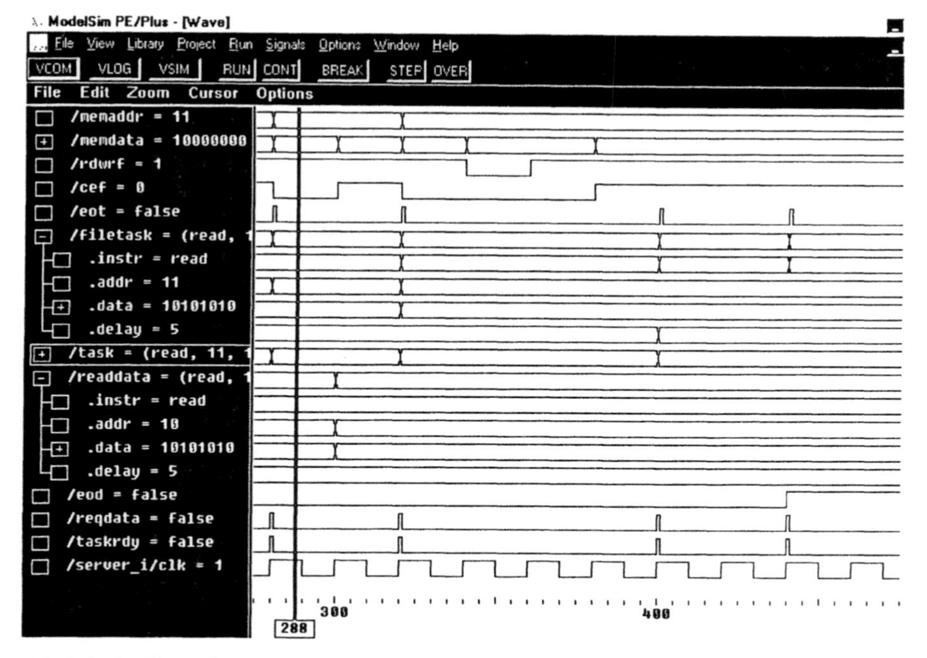

Figure 11.2.2-8 Simulation of the Memfile configuration*(with ModelSim from Model Technology).*

11.2.3 Scenario Generation Model : Binary Command File

The use of binary files is much faster than *ASCII* text files because conversion functions and *TextIO* routines are not required because the information is already in the proper data types. The same testbench architecture presented in Figure 11.2.2-3 can be used, but with different parser architecture of the same entity. The binary file represents a file of record type. In this application, the record type is defined in the *Client* package as:

```
type Task_Typ is record
     Instr : Instr_Typ; -- type Instr_Typ  is (READ, WRITE, IDLE);
     Addr  : Natural;
     Data  : Std_Logic_Vector(Atep_Lib.Mem_Pkg.DataWidth_c - 1 downto 0);
     Delay : Natural;                           -- in clock cycles
  end record;
```

Thus, in every binary file *READ* procedure call, the contents of a whole record is read. Since the information is already in the proper types no conversion functions are used.

Figure 11.2.3-1 provides an outline of the binary command file processing.

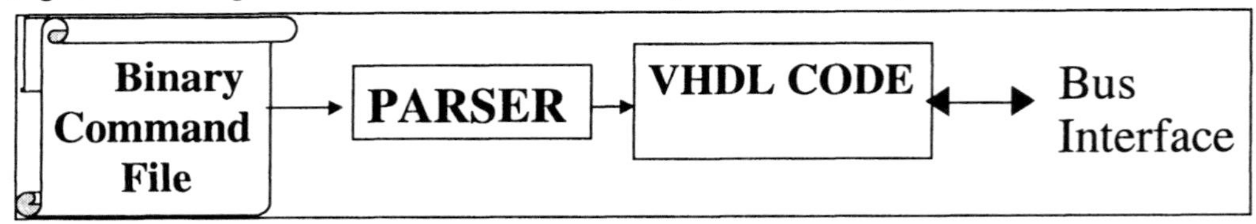

Figure 11.2.3-1 Binary Command File Processing

Figure 11.2.3-2 represents the parser that reads binary files.

```
architecture Parser_BinaryFile of Parser is
  type TaskFile_Typ is file of Client_Pkg.Task_Typ;  -- file data
begin  --  Parserory_a
  Parser_Proc : process
    -- file      ParserData_f : TaskFile_Typ is in FileName_g; -- VHDL'87
    file      ParserData_f  : TaskFile_Typ;
    variable fStatus_v   : File_Open_Status; -- New buitin type
      -- values: Open_OK, Status_Error, Name_Error
    variable Task_v      :   Client_Pkg.Task_Typ;
  begin  --  process Parser_Proc
    File_Open(Status        => fStatus_v,
              F             => ParserData_f,     -- implicit definition
              External_Name => FileName_g,
              Open_Kind     => Read_Mode);
    File_Lp: while not Endfile(ParserData_f) loop  -- is implicit built-in:
      -- Read 1 line from the input file
      Read(F        => ParserData_f,  -- Read a record of data
           Value    => Task_v);
      wait until ReqData;                   -- Wait for a request
       FileTask <= Task_v;
       TaskRdy <= True, False after 1 ns;
    end loop File_Lp;
    EOD <= True;
    wait;
  end process Parser_Proc;
end Parser_BinaryFile;
```

Figure 11.2.3-2 Binary File Parser (ch11_dir\parserbin.vhd)

Figure 11.2.2-5 represents the VHDL testbench model with the client handling commands defined in text or binary files. The binding of the parser architecture with the parser component determines the type of file handled. This is done in a configuration. For the binary files, the configuration is shown in Figure 11.2.3-3.

```
configuration MemfBin_cfg of MemFileTB is
  for MemFileTB_a
    for U1_Memory: Memory use
      entity ATEP_Lib.Memory(Memory_a)
        generic map(FileName_g    => "memdata1.txt");
    end for;

    for  ClientFile_I: ClientFile use
      entity ATEP_Lib.ClientFile(ClientFile_a);
    end for;

    for Server_I: Server use
      entity ATEP_Lib.Server(Server_a);
    end for;

    for Parser_I: Parser use
      entity ATEP_Lib.Parser(Parser_BinaryFile)
        generic map (FileName_g =>  "instr.bin");
    end for;
  end for;
end MemfBin_cfg;
```

Figure 11.2.3-3 Testbench Configuration Using Binary Files (ch11_dir\memfbin_c.vhd)

11.2.3.1 Generation of Binary Files

The simulation tool that reads those files must generate the binary files. Currently there are no standards on the definition of binary files; thus, these files are not necessarily compatible among vendors. Several techniques can be used to generate the files. Figure 11.2.3.1-1 makes use of modified parser architecture to create the binary file. Figure 11.2.3.1-2 represents the VHDL code. This code is a modification to the text *parser*. Once a text line is parsed into the proper record type, it is then written into a binary file.

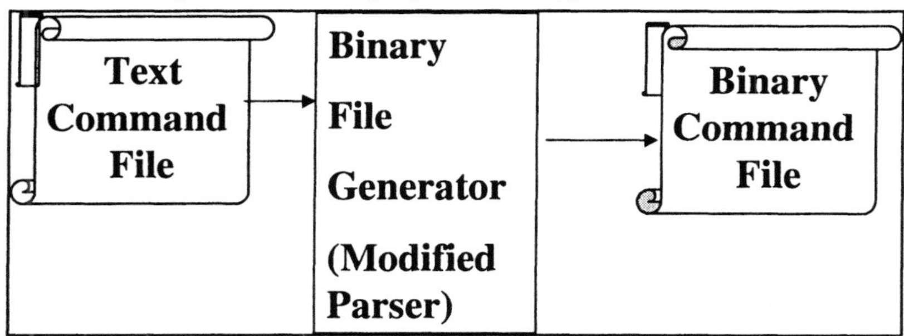

Figure 11.2.3.1-1 Modified Parser Model to Generate Binary File

```
library IEEE;
use IEEE.Std_Logic_1164.all;
  use IEEE.STD_LOGIC_TEXTIO.all; use IEEE.STD_LOGIC_TEXTIO;

library ATEP_Lib;
  use ATEP_Lib.Parser_Pkg.all; use ATEP_Lib.Parser_Pkg;
  use ATEP_Lib.Client_Pkg.all; use ATEP_Lib.Client_Pkg;
  use ATEP_Lib.Mem_Pkg.all;    use ATEP_Lib.Mem_Pkg;
entity GenBinary is
  generic (InstrFile_g  : String := "instr.txt";
           BinaryFile_g : String := "instr.bin");

end GenBinary;

architecture GenBinary_FromTextFile of GenBinary is

begin  --  GenBinaryory_a

  GenBinary_Proc : process
    use Std.TextIO.all; -- localize TextIO to this process
    use Std.TextIO;

    -- file     InstrFile_f : TextIO.text is in InstrFile_g; -- VHDL'87
    file InstrFile_f : TextIO.text open
                          Read_Mode is InstrFile_g; --VHDL'93
    variable InLine_v    :   TextIO.line;  -- TextIO line
    variable Task_v      :   Client_Pkg.Task_Typ;
    variable IsLegal_v   :   Boolean;
    variable EOD_v       :   Boolean;
    variable IsValidLine_v  : Boolean := True;

-- file  BinaryFile_f : Client_Pkg.Task_Typ is in BinaryFile_g; -- VHDL'87
    type TaskFile_Typ is file of Client_Pkg.Task_Typ;
    file BinDataOut_f : TaskFile_Typ open Write_Mode is BinaryFile_g;

    procedure ParseData -- Preload GenBinaryory from a file,
          (variable InLine_v   : inout TextIO.line;  -- TextIO line
           variable Task_v     : inout Client_Pkg.Task_Typ;
            variable IsLegal_v  : inout Boolean) is

      variable InstrStrng_v : Parser_Pkg.Strng6_Typ;  -- String of 6 char
     variable NumbParam_v  : Natural;  -- # of parameters in instruction
     variable Instr_v   : Instr_Typ;
    begin
        IsLegal_v := False;
        -- Read first string of 6 characters
       Std.TextIO.Read(InLine_v, InstrStrng_v);

        -- Test if read word is legal
        IsLegal_v := False;
        FindInstr_Lp : for I in Instr_Typ loop
          if InstrStrng_v = Instr2Strng_c(I) then  -- string  is legal?
            IsLegal_v := True;
            Instr_v := I;
            NumbParam_v := Instr2Param_c(I);
          end if;
        end loop FindInstr_Lp;


        -- Test for legal instruction and report error
        if not IsLegal_v then
          assert False report "Error in instruction" &
                              "Instruction = " & InstrStrng_v
               severity Warning;
          return;
        else
          -- Read other parameters
```

```
          -- Assume that all parameters do exist.  Otherwise,
          -- must do a test for legality.
          case Instr_v is
            when WRITE =>
              Task_v.Instr := WRITE;
              Std.TextIO.Read(InLine_v, Task_v.Addr);
              STD_LOGIC_TEXTIO.Read(InLine_v, Task_v.Data);

            when READ  =>
              Task_v.Instr := READ;
              Std.TextIO.Read(InLine_v, Task_v.Addr);

            when IDLE  =>
              Task_v.Instr := IDLE;
              Std.TextIO.Read(InLine_v, Task_v.Delay);
          end case;
          return;
        end if;
    end ParseData;

  begin  --  process GenBinary_Proc
    File_Lp : while not TextIO.Endfile(InstrFile_f) loop
    -- Read 1 line from the input file
    TextIO.ReadLine(InstrFile_f, InLine_v);
    -- Test if InLine_v is an empty line,
    if InLine_v'length = 0 then  -- null line then
      IsValidLine_v := False;
      next File_Lp;
    end if;

    -- Test for comment line
    if (InLine_v(1) = '-' and InLine_v(2) = '-') then
      IsValidLine_v := False;
      next File_Lp;
    end if;

    -- OK line
    IsValidLine_v := True;
    ParseData
      (InLine_v   => InLine_v,
       Task_v     => Task_v,
       IsLegal_v  => IsLegal_v);

     if IsLegal_v then
       Write(F         => BinDataOut_f,
             Value     => Task_v);
     end if;
    end loop File_Lp;

    assert False  report "End of Text File -- Done" severity Note;
    wait;
  end process GenBinary_Proc;

end GenBinary_FromTextFile;
```


Figure 11.2.3.1-2 represents the VHDL code (ch11_dir\genbinary.vhd)

Another method to generate the binary file is to reuse the parser that handles text files. This reusability concept is an essential element of object oriented languages. Figure 11.2.3.1-3 represents the architecture. The model consists of an interface *(storebin.vhd)* that handshakes with the text parser to generate the binary file. Figure 11.2.3.1-4 represents the *storebin* file, while figure 11.2.3.1-5 demonstrates the *genbinparser* environment.

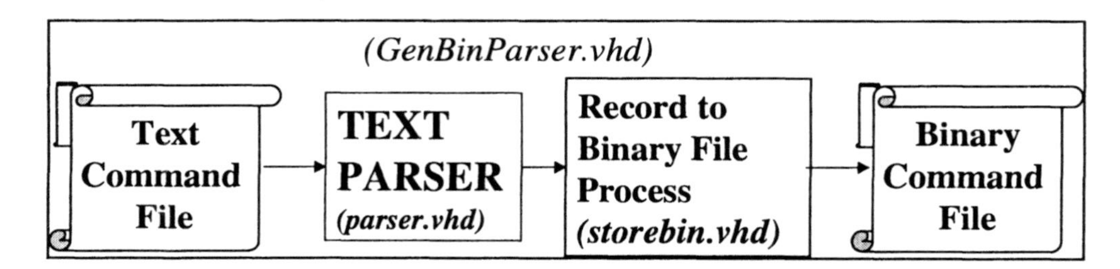

Figure 11.2.3.1-3 Generating Binary File with Text Parser

```
library IEEE;
  use IEEE.Std_Logic_1164.all;
  use IEEE.STD_LOGIC_TEXTIO.all; use IEEE.STD_LOGIC_TEXTIO;

library ATEP_Lib;
  use ATEP_Lib.Parser_Pkg.all; use ATEP_Lib.Parser_Pkg;
  use ATEP_Lib.Client_Pkg.all; use ATEP_Lib.Client_Pkg;
  use ATEP_Lib.Mem_Pkg.all;    use ATEP_Lib.Mem_Pkg;
entity StoreBin is
   generic (BinaryFile_g : String := "instr.bin");
   port(
     ReqData       : out  Boolean;              -- request for data
     FileTask      : in   Client_Pkg.Task_Typ;  -- file data
     TaskRdy       : in   Boolean;              -- task ready
     EOD           : in   Boolean         -- End of Data
     );
end StoreBin;

architecture StoreBin_a of StoreBin is
begin  --  GenBinaryory_a
  GenBinary_Proc : process
    -- file  BinaryFile_f : TextIO.text is in BinaryFile_g; -- VHDL'87
    type TaskFile_Typ is file of Client_Pkg.Task_Typ;
    file BinDataOut_f : TaskFile_Typ open Write_Mode is BinaryFile_g;

  begin  --  process GenBinary_Proc
     wait for 1 ns;                          -- End of initialization
    File_Lp : while not EOD loop
      ReqData <= True;
      wait until TaskRdy;
       ReqData <= False;
      Write(F        => BinDataOut_f,
            Value    => FileTask);
      wait for 10 ns;
    end loop File_Lp;

    assert False  report "End of Text File -- Done" severity Note;
    wait;
  end process GenBinary_Proc;
end StoreBin_a;
```

Callouts: Handshake with Text parser to get the task information — Write task data into binary file

Figure 11.2.3.1-4 represents the *storebin* file (ch11_dir\storebin.vhd)

```
library IEEE;
  use IEEE.Std_Logic_1164.all;
library ATEP_Lib;
  use ATEP_Lib.Mem_Pkg.all;
  use Atep_Lib.Client_Pkg.all;   use Atep_Lib.Client_Pkg;
entity GenBinParser is
end GenBinParser;

architecture GenBinParser_a of GenBinParser is
  signal ReqData            :  Boolean;
  signal FileTask           :  Client_Pkg.Task_Typ;
  signal TaskRdy            :  Boolean;
  signal EOD                :  Boolean;

  component StoreBin
    generic (BinaryFile_g : String := "instr.bin");
    port(
     ReqData       : out  Boolean;              -- request for data
     FileTask      : in   Client_Pkg.Task_Typ;  -- file data
     TaskRdy       : in   Boolean;              -- task ready
     EOD           : in   Boolean          -- End of Data
    );
  end component;

  component Parser
    generic (FileName_g : String := "instr.txt");
    port(
      ReqData       : in   Boolean;              -- request for data
      FileTask      : out  Client_Pkg.Task_Typ;  -- file data
      TaskRdy       : out  Boolean;              -- task ready
      EOD           : out  Boolean
      );
  end component;
begin  --  GenBinParser_a
  Parser_I: Parser
  generic map (FileName_g =>  "instr.txt")
  port map (
        ReqData        => ReqData,
        FileTask       => FileTask,
        TaskRdy        => TaskRdy,
        EOD            => EOD
    );

  StoreBin_I: StoreBin
    generic map(BinaryFile_g =>  "instr.bin")
    port map (
        ReqData        => ReqData,
        FileTask       => FileTask,
        TaskRdy        => TaskRdy,
        EOD            => EOD
    );
end GenBinParser_a;
```

Figure 11.2.3.1-5 Binary File Generation Environment (ch11_dir\genbinparser.vhd)

EXERCISES

1. What is the purpose of a testbench?

2. Modify the client/server/testbench model so that the client can request the writing or reading of 1 to 4 consecutive words starting at a specific address. You must modify the task package to accommodate the maximum number of data words, and add a field to identify how many words are to be written or fetched.

3. Define a component that performs this equation:

   Y = (A and B) or (A or C); -- Y, A, B, and C are ports.

Use 2 generics of type time to delay the **and** operation by 5 ns, and the **or** operation by 3 ns.

5. Define a testbench using the *Client/Server* format where a bus protocol component receives instructions to exhaustively test this logic component.

6. Define a verification process in the testbench that provides the following functions:
   - Logs, in a human readable form, the time and value of Y A B C data when their values change.
   - Logs an error (time and value of Y) when Y is not equal to the above equation.

6. How can the transition errors be ignored by the verifier?

12. UART PROJECT

A *UART* is a Universal Asynchronous Receiver Transmitter device utilizing an *RS232* serial protocol, and is in all personal computers to provide the interface between the *CPU* and the external devices through a serial port. This chapter presents the design of a simple ***UART*** using **synthesizable VHDL**. It also presents the design of a **complete testbench** to verify operation of the *UART*. The testbench employs the methods described in chapter 10.

12.1 *UART* ARCHITECTURE

A typical *UART* consists of a transmitter partition and a receiver partition. A commercial *UART* includes a set of programmable registers to characterize the device environment and error reporting features. The environment includes baud rate, number of STOP bits, and insertion of parity. The error reporting features include framing error (message with no STOP bit), overrun error (*UART* receives a new message prior to the *CPU* reading the old message), and parity error. For this project, a simple *UART* design is considered with NO personality registers, NO parity, and NO error detection logic.

12.1.1 UART Transmitter

12.1.1.1 General UART Concepts

A *CPU* typically loads an eight-bit word into a *UART*. The *UART* frames the word with a *START* bit (a logical 0) at the beginning and a *STOP* bit (a logical 1) at the end of the word. It sends the framing information along with the data in a serial manner from the Least Significant data Bit (*LSB*) to the Most Significant Bit (*MSB*). Figure 12.1.1-1 represents the timing waveform of a *UART* message issued by a *UART* transmitter and received by a *UART* receiver.

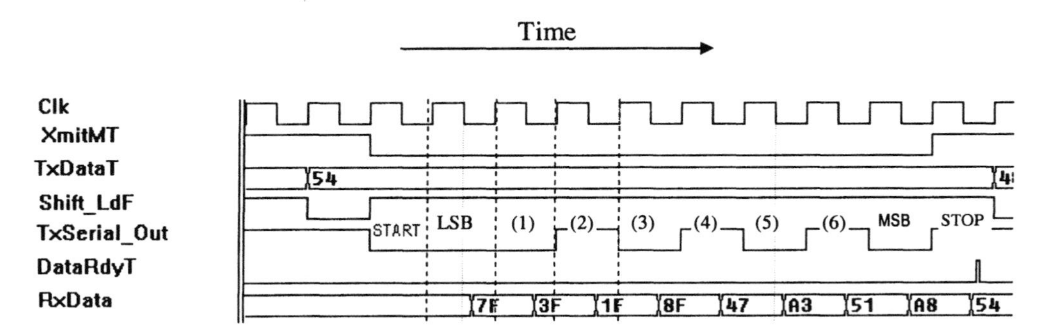

Figure 12.1.1.1 Timing of a *UART* Message

In this example, the *CPU* loads a value of 54 (in hex) to the *UART* transmitter. *TxDataT* represents the *CPU* parallel data, and *Shift_LdF* is the synchronous load, active low pulse to the *UART*. As soon as the data is loaded, the *UART* asserts on the *TxSerial_Out* signal the *START* bit (a logical 0), followed by the 8-bit data from *LSB* to *MSB*, followed by a *STOP* bit (a logical 1). In this example the data is 54 in hex or 0101_0100 in binary. Thus, the serial data sequence from *LSB* to *MSB* is 0010_1010, or 0_0010_1010_1 when the *START* and *STOP* bits are considered. The transmit *UART* deasserts a Transmitter Empty (*XmitMT*) signal when the data is sent. This flags the *CPU* that it cannot receive another character. When all bits are sent, it reasserts the *XmitMT* signal so that at the next clock cycle a new data can be loaded.

The serial data is received by a *UART* receiver. Synchronization is based on the negative transition of the *START* bit that resets a divide by 16 counter, which is clocked by a clock 16 times the bit clock. This counter is then used to detect mid-clock when it reaches a value of 7. The receive *UART* stores the serial data into a Receive Shift register. When all the data is framed, it alerts the receive *CPU* that data is ready (*DataRdyT* signal). The *RxData* signal represents the received 8-bit word.

12.1.1.2 UART Transmitter design

Figure 12.1.1.2-1 represents an architecture for a simple *UART* transmitter. It consists of a 10-bit transmit register that gets loaded with either a *STOP* bit concatenated with the 8-bit transmit data concatenated with the *START* bit during normal load cycle, or with a 10-bit value of all ones during a reset cycle. A state machine is used to control the data paths, and to identify the state of the transmit register, empty or in use. Figure 12.1.1.2-2 represents a synthesizable VHDL code for the transmit partition of a simple *UART*.

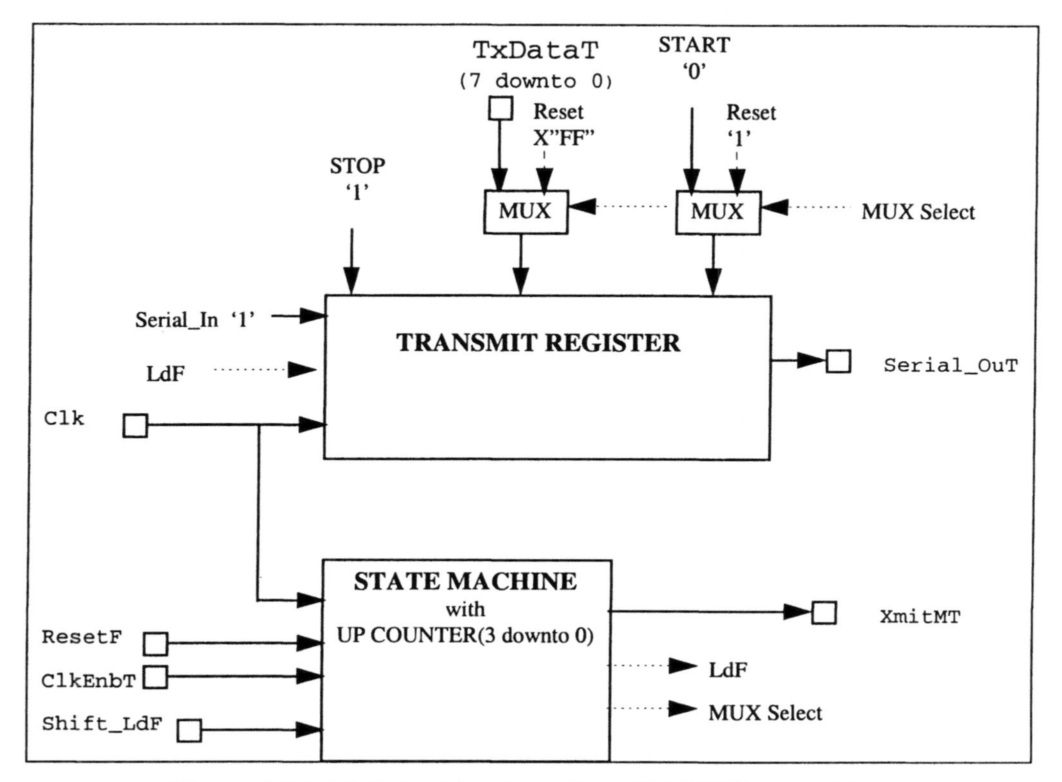

Figure 12.1.1.2-1 Architecture for a UART Transmitter

```
entity UartXmt is
  port (Shift_LdF      : in      bit;
        ClkEnbT        : in      bit;
        Clk            : in      bit;
        DataT          : in      Bit_Vector(7 downto 0);
        ResetF         : in      bit;
        Serial_OuT     : out     bit;
        XmitMT         : out     boolean);
end UartXmt;


architecture UartXmt_Beh of UartXmt is
  subtype Int0to9_Typ is integer range 0 to 9;
  signal Xmit_r : Bit_Vector(9 downto 0); -- the transmit register
  signal Count_r   : Int0to9_Typ;             --   # of serial bits sent
```

```
begin  --  UartXmt_Beh
  -----------------------------------------------------------------
  --  Process:   Xmit_Lbl
  --  Purpose:   Models the transmit register of a UART.
  --             Operation is as follows:
  --             . All operations occur on rising edge of CLK.
  --             . If ResetF = '0' then
  --                 Xmit_r is reset to "1111111111".
  --                 Count_r   is reset to 0.
  --             . If ClkEnbT = '1' and Shift_LdF = '0' and ResetF = '1' then
  --                 '1' & DataT & '0' get loaded into Xmit_r.
  --                 Count_r is reset to 0
  --             . If ClkEnbT = '1' and Shift_LdF = '1' and ResetF = '1' then
  --                 '1' & Xmit_r(9 downto 1) get loaded into Xmit_r
  --               (shift right with a '1' shifted in)
  --                 Count_r is incremented to no more then 10
  --                 (i.e. if it is 9, then it stays at 9)
  --
  ------------------------------------------------------------------------
  Xmit_Lbl : process
    variable Count_v : natural;
  begin  --  process Xmit_Lbl
    wait until Clk'event and Clk = '1'; --   rising edge of clock
    if ResetF = '0' then
      Xmit_r <= "1111111111";
      Count_r   <= 9;
    elsif ClkEnbT = '1' and Shift_LdF = '0' and ResetF = '1' then
      Xmit_r <=  '1' & DataT & '0';
      Count_r  <= 0;
    elsif ClkEnbT = '1' and Shift_LdF = '1' and ResetF = '1' then
      Xmit_r <= '1' & Xmit_r(9 downto 1);
      if Count_r /= 9 then
        Count_r <= Count_r + 1;
      end if;
    end if;
  end process Xmit_Lbl;

---------------------------------------------------------------------------
--  Concurrent signal assignment for Serial_Out
--  where Serial_Out is equal to Xmit_r(0)
---------------------------------------------------------------------------
  Serial_Out <= Xmit_r(0);

---------------------------------------------------------------------------
--  Concurrent signal assignment for XmitMT (transmitter eMpTy)
--  where XmitMT is true if COUNT_R is equal to 9.
---------------------------------------------------------------------------
  XmitMT <=       true when Count_r = 9
             else false;

end UartXmt_Beh;
```

Figure 12.1.1.2-2 Synthesizable VHDL Code for the Transmit Partition of a Simple UART (ch12_dir\UARTxmt.vhd)

Figure 12.1.1.2-3 represents the RTL view of the transmitter as synthesized with Synplicity.

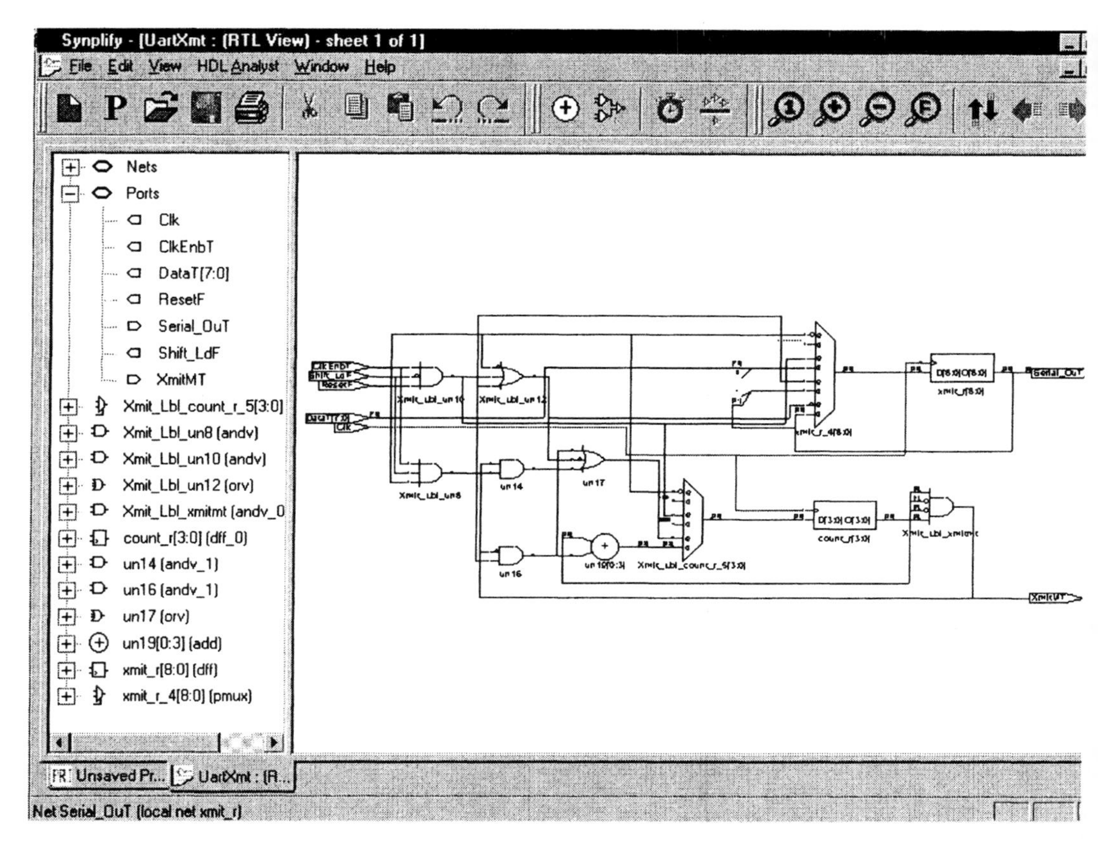

Figure 12.1.1.2-3 RTL View of the Transmitter *(with Synplicity from Synplicity).*

12.1.2 UART Receiver

Figure 12.1.2-1 represents a block diagram of a simple *UART* receiver. This architecture consists of a receive input register (*RxIn_s*) that reclocks the serial input data (*Serial_InT*) with a 16 times clock, so that the state machine operates in synchronous mode. The state machine remembers if the receive register is empty (*RxMT_s*) waiting for a new message. If it is empty and a *START* bit is detected, then the divide by 16 counter is reset to zero, thus synchronizing the start of a new bit. Also at this time, the *RxMT_s* register is reset to false since a new message is being received. When the counter reaches mid-bit count (*Count16_s* = 7) then the serial data is allowed to enter into the 10-bit receive register (*RxReg_s*). When a framed message is in the register, *RxReg_s*(9) shall be '1' and *RxReg_s*(0) shall be a '0'. At this point in time, the state machine sets the *RxMT_s* register and the Data Ready flag (*DataRdyT*) to alert the *CPU* that a new message is ready to be read.

Figure 12.1.2-2 represents a synthesizable VHDL code for the *UART* receiver. Figure 12.1.2-3 demonstrates the RTL view of the receiver as synthesized with Synplicity.

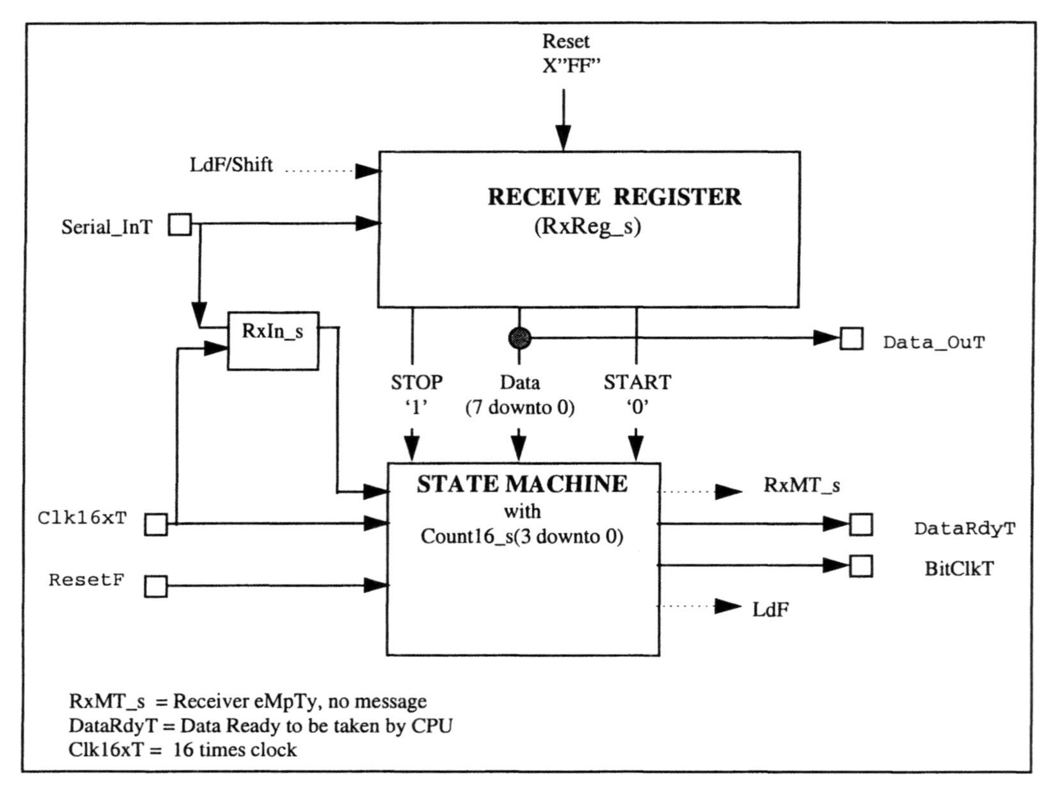

Figure 12.1.2-1 Block Diagram of a Simple *UART* Receiver

```
entity UartRx is
  port (Clk16xT      : in     bit;
        ResetF       : in     bit;
        Serial_InT   : in     bit;
        DataRdyT     : out    boolean;
        DataOuT      : out    Bit_Vector(7 downto 0);
        BitClkT      : out    bit);
end UartRx;

architecture UartRx_Beh of UartRx is
  subtype Int0to15_Typ is integer range 0 to 15;
  constant RxInit_c : Bit_Vector(9 downto 0) := "1111111111";
  signal RxReg_r   : Bit_Vector(9 downto 0); -- the receive  register
  signal Count16_r : Int0to15_Typ;             --  for divide by 16
  signal RxMT_r    : boolean;             --   Receive register empty
  signal RxIn_r    : bit;                 --   registered serial input
```

```
begin  --  UartRx_Beh
  -----------------------------------------------------------------
  --  Process:   Xmit_Lbl
  --  Purpose:   Models the receive portion of a UART.
  Rx_Lbl : process
  begin  --  process Rx_Lbl
     wait until Clk16xT'event and Clk16xT = '1';
     --   Clock serial input into RxIn_r
     RxIn_r <= Serial_InT;

     --   reset
     if (ResetF = '0') then
       Count16_r <= 0;                       --   reset divide by 16 counter
       RxMT_r <= true;                        --   new message starting
       RxReg_r <= RxInit_c;

     --   new bit start
     elsif (RxMT_r and RxIn_r = '0') then
       Count16_r <= 0;                       --   reset divide by 16 counter
       RxMT_r <= false;                        --   new message starting
       RxReg_r <= RxInit_c;

     --   If in a receive transaction mode
     --   if @ mid bit clock then clock data into register
     elsif Count16_r = 7  and not RxMT_r then --   mid clock
       RxReg_r <= RxIn_r & RxReg_r(9 downto 1);
       Count16_r <= Count16_r + 1;

     --   if @ 16X clock rollover
     elsif Count16_r = 15 then
       Count16_r <= 0;

     --   Normal count16 counter increment
     else
       Count16_r <= Count16_r + 1;
     end if;


     --   Check if a data word is received
     if not RxMT_r and RxReg_r(9) = '1' and RxReg_r(0) = '0' then
       DataRdyT <= true;
       RxMT_r <= true;
     else
       DataRdyT <= false;
     end if;
  end process Rx_Lbl;

-----------------------------------------------------------------
--  Concurrent signal assignment for BitClkT and DataOut
-----------------------------------------------------------------
  BitClkT <=       '1' when Count16_r = 9
              else '0';

  DataOuT <= RxReg_r(8 downto 1);

end UartRx_Beh;
```

Figure 12.1.2-2 Synthesizable VHDL Code for the UART Receiver (ch12_dir\UARTrx.vhd)

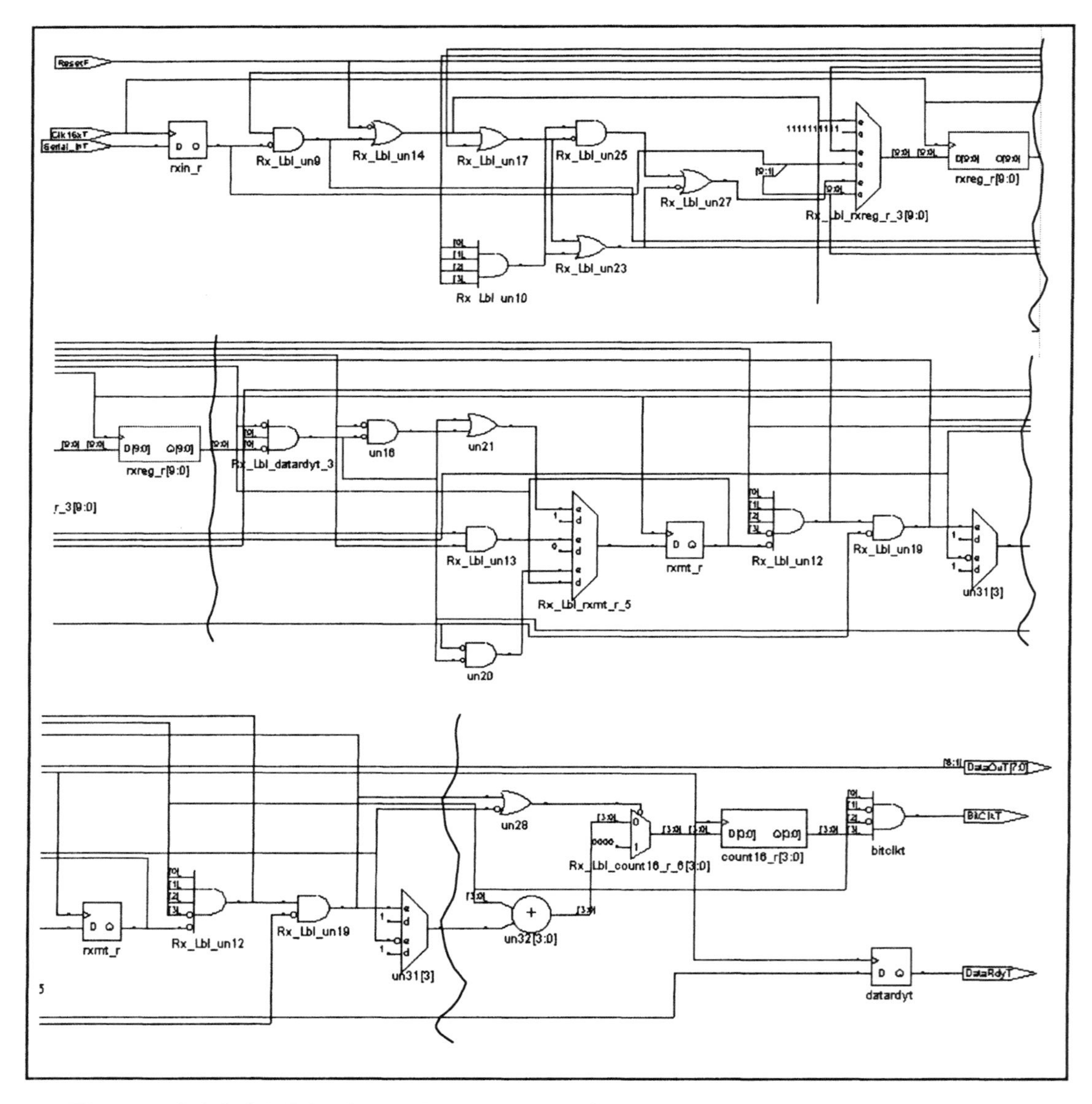

Figure 12.1.2-3 RTL View of Receiver as Synthesized with Synplicity.

12.2 *UART* TESTBENCH

The *UART* testbench provides the following functions:

1. Instantiates the *UART* transmitter and *UART* receiver into an architecture so that the serial data from the transmitter can be received by the receiver.
2. Emulates a *UART CPU* interface to load data into the transmit *UART*.
3. Sources the data to be sent to the *UART* from a text file to allow for easy modifications of the contents of the data.
4. Emulates a *UART CPU* interface to extract data received from the Receive *UART*.
5. Stores the receive data into an output file, and also displays received data onto the transcript window during simulation.
6. Emulates data link jitter between the transmitter and the receiver. This jitter represents long and random delays caused by switching networks.
7. Automatically verifies that the receiver properly receives the transmitted data.

Figure 12.2-1 represents a high level view of the *UART* testbench. A *CPU* transmit emulator reads characters from a file and transfers the binary encoding of that character to the transmit *UART*. It also sends through an abstract data path (global signal declared in a package) a copy of the character to alert the automatic verifier that a character was sent.

The serial output of the transmit *UART* is then passed through a link delay line that delays the data with a pseudorandom delay.

The *UART* Receiver decodes a framed word and passes that data to a *CPU* receive emulator. That *CPU* stores the received data into a file and passes the received information to the verifier. It also displays the received data onto the transcript window during simulation.

The verifier automatically verifies that the transmitted character matches the received character. It detects the case when a transmitter sent a message, but after a reasonable transmission delay the receiver never received the message. It also detects the case when the receiver detects a received message without the character being transmitted from the source.

Figure 12.2-2 is a more detailed representation of the testbench design that features the following methodology:

1. Instantiation of *UART* **transmitter** and **receiver** as **components.**
2. Instantiation of the **link delay line** as a **component**.
3. Partitioning the *CPU* transmit emulator into 2 sections:

- **Transmit Protocol component** that accepts high level instructions from a transmit process, and converts these directions using the appropriate interface protocol (see chapter 11 for more information on testbench designs). This component also transfers the sent character onto a global signal to be read by the verifier. A component is used here (instead of a process) to demonstrate the methodology that may be used for the modeling of complex transmit components (such as a processor interface).

- **Transmit process** that reads a test file, and instructs the transmit protocol component to send the desired character.

4. Partitioning the *CPU* Receive emulator into 2 sections:
   - **Receive Protocol component** that interfaces with the receiver component and converts the binary bits into characters. It then sends the received ASCII character to the receive process. This component also transfers the received character onto a global signal to be read by the verifier. Again a component is used here to demonstrate the methodology that may be used for the modeling of complex receive components

   - **Receive process** that stores the received characters into a file and onto the transcript display.

5. Instantiation of a **verifier** (Monitor component) that detects transmission errors.

6. Use of global signals to transfer information to be used by the verifier.

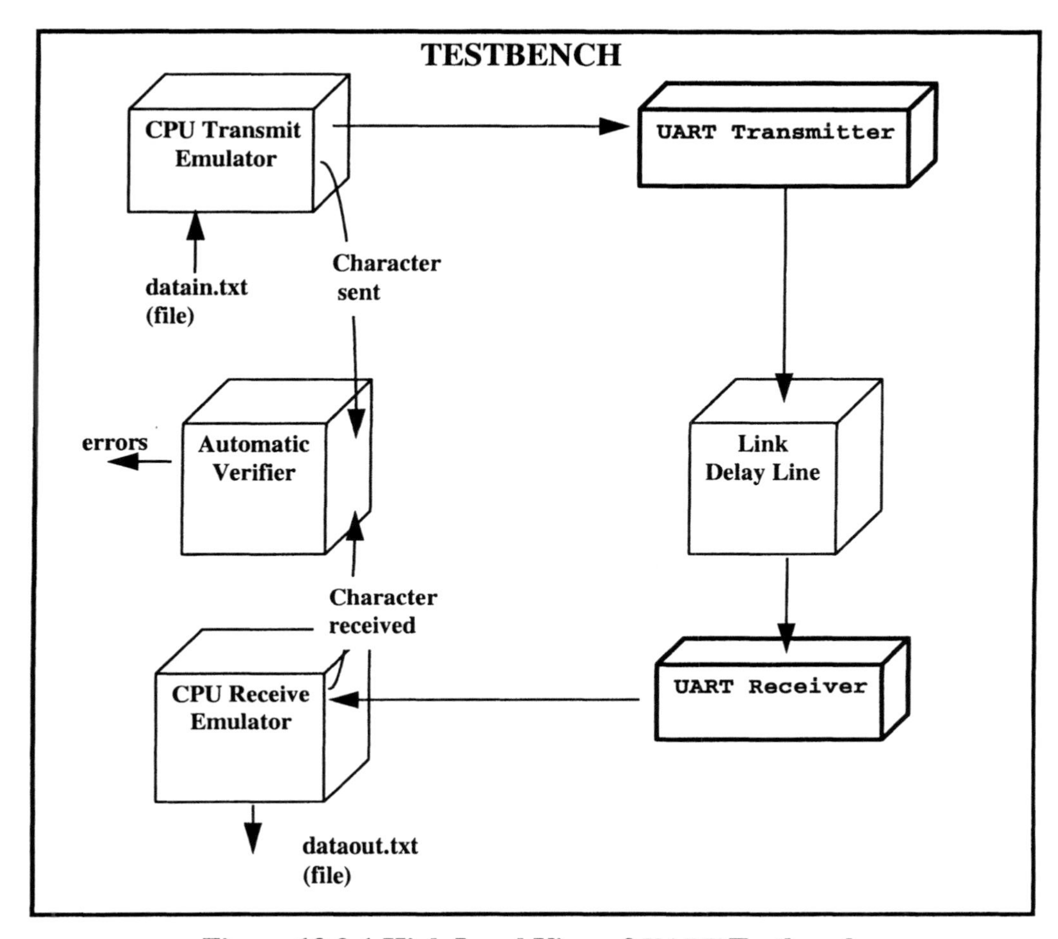

Figure 12.2-1 High Level View of *UART* Testbench

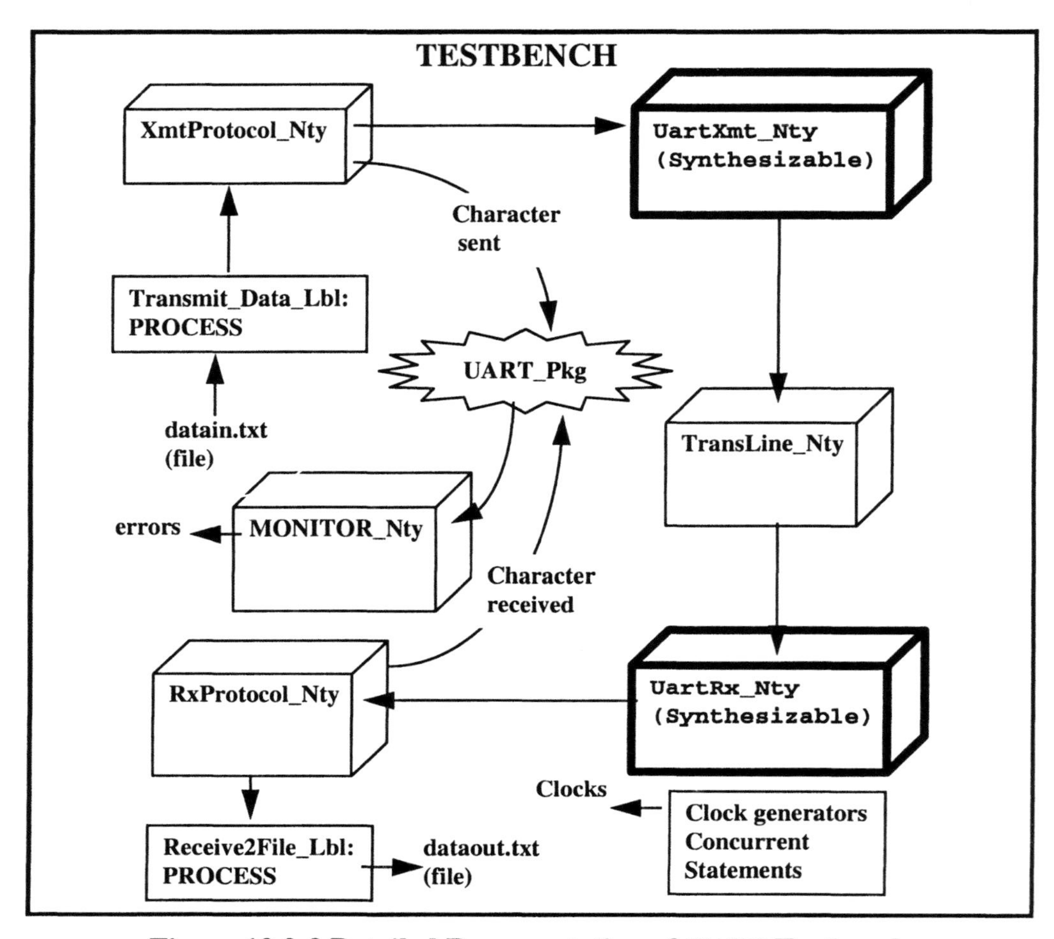

Figure 12.2-2 Detailed Representation of *UART* Testbench

The **compilation order** for the elements of the testbench is as follows:

1. *UART*xmt.vhd -- Synthesizable *UART* Transmitter
2. *UART*rx .vhd -- Synthesizable *UART* Receiver
3. *UART*_p.vhd -- *UART* Support package declaration for testbench
4. *UART*_b.vhd -- *UART* Support package body for testbench
5. xmtprtcl.vhd -- Transmit protocol component used by testbench
6. rcvprtcl.vhd -- Receive protocol component used by testbench
7. trnsline.vhd -- Transmission link component used by testbench
8. monitor.vhd -- Monitor or verifier component used by testbench
9. *UART*tb.vhd -- Testbench entity and architecture
10. *UART*ok_c.vhd -- Testbench with NO data link errors
11. *UART*tb_c.vhd -- Testbench with data link errors

All files are to be compiled in library "ATEP_Lib".

12.2.1 *UART* Package

The *UART* package declaration provides subtype declarations and constant declarations used by the testbench. To speedup the design of the testbench, a subset of the allowed characters was specified as legal characters to use for transmission. Specifically, the character 'A' through 'Z', the CR, and ' ' character are allowed. All other characters get translated to '?'. A constant *Asc2Bit_c* provides the type conversion between an ASCII character between 'A' through 'Z' to a bit vector equivalent. This package also defines two functions, one to determine if a character is in the range from 'A' to 'Z', and another function to convert a bit vector to its ASCII equivalent. Again, this last conversion functions limits the character sets to the one described above. A user may wish to enhance this package to include all the legal characters. The attributes 'POS and 'VAL could be used to provide for those conversions.

To support automatic verification of data transfer, this package defines two global signals:

1. **SentData_s** that is driven by the transmit protocol as it sends a character.
2. **ReceivedData_*s*** that is driven by the receive protocol as it receives a character.

A random number procedure provides a pseudo-random number given a seed and a modulus value. This procedure is used to generate the transmission line delay and the delay between characters. Figure 12.2.1-1 represents the package declaration while figure 12.2.1-2 represents the package body.

```
--   Title        :  ASCII conversions
--   Description :  Routines for UART
-- Compile into library ATEP_Lib
package UART_Pkg is
  subtype A2Z_Typ  is character range 'A' to 'Z';
  subtype Word_Typ is Bit_Vector(7 downto 0);
  subtype T16K_Typ is natural range 0 to 2**16;
  type Asc2Bit_Typ is array(A2Z_Typ) of Bit_Vector(7 downto 0);

  constant CR_c    : Bit_Vector(7 downto 0) := "00001010";
  constant SP_c    : Bit_Vector(7 downto 0) := "00100000";
  constant QMark_c : Bit_Vector(7 downto 0) := "00111111";   -- ? X"3F"
  constant LF_c    : Bit_Vector(7 downto 0) := "00001101";
  -- line feed, X"0D"

  constant Asc2Bit_c : Asc2Bit_Typ :=
     ('A' => "01000001",    -- 41
      'B' => "01000010",
      'C' => "01000011",
      'D' => "01000100",
      'E' => "01000101",
      'F' => "01000110",
      'G' => "01000111",
      'H' => "01001000",
      'I' => "01001001",
      'J' => "01001010",
      'K' => "01001011",
      'L' => "01001100",
      'M' => "01001101",
      'N' => "01001110",
      'O' => "01001111",
      'P' => "01010000",
```

```
      'Q' => "01010001",
      'R' => "01010010",
      'S' => "01010011",
      'T' => "01010100",
      'U' => "01010101",
      'V' => "01010110",


      'W' => "01010111",
      'X' => "01011000",
      'Y' => "01011001",
      'Z' => "01011010");

-------------------------------------------------------------------------
--  Global signals
-------------------------------------------------------------------------
  -- Signal driven by the TransmitData_Lbl process to monitor
  signal SentData_s      : character;

  -- signal driven by the Receive2File_Lbl process to monitor
  signal ReceivedData_s  : character;

-----------------------------------------------------------------------
--  Function: IsAtoZ determines if a character is in the range of 'A' to 'Z'
-----------------------------------------------------------------------
  function IsAtoZ(Cin_c    : Character) return boolean;

-------------------------------------------------------------------------
--  Function BV2AscII converts a bit vector to ASCII
--  Note: Not all the characters are implemented in this function
-------------------------------------------------------------------------
  function BV2AscII(BV8_c : Word_Typ) return Character;

-------------------------------------------------------------------------
--  Random number beteween 0 and 2^^16.  The Modulus determines the upper limit
-------------------------------------------------------------------------
  procedure Random_Number (variable Seed_v    : inout natural;
                           constant Modulus_c : in    T16K_Typ);

end UART_Pkg;
```

Figure 12.2.1-1 ***UART*** **Package Declaration (ch12_dir*UART*_p.vhd)**

```
--   Title         :  ASCII conversions
--   Description :  Routines for UART
-- Compile into library ATEP_Lib
package body UART_Pkg is
-----------------------------------------------------------------------------
--  Function: IsAtoZ determines if a character is in the range of 'A' to 'Z'
-----------------------------------------------------------------------------
  function IsAtoZ(Cin_c   : Character) return boolean is
    begin
      return (Cin_c > '@' and Cin_c < '[');
    end IsAtoZ;

  function BV2AscII(BV8_c : Word_Typ) return Character is
    begin
      case BV8_c is
        when CR_c         => return CR;
        when SP_c         => return ' ';
        when LF_c         => return LF;
        when "01000001"   => return 'A';    -- 41
        when "01000010"   => return 'B';
        when "01000011"   => return 'C';
        when "01000100"   => return 'D';
        when "01000101"   => return 'E';
        when "01000110"   => return 'F';
        when "01000111"   => return 'G';
        when "01001000"   => return 'H';
        when "01001001"   => return 'I';
        when "01001010"   => return 'J';
        when "01001011"   => return 'K';
        when "01001100"   => return 'L';
        when "01001101"   => return 'M';
        when "01001110"   => return 'N';
        when "01001111"   => return 'O';
        when "01010000"   => return 'P';
        when "01010001"   => return 'Q';
        when "01010010"   => return 'R';
        when "01010011"   => return 'S';
        when "01010100"   => return 'T';
        when "01010101"   => return 'U';
        when "01010110"   => return 'V';
        when "01010111"   => return 'W';
        when "01011000"   => return 'X';
        when "01011001"   => return 'Y';
        when "01011010"   => return 'Z';
        when others       => return '?';
      end case;                            --  BV8_c
    end BV2AscII;
-----------------------------------------------------------------------------
--  Random number beteween 0 and 2^^16
-----------------------------------------------------------------------------
  procedure Random_Number (variable Seed_v    : inout natural;
                           constant Modulus_c : in    T16K_Typ) is
    constant Multiplier_c  : integer := 25173;
    constant Increment_c   : integer := 13849;
  begin
    Seed_v :=  (Multiplier_c * Seed_v + Increment_c) Mod Modulus_c;
  end  Random_Number;
end UART_Pkg;
```

Figure 12.2.1-2 *UART* Package Body (ch12_dir*UART*_b.vhd)

12.2.2 Transmit Protocol

The transmit protocol is accomplished in a component that accepts a character from the Command process, and is responsible for transferring that character to the Transmit *UART* component using the *UART* processor interface protocol. This component also alerts the Command process when the *UART* has completed the transaction. Figure 12.2.2 represents the VHDL code for the transmit protocol component.

```
--   File name   :  xmtprtcl.vhd
--   Title       :  Transmit Bus Protocol Entity
--   Description :  Component accepts a character from the Command process
--               :  and is responsible for transferring that character to the
--               :  Transmit UART using the UART processor interface protocol.
--               :  Component also alerts the Command process when the UART has
--               : completed the transaction.
---------------------------------------------------------------------------------
library ATEP_Lib;
  use ATEP_Lib.UART_Pkg.all;   use ATEP_Lib.UART_Pkg;
entity XmtProtocol is
  generic (ProcDelay_g:          time := 100 ns);
  port (Shift_LdF      : out     bit;  -- Transmit UART interface (I/F)
        ClkEnbT        : out     bit;  -- Transmit UART I/F
        DataT          : out     Bit_Vector(7 downto 0); -- to Transmit UART
        ResetF         : out     bit;  -- Transmit UART I/F
        XmitMT         : in      boolean; -- Transmit UART I/F

        Clk            : in      bit;  -- Testbench I/F
        ReseT_s        : in      boolean;  -- Command Process I/F
        Char2Bus_s     : in      Character; -- Command Process I/F
        DoneXmit_s     : out     boolean); -- Command Process I/F
end XmtProtocol;

architecture XmtProtocol_a of XmtProtocol is
begin  --  XmtProtocol
  -------------------------------------------------------------------------------
  --  Process:    TransmitData_Lbl
  --  Purpose:    Read character from Command Process and send the converted
  --              data (ASCII to binary) to the UART Transmitter.
  --              Note that a subset of the Alphabet is acceptable
  --              is this model (A to Z,
  --              ' ', and CR.  All other characters are to be converted to '?'
  -------------------------------------------------------------------------------
  TransmitData_Lbl : process
    -----------------------------------------------------------------------------
    --  Procedure: SendChar
    --  Purpose:   Sends a converted character to the transmit UART
    --             Procedure provides the control signals.
    --             This procedure has NO side effects.
    --  Inputs:    BV8In_c       :  The input vector to be sent
    --             Delay_c       :  The data is the data and controls
    --             Ready2Xmit_s  :  The Ready to transmit signal
    --             Clk_s         :  The 1 X clock
    --  Outputs:   Data_s        :  The data signal to the transmit UART
    --             LdF_s         :  The load control signal to the UART
    -----------------------------------------------------------------------------
    procedure SendChar
           (constant BV8In_c      : in    character;
            constant Delay_c      : in    time;
            signal   Ready2Xmit_s : in    boolean;
            signal   Clk_s        : in    Bit;
            signal   Data_s       : out   Bit_Vector(7 downto 0);
            signal   LdF_s        : out   Bit) is
    begin  --  procedure SendChar
        -- Wait unitl UART is ready to send
```

```
        -- (required for startup or if UART conditions change)
        if not Ready2Xmit_s then
          wait until Ready2Xmit_s;
        end if;

        -- Send character to UART
        LdF_s <= '0' after Delay_c;
        case BV8In_c is
          when CR           =>
            Data_s  <= UART_Pkg.CR_c after Delay_c;
          when ' '          =>
            Data_s  <= UART_Pkg.SP_c after Delay_c;
          when 'A' to 'Z'   =>
            Data_s  <= UART_Pkg.Asc2Bit_c(BV8In_c) after Delay_c;
          when others       =>
            Data_s  <= QMark_c after Delay_c;
        end case;                             -- BV8In_c

        wait until Clk_s'event and Clk_s = '1';
        LdF_s <= '1' after Delay_c;

        -- Alert the Verifier that a character was sent
        UART_Pkg.SentData_s <= BV8In_c;
        wait until Ready2Xmit_s; -- Allow LdF_s to propagate
    end SendChar;

  begin  -- TransmitData_Lbl
    -- Assume that command process is synchronous to the
    -- the rising edge of clock
    -- Wait for a new character from the command process
    Shift_LdF <= '1'; -- for proper startup


    wait on Char2Bus_s'transaction; -- Wait for a new character to send

    SendChar (BV8In_c        => Char2Bus_s,  -- the character to send
              Delay_c        => ProcDelay_g, -- processor delay
              Ready2Xmit_s => XmitMT,        -- UART Transmitter eMpTy flag
              Clk_s          => Clk,         -- clock
              Data_s         => DataT,       -- UART DATA input from processor
              LdF_s          => Shift_LdF);  -- UART Load control

    -- Create a short pulse for the Done with the transmission.
    -- Short pulse is easy to debug with timing waveforms.
    DoneXmit_s  <= true, false after 1 ns;
  end process TransmitData_Lbl;

  -- Concurrent signal assignment for the reset signal
  ResetF <=          '0' after ProcDelay_g when ReseT_s
              else '1' after ProcDelay_g;

  -- Concurrent signal assignment for clock enable
  ClkEnbT  <= '0' after 100 ns,
              '1' after 200 ns;
end XmtProtocol_a;
```

Figure 12.2.2 VHDL Code for the Transmit Protocol Component, (ch12_dir\xmtprtcl.vhd)

12.2.3 Receive Protocol Component

The Receive protocol component interfaces with the *UART* receiver, thus emulating a processor interface to the receive *UART*. This component detects the receipt of data and passes the converted value to a receive process located in the testbench. Figure 12.2.3 represents the VHDL code for the receive protocol component.

```
--    Title         :  Receiver Protocol
--    Description :  This model interfaces with the UART receiver
--                  :  thus emulating a processor interface.
--                  :  This component detects the receipt of data
--                  :  and passes the converted value to a receive process
--                  :  located in the testbench
-------------------------------------------------------------------------------
library ATEP_Lib;
  use ATEP_Lib.UART_Pkg.all;
  use ATEP_Lib.UART_Pkg;

entity RxProtocol is
  generic (RxDelay_g      : time := 100 ns);
  port (ReseT_s          : in      boolean; -- Reset command to be passed out
        RxChar_s         : out     character; -- Received character from RxUART

        DataRdyT         : in      boolean;  -- Data Ready from RxUART
        RxData           : in      Bit_Vector(7 downto 0); -- Data from RxUART
        ResetF           : out     bit);  -- Reset to RxUART
end RxProtocol;


architecture RxProtocol_a of RxProtocol is

begin
  ------------------------------------------------------------------------------
  --  Process:    ReceiveChar_Lbl
  --  Purpose:    Receive a character and pass it to the receive process
  --          :   of the testbench and to a global signal for
  --          :   verification of proper reception.
  ------------------------------------------------------------------------------
  ReceiveChar_Lbl : process
-------------------------------------------------------------------------
  --  Procedure: ReceiveChar
  --  Purpose:    Receive a Character and send it to the component interface.
  --              When the receive UART indicates a data ready, this
  --              procedure converts the received bit vector data to ASCII
  --              and sends it to the component interface.
  --              It also sends the character to a global signal to be
  --              processed by a verifier.
  --  Inputs:     BV8In_c       : The 8 bit data from the Receive UART
  --              DataRdyT_s    : The ready signal from the Receive UART
  --  Outputs:    Char_s        : To port or component
  --              RcvData_s     : To Global signal for verifier
  ---------------------------------------------------------------------------
```

```
    procedure ReceiveChar
           (signal    BV8In_s       : in    Bit_Vector(7 downto 0); -- Data in
            signal    DataRdyT_s    : in    boolean;  -- Data ready handshake
            signal    Char_s        : out   character;  -- To port or component
            signal    RcvData_s     : out   character) is -- to Global signal

      variable Char_v : Character;
    begin
      -- wait for a character to be received at the UART
      wait until DataRdyT_s;
      Char_v := UART_Pkg.BV2AscII(BV8In_s); -- convert received
                                             -- Bit vector to ASCII
      Char_s <= Char_v;  -- send character to output of procedure
      RcvData_s  <= Char_v; -- send character (to verifier)
    end ReceiveChar;

  begin  --  process ReceiveChar_Lbl
      ReceiveChar
           (BV8In_s        => RxData,    -- Receive data from UART
            DataRdyT_s     => DataRdyT, -- Data ready from UART
            Char_s         => RxChar_s,    -- Received character
            RcvData_s      => UART_Pkg.ReceivedData_s); -- Char to verifier

  end process ReceiveChar_Lbl;

  -- Concurrent signal assignment for the reset signal
  ResetF <=          '0' after RxDelay_g when ReseT_s
              else '1' after RxDelay_g;

end RxProtocol_a;
```

Figure 12.2.3 VHDL code for the Receive Protocol Component (ch12_dir\rcvprtcl.vhd)

12.2.4 Transmission Line Component

This component models the transmission line jitter. This component delays the serial input transitions by a pseudorandom delay defined in a generic. A configuration declaration can be used to modify the value of this generic. A transport delay is used to emulate the transmission line. Figure 12.2.4 represents the VHDL code for transmission line component.

```
--   Title        :  Transmission line
--   Description :  Model of transmission line data jitter

library ATEP_Lib;
  use ATEP_Lib.UART_Pkg.all;
  use ATEP_Lib.UART_Pkg;


entity TransLine is
  generic (LinkJitter_g : natural := 10); -- transmission link jitter (in us)
  port (SerialIn       : in      bit;
        SerialOut      : out     bit);
end TransLine;
```

```
architecture TransLine_a of TransLine is

begin  --  TransLine_a
-------------------------------------------------------------------------------
  --  Process:   DelayLine
  --  Purpose:   Provides a random delay on the serial Output of the
  --             UART transmitter.
  --  Inputs:    Serial_Ou
  --  Outputs:   SerialDlyOut
  -------------------------------------------------------------------------------
  DelayLine : process (SerialIn)
    variable Seed_v      : natural := 101;
  begin  --  process DelayLine
      -- Compute new delay
      UART_Pkg.Random_Number
        (Seed_v    => Seed_v,
         Modulus_c => LinkJitter_g);

      SerialOut <= transport SerialIn after Seed_v * 1 us;
  end process DelayLine;
end TransLine_a;
```

Figure 12.2.4 VHDL Code for Transmission Link Component (ch12_dir\trnsline.vhd)

12.2.5 Monitor or Verifier Component

The verifier is used to detect errors in the data transmission between the transmitter and receiver. The verifier performs the following functions:

1. Automatically verifies that the transmitted character matches the received character.
2. Detects that a message is received after it is sent.
3. Detects when a message is received without being sent.

Figure 12.2.5 represents the VHDL code for the verifier.

```
--   Title       :  Monitor for UART project
--   Description :  Alerts if transmitted data is different than received data
entity Monitor is
  generic (WordDelay_g:          time := 100 us * 11); -- 1 us of extra margin
                                                        -- 10 bits/word + 1 margin

end Monitor;

library ATEP_Lib;
  use ATEP_Lib.UART_Pkg.all;
  use ATEP_Lib.UART_Pkg;

architecture Monitor_a of Monitor is
  signal DataTransmitted_s : boolean := false;
begin  --  Monitor_a
  ------------------------------------------------------------------------------
  --  Process:   Test_Lbl
  --  Purpose:   Verifies that transmitted data is same as received data
  --  Inputs:    UART_Pkg.SentData_s,  UART_Pkg.ReceivedData_s
  --  Outputs:   Assert statement
  ------------------------------------------------------------------------------
  Test_Lbl : process
    variable SentData_v : character;
  begin  --  process Test_Lbl
    wait on UART_Pkg.SentData_s'transaction; -- new character sent
    SentData_v := UART_Pkg.SentData_S; -- store sent character
    DataTransmitted_s <= true;   -- flag to other process

    wait on UART_Pkg.ReceivedData_s'transaction for WordDelay_g;
    DataTransmitted_s <= false;
    if (UART_Pkg.ReceivedData_s'active) then
      assert SentData_v = ReceivedData_s
        report "Sent Data is /= Received data"
        severity warning;
    else
      assert false
        report "Sent data was not received"
        severity warning;
    end if;
  end process Test_Lbl;

  ------------------------------------------------------------------------------
  --  Process:   Test2_Lbl
  --  Purpose:   Verifies that Data was not received without
  --              being transmitted.
  --  Inputs:    UART_Pkg.SentData_s,  UART_Pkg.ReceivedData_s
  --  Outputs:   Assert statement
  ------------------------------------------------------------------------------
  Test2_Lbl : process
    variable SentData_v : character;
  begin  --  process Test_Lbl
    wait on UART_Pkg.ReceivedData_s'transaction;
    if not DataTransmitted_s then
      assert false
        report "Received Data without the character being transmitted"
        severity warning;
    end if;
  end process Test2_Lbl;
end Monitor_a;
```

Figure 12.2.5 VHDL Code for the Verifier (ch12_dir\monitor.vhd)

12.2.6 Testbench Entity and Architecture

The testbench components incorporates all the elements required to verify the operation of the *UART* transmitter and receiver. Figure 12.2.6-1 represents the VHDL code for the *UART* testbench. Figures 12.2.6-2 through 12.2.6-5 represent some simulation results.

```
--   Title        :  TestBench for UART Transmitter & receiver
--   Description :  Testbench with File IO fro data sent
library ATEP_Lib;
  use ATEP_Lib.UART_Pkg.all;
  use ATEP_Lib.UART_Pkg;

  use Std.TextIO.all;
  use Std.TextIO;

entity UARTBench is
  generic (Ck16Jitter_g     : time := 0 ns);    -- 16X clock jitter
end UARTBench;


architecture UARTBench_a of UARTBench is
  constant FastDelay_c : time := 100 ns;
------------------------------------------------------------------------------
--  Signal names are same as component port name to ease readability
--  (1 to 1 name association)
------------------------------------------------------------------------------
  signal BitClkT       :  bit;
  signal Char2Bus_s    :  character;
  signal ClkEnbT       :  bit := '1';
  signal Clk           :  bit;
  signal Clk16xT       :  bit;
  signal DataRdyT      :  boolean;
  signal DoneXmit_s    :  boolean;
  signal RxData        :  Bit_Vector(7 downto 0);
  signal RxResetF      :  bit;  -- receive Reset control
  signal RxReset_s     :  boolean;
  signal RxChar_s      :  character;
  signal SerialDlyOut  :  bit;
  signal Shift_LdF     :  bit;
  signal TxDataT       :  Bit_Vector(7 downto 0);
  signal TxResetF      :  bit;  -- transmit Reset control
  signal TxReset_s     :  boolean;
  signal TxSerial_OuT  :  bit;
  signal XmitMT        :  boolean;

component UARTXmt
  port (Shift_LdF      : in      bit;
        ClkEnbT        : in      bit;
        Clk            : in      bit;
        DataT          : in      Bit_Vector(7 downto 0);
        ResetF         : in      bit;
        Serial_OuT     : out     bit;
        XmitMT         : out     boolean);
end component;

component UARTRx
  port (Clk16xT        : in      bit;
        ResetF         : in      bit;
        Serial_InT     : in      bit;
        DataRdyT       : out     boolean;
        DataOuT        : out     Bit_Vector(7 downto 0);
        BitClkT        : out     bit);
end component;
```

Signal declarations

Component declarations. If there are too many components, they could be declared in a package. A **use** of that package will then be required.

```
component Monitor
end component ;                     <---- This component has no ports.
                                          It uses global signals.

component TransLine
  port(
        SerialIn              : in        bit;
        SerialOut             : out       bit);
end component;

component XmtProtocol
  port(
        Shift_LdF             : out       bit;
        ClkEnbT               : out       bit;
        DataT                 : out       Bit_Vector(7 downto 0);
        ResetF                : out       bit;
        XmitMT                : in        boolean;

        Clk                   : in        bit;
        ReseT_s               : in        boolean;
        Char2Bus_s            : in        Character;
        DoneXmit_s            : out       boolean);
end component;

component RxProtocol
  port(
        ReseT_s               : in        boolean;
        RxChar_s              : out       character;

        DataRdyT              : in        boolean;
        RxData                : in        Bit_Vector(7 downto 0);
        ResetF                : out       bit);
end component;

begin  -- UARTBench
  ------------------------------------------------------------------------
  -- Component Instantiation
  ------------------------------------------------------------------------
  U1_UARTXmt: UARTXmt
  port map
    (Shift_LdF      => Shift_LdF,
     ClkEnbT        => ClkEnbT,
     Clk            => Clk,
     DataT          => TxDataT,
     ResetF         => TxResetF,
     Serial_OuT     => TxSerial_OuT,
     XmitMT         => XmitMT);

  U1_UARTRx: UARTRx
  port map
    (Clk16xT        => Clk16xT,
     ResetF         => RxResetF,
     Serial_InT     => SerialDlyOut, -- Random delayed Serial_Out
     DataRdyT       => DataRdyT,
     DataOuT        => RxData,
     BitClkT        => BitClkT);

  U1_Monitor: Monitor;

  U1_TransLine: TransLine
  port map (
        SerialIn                => TxSerial_Out,
        SerialOut               => SerialDlyOut);
  U1_XmtProtocol: XmtProtocol
  port map (
        Shift_LdF               => Shift_LdF,
        ClkEnbT                 => ClkEnbT,
        DataT                   => TxDataT,
```

```
      ResetF                  => TxResetF,
      XmitMT                  => XmitMT,

      Clk                     => Clk,
      ReseT_s                 => TxReset_s,
      Char2Bus_s              => Char2Bus_s,
      DoneXmit_s              => DoneXmit_s);


U1_RxProtocol: RxProtocol
port map (
      ReseT_s                 => RxReset_s,
      RxChar_s                => RxChar_s,

      DataRdyT                => DataRdyT,
      RxData                  => RxData,
      ResetF                  => RxResetF);


-------------------------------------------------------------------------------
--  Clock definition and Clock Enable
-------------------------------------------------------------------------------
Clk <= not Clk after 50 us;
Clk16xT  <= not Clk16xT after (50 us + Ck16Jitter_g) / 16;


-------------------------------------------------------------------------------
--  Process:   TransmitData_Lbl
--  Purpose:   Read data from file "datain.txt" and send the character
--             to the UART Transmit Protocol component.
--             Note that a subset of the Alphabet is acceptable (A to Z,
--             ' ', and CR.  All other characters are to be converted to '?'
-------------------------------------------------------------------------------
TransmitData_Lbl : process
  file DataIn_f        : TextIO.text is in  "datain.txt";
  variable InLine_v    : TextIO.line;  -- pointer to string
  variable ClkBetweenChar_v : natural;
begin
  -- Force  TxResetF = '1';
  TxReset_s  <= true after FastDelay_c;
  Reset_Lbl  : for Lp_i in 1 to 2 loop
    wait until Clk'event and Clk = '1';
  end loop Reset_Lbl;
  TxReset_s  <= false after FastDelay_c;

  wait until Clk'event and Clk = '1';  -- wait for 1 cycle
  -- Read data from file
  File_Lbl  : while not TextIO.Endfile(DataIn_f) loop
    -- Read 1 line from the input file
    TextIO.ReadLine(DataIn_f, InLine_v);

    -- Test if InLine_v is an empty line, then write it (send a CR)
    if InLine_v'length = 0 then                -- null line
      -- Send to UART CR_c
      Char2Bus_s  <= CR;
      wait on DoneXmit_s'transaction until DoneXmit_s;

    else      -- Line is not empty
      -- Scan the characters in the line
      -- Check if character is A to Z, or ' ' else send a ?
      Line_Lbl  : for Index_i in InLine_v'low to InLine_v'high loop
        if UART_Pkg.IsAtoZ(InLine_v(Index_i))  then
          Char2Bus_s <= InLine_v(Index_i);
        elsif InLine_v(Index_i) = ' ' then -- space
          Char2Bus_s  <= ' ';
        else
          Char2Bus_s <= '?';
        end if;
```

```
          wait on DoneXmit_s'transaction until DoneXmit_s;
          UART_Pkg.Random_Number
            (Seed_v    =>  ClkBetweenChar_v,
             Modulus_c =>  6);
          DeadTime_Lbl: for T_i in 1 to ClkBetweenChar_v loop
            wait until Clk'event and Clk = '1';
          end loop DeadTime_Lbl;
        end loop Line_Lbl;

        -- Now send a Carriage return for end of line
        Char2Bus_s <= Cr;
        wait on DoneXmit_s'transaction until DoneXmit_s;
      end if;
    end loop File_Lbl;

    assert false
      report "end of file reached, No more data"
      severity note;

    wait; -- to prevent process from repeating
  end process TransmitData_Lbl;

  ------------------------------------------------------------------------------
  --   Process:  Receive2File_Lbl
  --   Purpose:  This process emulates the interface to the RECEIVE bus
  --             protocol.  It accepts a  character from the receive
  --             protocol component and sends it to output
  --             file "dataout.txt".
  ------------------------------------------------------------------------------
  Receive2File_Lbl : process
    file      DataOut_f   : TextIO.text is out   "dataout.txt";
    variable OutLine_v    : TextIO.line;  -- pointer to string
    variable OutLine2_v  : TextIO.line;  -- for display to transcript
  begin
    RxReset_s   <= true after FastDelay_c;
    Reset_Lbl  : for Lp_i in 1 to 2 loop
      wait until Clk'event and Clk = '1';
    end loop Reset_Lbl;
    RxReset_s   <= false after FastDelay_c;

    AfterReset_Lbl : loop
      wait on RxChar_s'transaction; -- receipt of a new character.
      if RxChar_s = CR then
        -- Create a new pointer to point to a new string
        OutLine2_v := new string(OutLine_v'low to OutLine_v'high);

        -- Must copy the data, but not the pointers
        OutLine2_v.all := OutLine_v.all;

        -- OutLine2_v is deallocated after the write of the line
        TextIO.WriteLine(output, OutLine2_v);      -- to screen

        -- Write Outline to file
        TextIO.Writeline(DataOut_f, OutLine_v);
      else
        TextIO.Write(Outline_v, RxChar_s);
      end if;
    end loop AfterReset_Lbl;
  end process Receive2File_Lbl;
end UARTBench_a;
```

Figure 12.2.6-1 VHDL Code for the *UART* Testbench (ch12_dir*UART*tb.vhd)

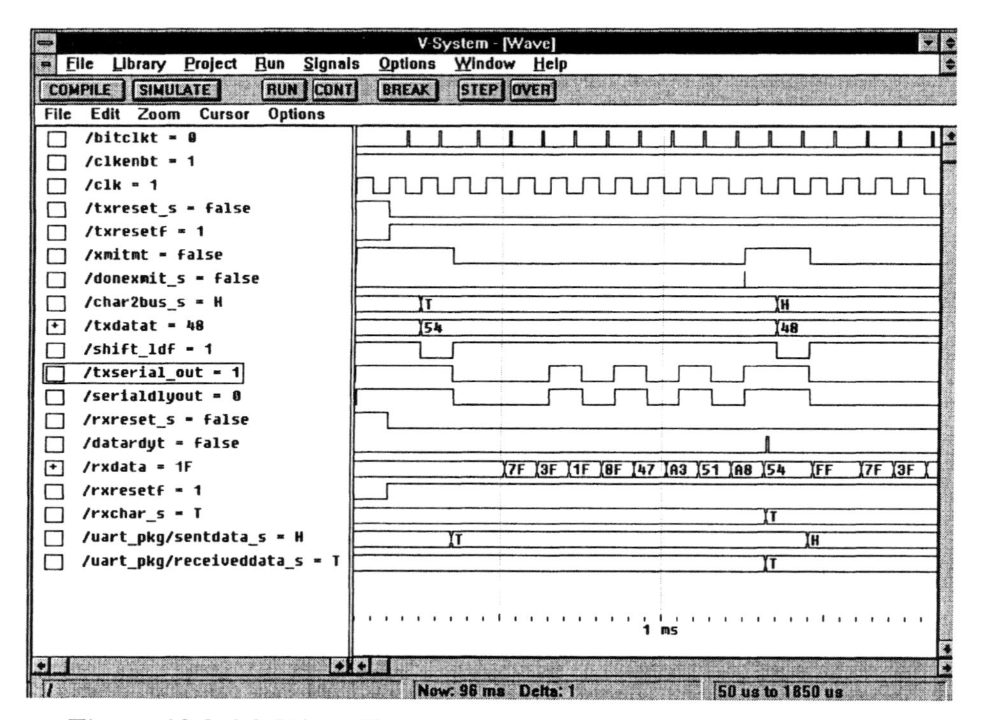

Figure 12.2.6-2 ***UART*** **Testbench Simulation -- Transfer of 1 Character**

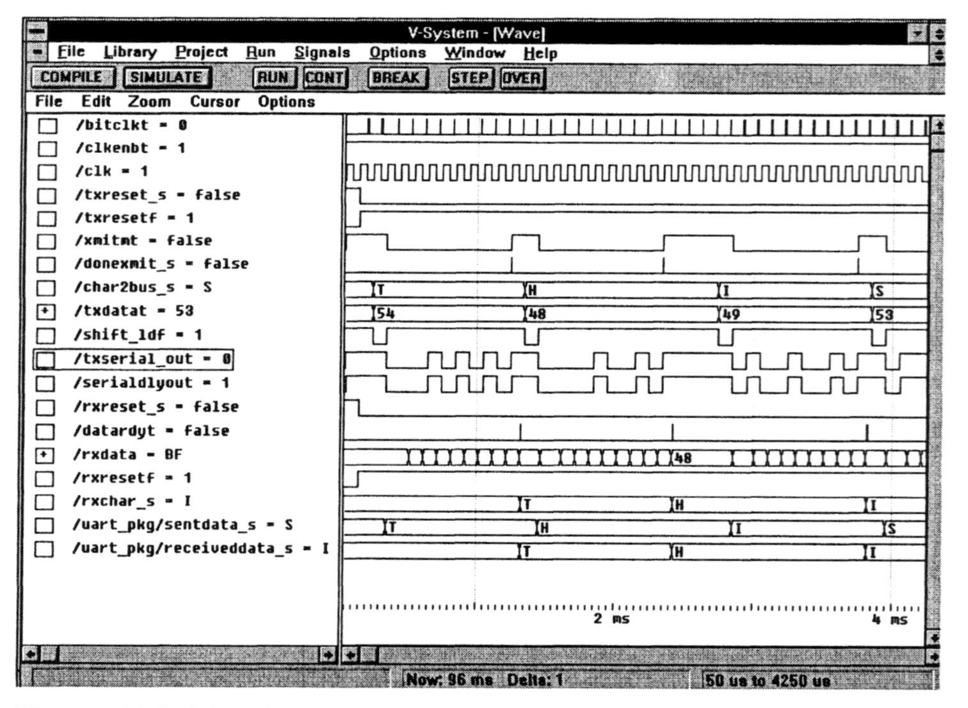

Figure 12.2.6-3 ***UART*** **Testbench Simulation -- Transfer of 3 Characters**

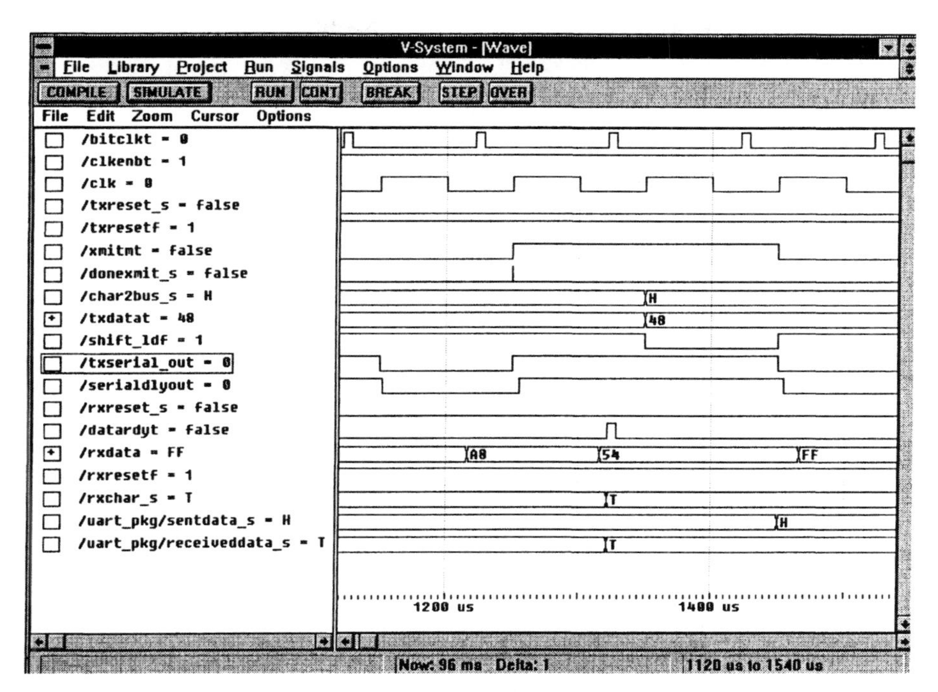

Figure 12.2.6-4 *UART* Testbench Simulation -- Receipt of 1 Character

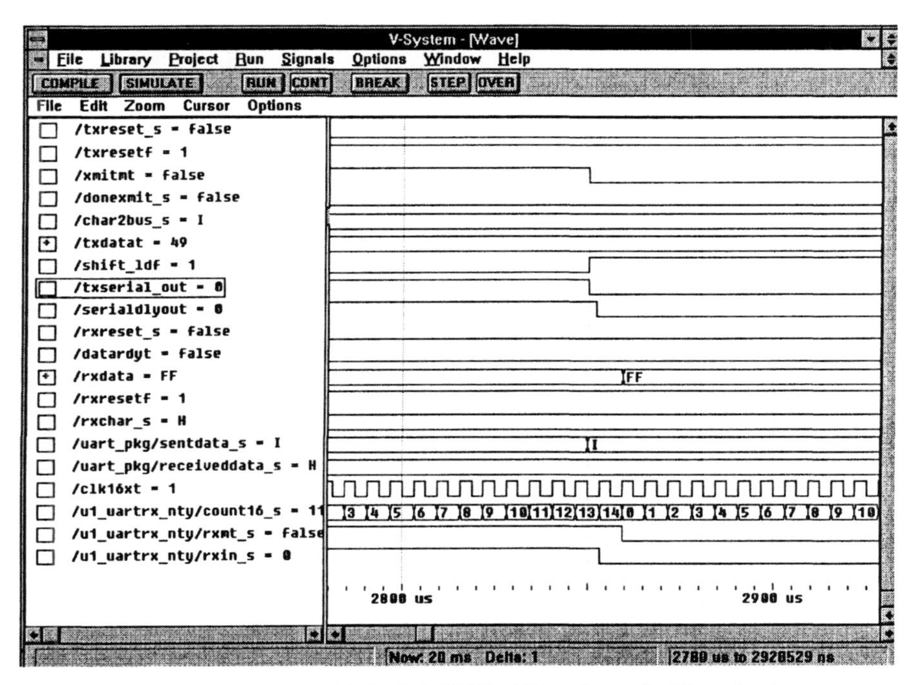

Figure 12.2.6-4 *UART* Testbench Simulation

-- Resetting of Receive 16X Counter

12.2.7 Configuration

A configuration can be used to modify the value of generics or to modify the selection of an architecture. Figure 12.2.7-1 represents a configuration of the testbench to modify the value of the delay generic in the transmission line jitter. Figure 12.2.7-2 represents the simulation for this configuration. Note the transmission errors caused by this excessive data line jitters.

```
--   Title       :  UART Testbench Configuration
--   Description :  Configuration for transmission line delay

library ATEP_lib;

configuration UARTLinkError_Cfg of UARTBench is
  for UARTBench_a
    for U1_TransLine: TransLine use
      entity ATEP_lib.TransLine(TransLine_a)
        generic map
          (LinkJitter_g  => 75);  -- set to 1 for normal (see file UARTok_c.vhd)
    end for;
  end for;
end UARTLinkError_Cfg;
```

Figure 12.2.7-1 Configuration of the Testbench (ch12_dir\UARTtb_c.vhd)

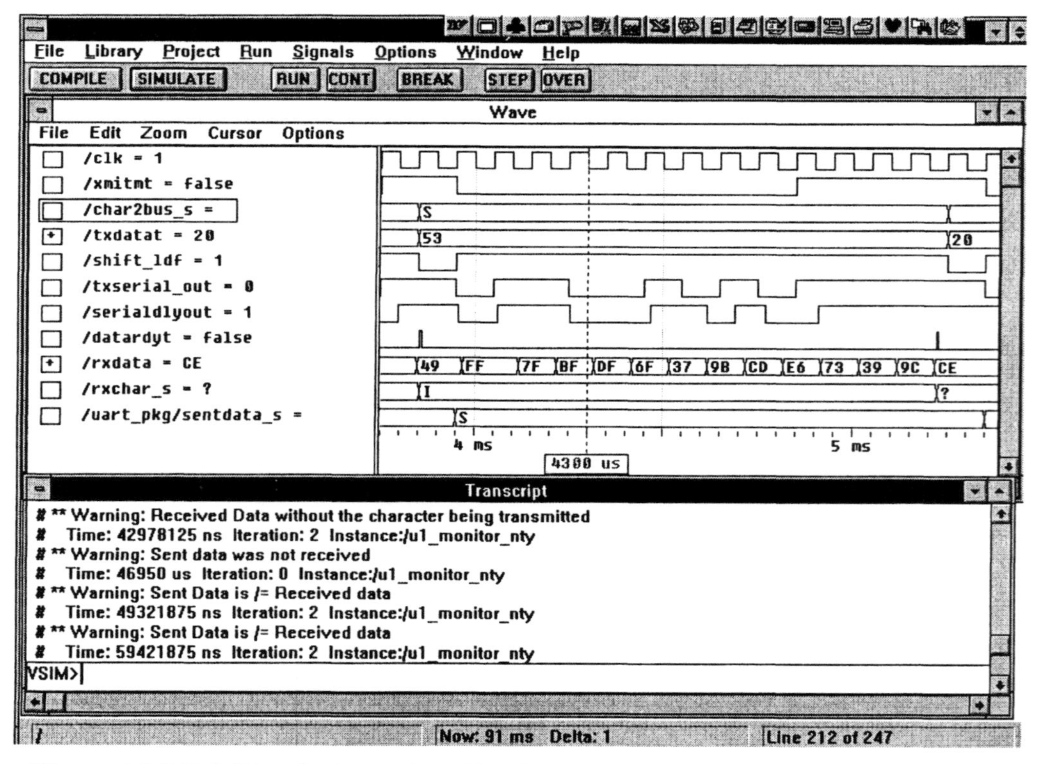

Figure 12.2.7-2 Simulation using Configuration for Excessive Transmission Jitter.

EXERCISES

1. Modify the transmit and receive *UART* components to include 2 STOP bits. Rerun testbench and verify your results.

2. Modify the transmit and receive *UART* components to include odd parity on the data. Modify the testbench. Rerun the testbench and verify your results.

3. A real *UART* device includes a transmit Hold register that interfaces with the *CPU* and loads an 8-bit word under the control of the processor. When the transmit shift register is empty, it loads data from the transmit hold register if that register has a value. Modify the architecture of the transmit *UART* to incorporate this transmit HOLD register. Rerun the testbench and verify your results.

4. A real *UART* device includes a receive HOLD register to hold the value of the received data. Data passed to the *CPU* is read from this receive HOLD register. Modify the architecture of the receive *UART* to incorporate this receive HOLD register. Rerun the testbench and verify your results.

13. VITAL

The purpose of this chapter is to introduce the reader to the concepts of VHDL Initiative Toward *ASIC* Libraries (VITAL). It is not intended as a coding guide for *VITAL*. *VITAL* coding styles, package descriptions, and specifications are described in a document entitled *VITAL Model Development Specification.* This chapter is based on *VITAL* Version 2.2b of the specification. The specification was updated in 1995. The latest version of the specification and packages can be downloaded from the Internet site *http://www.vhdl.org/vi/vital/.* The documents supporting *VITAL* version 2.2b include:

vital22b.ps	VITAL Model Development Specification, Version 2.2b
coverltr.ps	Cover letter
appndx_a.ps	Appendix A: Vital_Timing Package
appndx_a.ps	Appendix B: Vital_Primitives Package
appndx_c.ps	Appendix C: Vital-SDF Mapping
appndx_d.ps	Appendix D: Unresolved Issued
timing_p.vhd	VHDL Vital_Timing Package Declaration
timing_b.vhd	VHDL Vital_Timing Package Body
prmtvs_p.vhd	VHDL Vital_Primitives Package Declaration
prmtvs_b.vhd	VHDL Vital_Primitives Package Body

Version 2.2b packages are supplied on disk in the subdirectory "vital" to enable the compilation of the exercises. However, the user is encouraged to download the latest version.

13.1 VITAL

13.1.1 Overview

The VHDL Initiative Towards *ASIC* Libraries (*VITAL*) is an industry-based, informal consortium formed with the following in mind:

- **Charter**: Accelerate the availability of *ASIC* libraries across industry VHDL simulators.

- **Objective**: High-performance, accurate (sign-off quality) *ASIC* simulation across VITAL-compliant Electronic Design Automation (EDA) tools from a single *ASIC* vendor description.

- **Approach**: Define a modeling specification (in conjunction with VHDL packages) that leverages existing practices and techniques, is compliant with IEEE Standards 1076 and 1164, and utilizes Open Verilog International's (*OVI*) Standard Delay Format (*SDF*) timing format. Standardize the approved result through the IEEE.

- **End-Product**:
  (1) *VITAL_Timing* VHDL package defining standard, acceleratable timing procedures for delay value selection, timing checks, and timing error reporting;

  (2) *VITAL_Primitives* VHDL package defining standard, acceleratable primitives for Boolean and table-based functional description;

  (3) Specification of *OVI*'s Standard Delay Format (*SDF*) for communication of instance delay values; and,

  (4) Model Development Specification documents defining utilization of *VITAL* and VHDL elements for *ASIC* library development.

13.2 VITAL FEATURES

The main features of VITAL (based on version 2.2B) include the following:

1. **Modeling specification**. This specification covers:

- **Naming conventions** for the timing parameters and internal signals.
- **Use of types** defined in the **timing package** to specify timing parameters.
- **Coding style methodologies to define and back annotate** timing parameters.
- **Two Coding styles** that accommodate acceleratable models:
  . **Pin-to-pin delay** style with a single process to describe the behavior.
  . **Distributed delay** style with use of predefined concurrent procedures.
- **Two levels of compliance**:
  - Level 0: Complex models described at higher level,
  - Level 1: Model acceleration permitted.

2. **VITAL timing package**. This package contains data types and subprograms to support development of macrocell timing models. Included in this package are routines for delay selections, timing violations checking and reporting, and glitch detection.

3. **VITAL_Primitives package**. This package includes a set of commonly used combinatorial primitives defined as functions and concurrent procedures to support behavioral or structural modeling styles. Examples of *VITAL* primitives include *VitalAND, VitalOR, Vital AND2, VitalMux4*, etc. The procedure primitives support separate pin-to-pin delay path and *GlitchOnEvent* glitch detection. In addition, this package contains general purpose Truth tables and state tables that are very useful in defining state machines and registers.

4. **VITAL_SDFmap**. The Standard Delay File (*SDF*) to VHDL mapping specification defines the mapping/translation of *SDF* data to Generic Parameter values on *VITAL* models. Simulator vendors use this mapping standard to build tools that automatically back-annotate the VHDL circuit and timing data values.

13.3 VITAL MODEL

A *VITAL* compliant representation of level 0 and level 1 models consists of an entity with generics defining the timing parameters of the ports. The types of the generics are defined in *VITAL_Timing* package and can be any of the following types and subtypes shown in Figure 13.3.

```
type     TransitionType           is ( tr01, tr10, tr0z, trz1, tr1z, trz0 ); -- {TVTG}
type     TransitionIOType         is ( trll, trlh, trhl, trhh );         -- (R&R)
type     TransitionArrayType      is array (TransitionType range <>) of time;
type     TransitionIOArrayType    is array (TransitionIOType range <>) of time;
subtype  DelayTypeXX              is time;
subtype  DelayType01              is TransitionArrayType (tr01 to tr10);
subtype  DelayType01Z             is TransitionArrayType (tr01 to trz0);
subtype  DelayTypeIO              is TransitionIOArrayType (trll to trhh);
type     DelayArrayTypeXX         is array (natural range <>) of DelayTypeXX;
type     DelayArrayType01         is array (natural range <>) of DelayType01;
type     DelayArrayType01Z        is array (natural range <>) of DelayType01Z;
type     DelayArrayTypeIO         is array (natural range <>) of DelayTypeIO ;
type     TimeArray                is array ( natural range <>) of time;
```

Figure 13.3 Vital (Version 2.2B) Types for Timing Parameters

The *VITAL* generic timing parameters are one of the following forms:

VITAL_Prefix
VITAL_Prefix_PortName
VITAL_Prefix_Port1Name_Port2Name_Condition_Edge1_edge2

The *VITAL* timing parameter prefixes are defined in table 13.3-1.

Table 13.3 VITAL Timing Parameter Prefixes

VITAL_Prefix	SIGNIFICANCE
tipd	**Interconnect Path Delay (IPD).** This is the delay from the input port to an internal signal that represents the input to the internal hardware.
tpd	**Propagation Delay.** This is the propagation delay between the output of a functional block and its interconnect signal or port.
tsetup	**Setup constraint.** This timing parameter is used to verify timing violations.
thold	**Hold constraint.** This timing parameter is used to verify timing violations.
trelease	**Release constraint** (like setup for asynchronous signals).
trecovery	**Recovery constraints** (like hold for asynchronous signals).
tperiod	**Period** where Min or Max is not specified.
tperiod_min	**Minimum period**
tperiod_max	**Maximum period**
tpw	**Minimum pulse width**
tdevice	subcomponent to which the specification applies
tskew	subcomponent to which the specification applies
tpulse	**Path pulse delay**

The ports of the entity use the *Std_Logic_1164* data type or subtypes. Array ports use the *Std_Logic_Vector* data types and may be constrained,
(e.g., x : Std_Logic_Vector(31 downto 0);)

The *VITAL* architecture for level 1 models can be written in one of two styles, but not both, including:

1. Pin-to-pin delay style
2. Distributed delay style

13.3.1 Pin-to-Pin Delay Modeling Style.

Figure 13.3.1 is a representation of VITAL pin-to-pin architecture, and includes the following:

1. **"Wire_Delay" Input path delay block.** A block called *Wire_Delay* is a requirement if the model used the *tipd* generics to define the input propagation delays between the input ports and internal signals. Those internal architectural signals are named *PortName_tipd.* The *Wire_Delay* block calls concurrent procedure *VitalPropagateWireDelay* defined in package *Vital_Timing*.

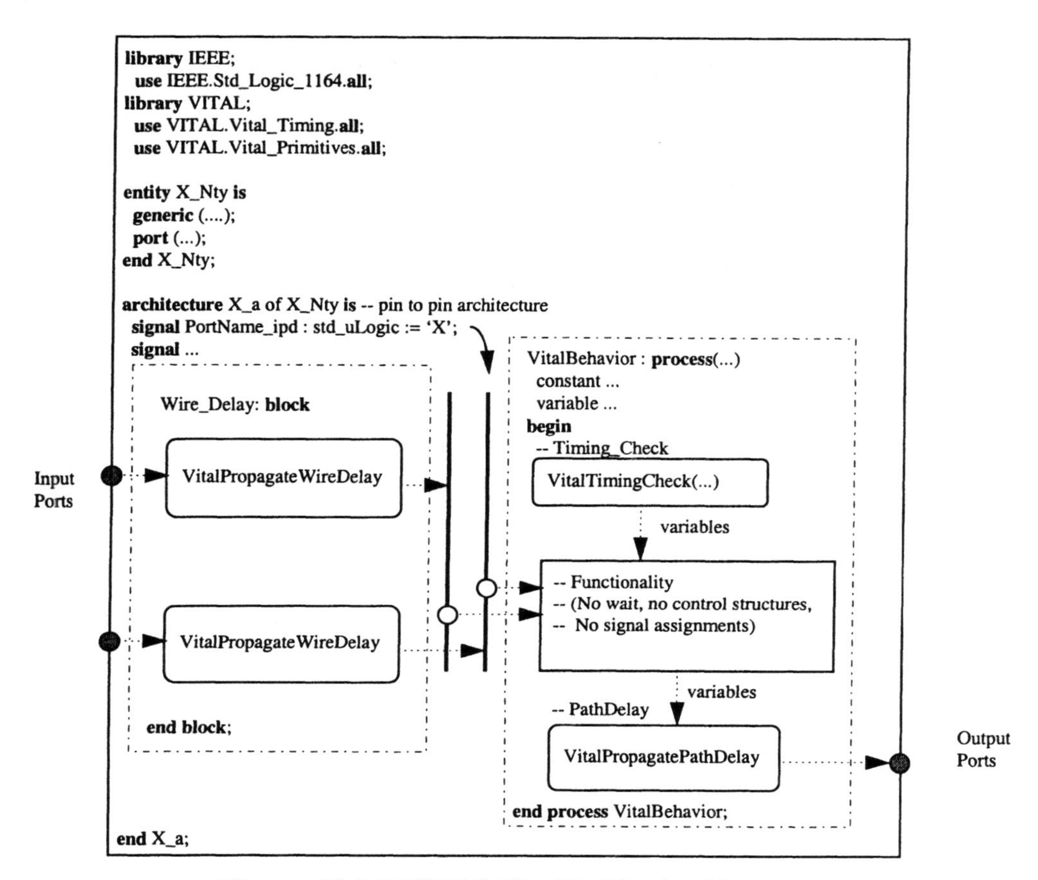

Figure 13.3.1 VITAL Pin-To-Pin Architecture

2. **"VitalBehavior" process.** This process, like any other VHDL process, includes a declaration section where constants, variables, and aliases are declared. Constants of type *VitalTruthTableType* or *VitalStateTableType* can be used to define truth tables for Boolean logic, or state tables for state machine definitions. Those constants are used as actual parameters to functions and procedures to compute the results based on the inputs and the previous states (for state machines). The *VitalBehavior* process structure is rigid, and is divided into three parts:

   1. Timing checks. These are procedure calls to the *VitalTimingCheck* procedure defined in package VITAL_Timing. A check on a user defined generic *TimingCheckOn* of type Boolean can be made with an *if* statement prior to calling the *VitalTimingCheck* procedure.

   2. Functionality. This section may contain one or more calls to subprograms contained in the *Vital_Primitives* package. It may also contain assignments to internal **temporary variables**. This process has **no wait statements**, and the sensitivity list includes a reference to all the inputs ports or their **_ipd* signal counterparts, but not both. **No control structures are allowed** (if, case, loop) in the process (except for one *if* statement in the timing checks section).

No signal assignments are allowed, outside of the path delay section. Data is passed between sections in the process using variables. Specifically, the violation section is only allowed to make assignments to violation flags.

3. Path Delay section. Each output signal is driven by a call to *VitalPropagatePathDelay* procedure. The *VitalPropagatePathDelay* routine determines the appropriate delay to use given that an output signal has changed value. It accomplishes this task by knowing when the input paths to the routine last changed their values. From that information, its picks the minimum delay of all possible active stimulus-response paths.

Figure 13.3.1 represents an example of a two-bit counter with synchronous reset using the pin-to-pin modeling style.

```
--   Title       :  VITAL PinToPin Modeling Style
--   Description :  Modeling of a synchronous 2 bit counter with a reset
--
-------------------------------------------------------------------------
library IEEE;
  use IEEE.Std_Logic_1164.all;

library VITAL;
  use VITAL.Vital_Timing.all;  -- VITAL packages compiled in library VITAL
  use VITAL.Vital_Timing;
  use VITAL.Vital_Primitives.all;
  use VITAL.Vital_Primitives;

entity Counter_Nty is
  generic(tipd_ResetF        : DelayType01 := (2 ns, 2 ns);
          tipd_Clk           : DelayType01 := (2 ns, 2 ns);
          tsetup_ResetF_Clk  : DelayType01 := (15 ns, 15 ns);
          thold_ResetF_Clk   : DelayType01 := (10 ns, 10 ns);
          tpd_Clk_Q0         : DelayType01 := (10 ns, 10 ns);
          tpd_Clk_Q1         : DelayType01 := (10 ns, 10 ns);
          TimingCheckOn      : boolean := true);
  port   (ResetF    : in    Std_Logic := 'U';
          Clk       : in    Std_logic := 'U';
          Count     : out   Std_logic_Vector(1 downto 0));

end Counter_Nty;



architecture PinToPin_a of Counter_Nty is
  attribute Vital_level1 of PinToPin_a : architecture is true;
  signal ResetF_ipd   : Std_uLogic := 'X';
  signal Clk_ipd      : Std_uLogic := 'X';

begin
  ---------------------------------------------------------------
  -- Input path delay
  ---------------------------------------------------------------
  Wire_Delay: block
  begin
   Vital_Timing.VitalPropagateWireDelay
      (OutSig     =>  ResetF_ipd,
       InSig      =>  ResetF,
       tWire      =>  VitalExtendToFillDelay(tipd_ResetF));


   Vital_Timing.VitalPropagateWireDelay
```

```
    (OutSig      =>  Clk_ipd,
     InSig       =>  Clk,
     tWire       =>  VitalExtendToFillDelay(tipd_Clk)
    );
end block;

----------------------------------
-- Behavior section
----------------------------------
VitalBehavior: process(Clk_ipd, ResetF_ipd)
  -- Timinf checks reseults
  variable Tviol_ResetF_Clk   : X01 := '0';
  variable violation          : X01 := '0';
  variable TimeMarkerResetfClk : Vital_Timing.TimeMarkerType
             := (RefTimeMarker    => - 500 ns,
                 HoldCheckPassed => null,
                 LockOutCheck     => null);

  variable Results   : Std_Logic_Vector(1 to 2);
  alias    Count1_zd : Std_Logic is Results(1);
  alias    Count0_zd : Std_Logic is Results(2);
  variable PrevData : Std_Logic_Vector(1 to 3);
  variable Count0_GlitchData   : VITAL_Timing.GlitchDataType;
  variable Count1_GlitchData   : VITAL_Timing.GlitchDataType;

  constant H : Vital_Primitives.VitalTableSymbolType := '1';
  constant L : Vital_Primitives.VitalTableSymbolType := '0';
  constant x : Vital_Primitives.VitalTableSymbolType := '-'; -- any match
  constant S : Vital_Primitives.VitalTableSymbolType := 'S'; -- same
  constant R : Vital_Primitives.VitalTableSymbolType := '/'; -- rising
  constant U : Vital_Primitives.VitalTableSymbolType := 'X';
  constant V : Vital_Primitives.VitalTableSymbolType := 'B'; -- valid CLK


  constant CounterState :
     Vital_Primitives.VitalStateTableType(1 to 8, 1 to 7) := (
    -- previous_data
    --               States
    --                      Results(next state)
    -- Viol Clk RSF  Q1 Q0  Q1 Q0
     ( U,    x,  x,  x, x,  U, U),  -- Timing Violation
     ( x,    R,  L,  x, x,  L, L),  -- Synchronous Reset
     ( x,    R,  H,  L, L,  L, H),  -- Count
     ( x,    R,  H,  L, H,  H, L),  -- Count
     ( x,    R,  H,  H, L,  H, H),  -- Count
     ( x,    R,  H,  H, H,  L, L),  -- Count
     ( x,    R,  H,  U, U,  U, U),  -- Count
     ( x,    V,  x,  x, x,  S, S)); -- Same, no change
```

```
  begin
    -- TimingCheck:
    if TimingCheckOn then
      Vital_Timing.VitalTimingCheck
         (TestPort        =>  ResetF_ipd,
          TestPortName    =>  "ResetF",
          RefPort         =>  Clk_ipd,
          RefPortName     =>  "Clk",
          t_setup_hi      =>  tsetup_ResetF_Clk(tr01),
          t_setup_lo      =>  tsetup_ResetF_Clk(tr10),
          t_hold_hi       =>  thold_ResetF_Clk(tr01),
          t_hold_lo       =>  thold_ResetF_Clk(tr10),
          CheckEnabled    =>  true,
          RefTransition   =>  Clk_Ipd = '1',
          HeaderMsg       =>  "Counter",
          TimeMarker      =>  TimeMarkerResetfClk,
          Violation       =>  Tviol_ResetF_Clk);
    end if;

  -- Functionality
  violation  := Tviol_ResetF_Clk;  -- Usually ORed with other violations
  Vital_Primitives.VitalStateTable
    (StateTable     => CounterState,
     DataIn         => (violation, Clk_ipd, ResetF_ipd),  -- inputs
     NumStates      => 2,  -- Number state variables (FF)
     Result         => Results,
     PreviousDataIn => PrevData);

  -----------------
  -- Path Delay Section
  -- PathDelay:
  Vital_Timing.VitalPropagatePathDelay
    (OutSignal      => Count(1),
     OutSignalName  => "Count_MSB",
     OutTemp        => Count1_zd, -- intermediate 0 delay (input)
     Paths(0).InputChangeTime => Clk_ipd'last_event,
     Paths(0).PathDelay       => VitalExtendToFillDelay(tpd_Clk_Q1),
     Paths(0).PathCondition   => true,
     GlitchData     => Count1_GlitchData,
     GlitchMode     => Xonly,
     GlitchKind     => OnEvent);

  Vital_Timing.VitalPropagatePathDelay
    (OutSignal      => Count(0),
     OutSignalName  => "Count_LSB",
     OutTemp        => Count0_zd, -- intermediate 0 delay (input)
     Paths(0).InputChangeTime => Clk_ipd'last_event,
     Paths(0).PathDelay       => VitalExtendToFillDelay(tpd_Clk_Q0),
     Paths(0).PathCondition   => true,
     GlitchData     => Count0_GlitchData,
     GlitchMode     => Xonly,
     GlitchKind     => OnEvent);

  end process VitalBehavior;
end PinToPin_a;
```

Alias of Result (count1_zd, Count0_zd) used in the `VitalPropagatePathDelay procedure`

Figure 13.3.1 Pin-To-Pin Modeling Style of a Two-Bit Counter (ch13_dir\counter.vhd)

13.3.2 Distributed Delay Modeling Style

Distributed delay is a style of delay modeling where a cell is actually composed of structural portions, each is left to account for its own delay. The output of the cell is an artifact of the structure, event profile and actual delays that are present in the underlying implementation of this cell. Figure 13.3.2-1 represents *VITAL* distributed delay architecture.

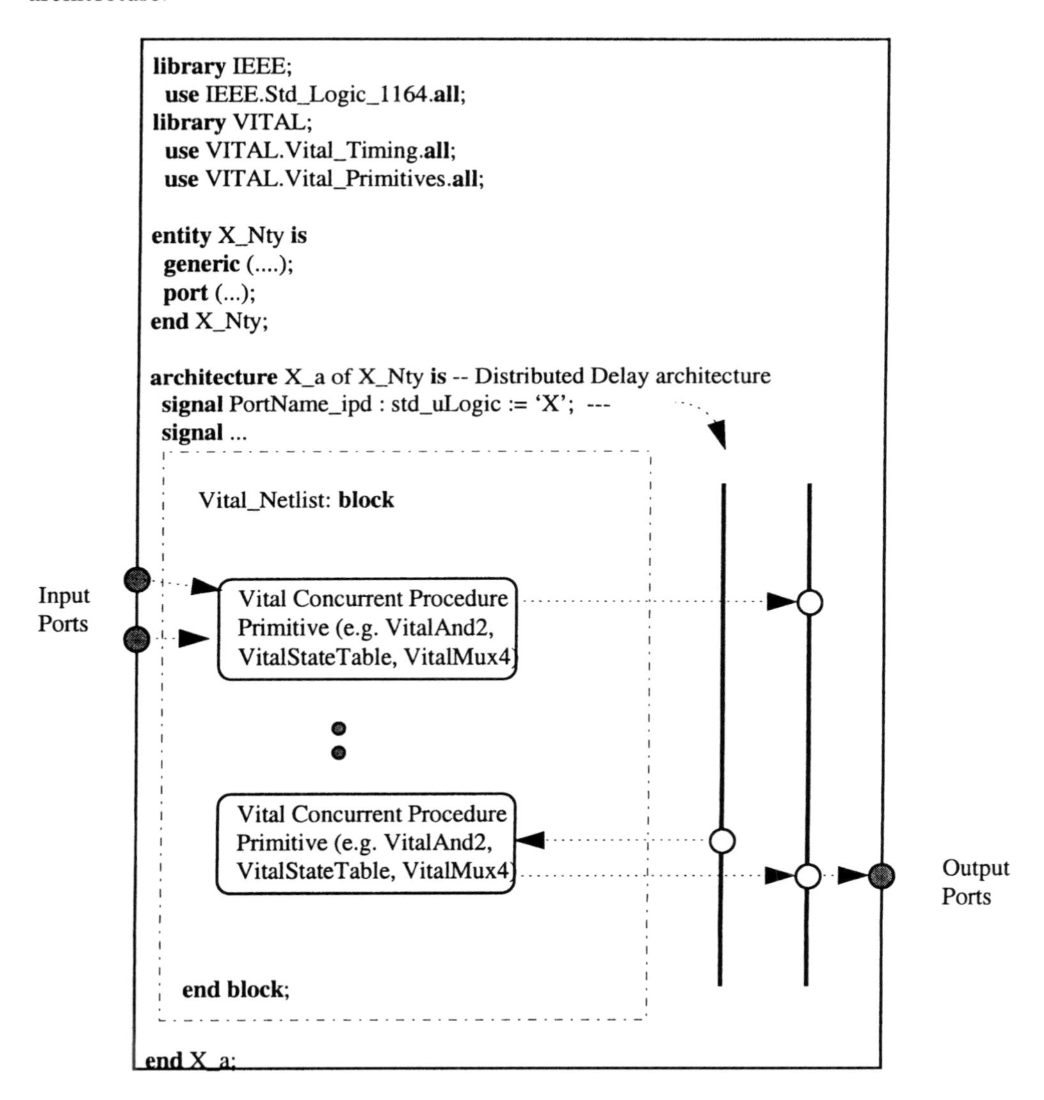

Figure 13.3.2-1 VITAL Distributed Delay Architecture

Functionality for the distributed delay model is contained within the *Vital_Netlist* block. Only calls to *VITAL_Primitives* concurrent procedures may be made within this block. Local signals may be declared solely to support the interconnection of the concurrent procedures. Timing information may be passed into the concurrent procedures from the generic parameter list. Figure 13.3.2-2 represents the high level architecture of a model using the distributed VITAL modeling style. Figure 13.3.2-3 represents the VITAL code for this model using this distributed modeling style.

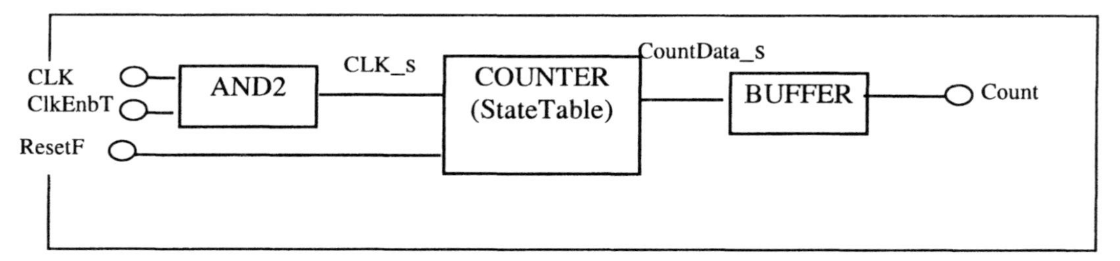

Figure 13.3.2-2 High Level Architecture of a Model Using the Distributed Vital Modeling style

```
--    Title        :  VITAL Distributed Modeling Style
--    Description :  Modeling ofa synchronous 2 bit counter with a reset
--
--------------------------------------------------------------------------
library IEEE;
  use IEEE.Std_Logic_1164.all;

library VITAL;
  use VITAL.Vital_Timing.all;
  use VITAL.Vital_Timing;
  use VITAL.Vital_Primitives.all;
  use VITAL.Vital_Primitives;

entity Counter_Nty is
  generic(tpd_And          : DelayType01 := (5 ns, 5 ns);
          tpd_Clk_Q0       : DelayType01 := (10 ns, 10 ns);
          tpd_Clk_Q1       : DelayType01 := (10 ns, 10 ns));
  port    (ResetF     : in     Std_Logic := 'U';
           Clk        : in     Std_logic := 'U';
           ClkEnbT    : in     Std_logic := 'U';
           Count      : out    Std_logic_Vector(1 downto 0));

end Counter_Nty;

architecture Count_Distrib of Counter_Nty is
  attribute Vital_level1 of Count_Distrib : architecture is true;
  signal Clk_s           : Std_Logic := 'X';
  signal CountData_s    : Std_Logic_Vector(1 downto 0) := "XX";

begin
  ---------------------------------------------------------------
  -- Input path delay
  ---------------------------------------------------------------
  Vital_Netlist: block
    constant H : Vital_Primitives.VitalTableSymbolType := '1';
    constant L : Vital_Primitives.VitalTableSymbolType := '0';
    constant x : Vital_Primitives.VitalTableSymbolType := '-'; -- any match
    constant S : Vital_Primitives.VitalTableSymbolType := 'S'; -- same
    constant R : Vital_Primitives.VitalTableSymbolType := '/'; -- rising
    constant U : Vital_Primitives.VitalTableSymbolType := 'X';
```

```
    constant V : Vital_Primitives.VitalTableSymbolType := 'B'; -- valid CLK

    constant CounterState :
      Vital_Primitives.VitalStateTableType(1 to 9, 1 to 6) := (
     --<dataIn ->
     --  CLk  Reset
     --          Present
     --          States
     --                Results
     --                (Next state)
     --    1  2  3  4   5  6
     --   Clk R   Q1 Q0 O1 O0
      (    x, U,  U, U, U, U),  --  Violation
      (    R, L,  x, x, L, L),  -- Reset
      (    R, H,  L, L, L, H),  -- Count
      (    R, H,  L, H, H, L),  -- Count
      (    R, H,  H, L, H, H),  -- Count
      (    R, H,  H, H, L, L),  -- Count
      (    R, H,  U, x, U, U),  -- Count
      (    R, H,  x, U, U, U),  -- Count
      (    V, x,  x, x, S, S)); -- Same, no change

  begin
    -- For demonstration, the cicuit includes an AND gate followed by
    -- the state table
    Vital_Primitives.VitalAND2(q       => Clk_s, -- Internal clock signal
                               a       => Clk,
                               b       => ClkEnbT,
                               tpd_a_q => tpd_And,
                               tpd_b_q => tpd_And);

    Vital_Primitives.VitalStateTable
         (StateTable    =>  CounterState,
          DataIn(1)     =>  Clk_s,    -- sub-element association
          DataIn(2)     =>  ResetF,
          NumStates     =>  2,
          Result(1)     =>  CountData_s(1),
          Result(2)     =>  CountData_s(0));

     Vital_Primitives.VitalBuf(q       => Count(1),
                               a       => CountData_s(1),
                               tpd_a_q => tpd_Clk_Q1);

     Vital_Primitives.VitalBuf(q       => Count(0),
                               a       => CountData_s(0),
                               tpd_a_q => tpd_Clk_Q0);
  end block;
end Count_Distrib;
```

Figure 13.3.2-3 VITAL Model Using the Distributed Modeling Style (ch13_dir\countdd.vhd)

EXERCISES

1. Define a testbench to test the pin-to-pin architecture for this counter. Use the timing results shown below as guidelines for the input stimuli and the expected results.

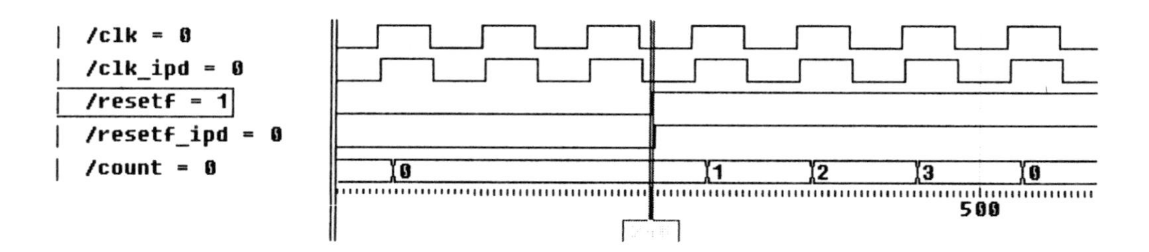

2. Define a testbench to test the distributed architecture for this counter. Use the timing results shown below as guidelines for the input stimuli and the expected results.

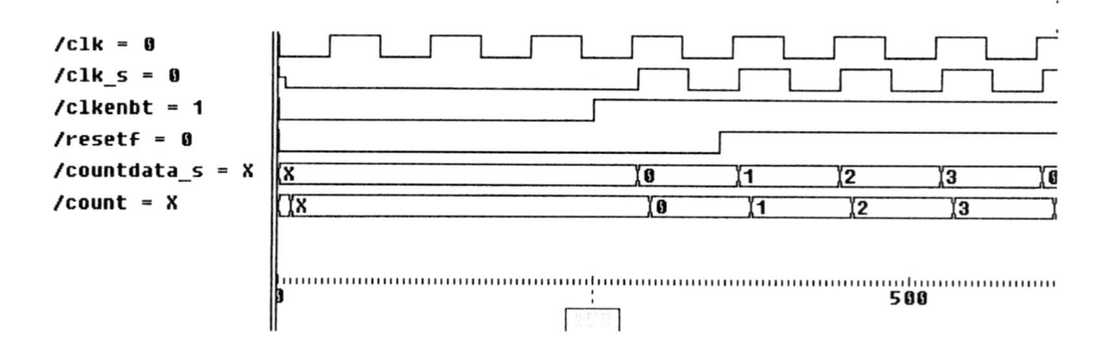

APPENDIX A : VHDL'93 AND VHDL'87 SYNTAX SUMMARY

abstract_literal ::= decimal_literal | based_literal

access_type_definition ::= **access** subtype_indication

actual_designator ::=
expression
| *signal*_name
| *variable*_name
| *file*_name
| **open**

actual_parameter_part ::= *parameter*_association_list

actual_part ::=
actual_designator
| *function*_name (actual_designator)
| type_mark (actual_designator)

adding_operator ::= + | - | &

aggregate ::=
(element_association { , element_association })

alias_declaration ::=
alias alias_designator [: subtype_indication] **is**
name [signature] ;

alias_designator ::= identifier | character_literal |
operator_symbol

allocator ::=
new subtype_indication
| **new** qualified_expression

architecture_body ::=
architecture identifier **of** *entity*_name **is**
architecture_declarative_part
begin
architecture_statement_part
end [**architecture**] [*architecture*_simple_name] ;

architecture_declarative_part ::=
{ block_declarative_item }

architecture_statement_part ::=
{ concurrent_statement }

array_type_definition ::=
unconstrained_array_definition |
constrained_array_definition

assertion ::=
assert condition
[**report** expression]
[**severity** expression]

assertion_statement ::= [label :] assertion ;

association_element ::=
[formal_part =>] actual_part

association_list ::=
association_element { , association_element }

attribute_declaration ::=
attribute identifier : type_mark ;

attribute_designator ::= *attribute*_simple_name

attribute_name ::=
prefix [signature] ' attribute_designator
[(expression)]

attribute_specification ::=
attribute attribute_designator **of**
entity_specification **is** expression ;

base ::= integer

base_specifier ::= B | O | X

base_unit_declaration ::= identifier ;

based_integer ::=
extended_digit { [underline] extended_digit }

based_literal ::=
base # based_integer [. based_integer] #
[exponent]

basic_character ::=
basic_graphic_character | format_effector

basic_graphic_character ::=
upper_case_letter | digit | special_character|
space_character

basic_identifier ::=
letter { [underline] letter_or_digit }

binding_indication ::=
[**use** entity_aspect]
[generic_map_aspect]
[port_map_aspect]

bit_string_literal ::= base_specifier " bit_value "

bit_value ::= extended_digit { [underline]
extended_digit }

block_configuration ::=
for block_specification
{ use_clause }
{ configuration_item }
end for ;

block_declarative_item ::=
subprogram_declaration
| subprogram_body
| type_declaration
| subtype_declaration
| constant_declaration
| signal_declaration
| shared_variable_declaration
| file_declaration
| alias_declaration
| component_declaration
| attribute_declaration
| attribute_specification
| configuration_specification
| disconnection_specification
| use_clause
| group_template_declaration
| group_declaration

block_declarative_part ::=
{ block_declarative_item }

block_header ::=
[generic_clause
[generic_map_aspect ;]]
[port_clause
[port_map_aspect ;]]

block_specification ::=
*architecture*_name
| *block_statement*_label
| *generate_statement*_label
[(index_specification)]

block_statement ::=
block_label :
block [(guard_expression)] [**is**]
block_header
block_declarative_part
begin
block_statement_part
end block [block_label] ;

block_statement_part ::=
{ concurrent_statement }

case_statement ::=
[case_label :]
case expression **is**
case_statement_alternative
{ case_statement_alternative }
end case [case_label] ;

case_statement_alternative ::=
when choices =>

sequence_of_statements

character_literal ::= ' graphic_character '

choice ::=
simple_expression
| discrete_range
| *element*_simple_name
| **others**

choices ::= choice { | choice }

component_configuration ::= -- VHDL'87
for component_specification
[**use** binding_indication ;]
[block_configuration]
end for ;

component_configuration ::= -- VHDL'93
for component_specification
[binding_indication ;]
[block_configuration]
end for ;

component_declaration ::=
component identifier [**is**]
[*local*_generic_clause]
[*local*_port_clause]
end component [component_simple_name] ;

component_instantiation_statement ::= -- VHDL'87
*instantiation*_label :
component_name
[generic_map_aspect]
[port_map_aspect] ;

component_instantiation_statement ::= -- VHDL'93
*instantiation*_label :
instantiated_unit
[generic_map_aspect]
[port_map_aspect] ;

component_specification ::=
instantiation_list : *component*_name

composite_type_definition ::=
array_type_definition
| record_type_definition

concurrent_assertion_statement ::=
[label :] [**postponed**] assertion ;

concurrent_procedure_call_statement ::=
[label :] [**postponed**] procedure_call ;

concurrent_signal_assignment_statement ::=
[label :] [**postponed**]
conditional_signal_assignment
| [label :] [**postponed**]
selected_signal_assignment

concurrent_statement ::=
block_statement
| process_statement
| concurrent_procedure_call_statement
| concurrent_assertion_statement
| concurrent_signal_assignment_statement
| component_instantiation_statement
| generate_statement

condition ::= *boolean*_expression

condition_clause ::= **until** condition

conditional_signal_assignment ::=
target <= options
conditional_waveforms ;

conditional_waveforms ::=
{ waveform **when** condition **else** }
waveform [**when** condition]

configuration_declaration ::=
configuration identifier **of** *entity*_name **is**
configuration_declarative_part
block_configuration
end [**configuration**]
[configuration_simple_name] ;

configuration_declarative_item ::=
use_clause
| attribute_specification
| group_declaration

configuration_declarative_part ::=
{ configuration_declarative_item }

configuration_item ::=
block_configuration
| component_configuration

configuration_specification ::= -- VHDL'87
for component_specification **use**
binding_indication ;

configuration_specification ::= -- VHDL'93
for component_specification binding_indication ;

constant_declaration ::=

constant identifier_list : subtype_indication [:= expression] ;

constrained_array_definition ::=
array index_constraint **of**
*element*_subtype_indication

constraint ::=
range_constraint
| index_constraint

context_clause ::= { context_item }

context_item ::=
library_clause
| use_clause

decimal_literal ::= integer [. integer] [exponent]

declaration ::=
type_declaration
| subtype_declaration
| object_declaration
| interface_declaration
| alias_declaration
| attribute_declaration
| component_declaration
| group_template_declaration
| group_declaration
| entity_declaration
| configuration_declaration
| subprogram_declaration
| package_declaration

delay_mechanism ::= -- VHDL'93
transport
| [**reject** time_expression] **inertial**

design_file ::= design_unit { design_unit }

design_unit ::= context_clause library_unit

designator ::= identifier | operator_symbol

direction ::= **to** | **downto**

disconnection_specification ::=
disconnect guarded_signal_specification **after**
*time*_expression ;

discrete_range ::= *discrete*_subtype_indication | range

element_association ::=
[choices =>] expression

element_declaration ::=
identifier_list : element_subtype_definition ;

element_subtype_definition ::= subtype_indication

entity_aspect ::=
entity *entity*_name [(*architecture*_identifier)]
| **configuration** configuration_name
| **open**

entity_class ::=
entity | **architecture** | **configuration**
| **procedure** | **function** | **package**
| **type** | **subtype** | **constant**
| **signal** | **variable** | **component**
| **label** | **literal** | **units**
| **group** | **file**

entity_class_entry ::= entity_class [<>]

entity_class_entry_list ::=
entity_class_entry { , entity_class_entry }

entity_declaration ::=
entity identifier **is**
entity_header
entity_declarative_part
[**begin**
entity_statement_part]
end [**entity**] [*entity*_simple_name] ;

entity_declarative_item ::=
subprogram_declaration
| subprogram_body
| type_declaration
| subtype_declaration
| constant_declaration
| signal_declaration
| shared_variable_declaration
| file_declaration
| alias_declaration
| attribute_declaration
| attribute_specification
| disconnection_specification
| use_clause
| group_template_declaration
| group_declaration

entity_declarative_part ::=
{ entity_declarative_item }

-- VHDL'87
entity_designator ::= simple_name | operator_symbol

-- VHDL'93

entity_designator ::= entity_tag [signature]

entity_header ::=
    [*formal*_generic_clause]
    [*formal*_port_clause]

entity_name_list ::=
    entity_designator { , entity_designator }
    | **others**
    | **all**

entity_specification ::=
    entity_name_list : entity_class

entity_statement ::=
    concurrent_assertion_statement
    | *passive*_concurrent_procedure_call_statement
    | *passive*_process_statement

entity_statement_part ::=
    { entity_statement }

-- VHDL'93
entity_tag :: simple_name | character_literal |
  operator_symbol

enumeration_literal ::= identifier | character_literal

enumeration_type_definition ::=
    (enumeration_literal { , enumeration_literal })

exit_statement ::=
    [label :] **exit** [*loop*_label] [**when** condition] ;

exponent ::= E [+] integer | E - integer

expression ::=
    relation { **and** relation }
    | relation { **or** relation }
    | relation { **xor** relation }
    | relation [**nand** relation]
    | relation [**nor** relation]
    | relation { **xnor** relation }

extended_digit ::= digit | letter

extended_identifier ::=
    \ graphic_character { graphic_character } \

factor ::=
    primary [** primary]
    | **abs** primary
    | **not** primary

-- VHDL'87
file_declaration ::=
    file identifier : subtype_indication **is** [mode]
    file_logical_name ;
-- VHDL'93
file_declaration ::=
    file identifier_list : subtype_indication
    file_open_information] ;

file_logical_name ::= string_expression

-- VHDL'93
file_open_information ::=
    [**open** *file_open_kind*_expression] **is**
    file_logical_name

file_type_definition ::=
    file of type_mark

floating_type_definition := range_constraint

formal_designator ::=
    *generic*_name
    | *port*_name
    | *parameter*_name

formal_parameter_list ::= *parameter*_interface_list

formal_part ::=
    formal_designator
    | *function*_name (formal_designator)
    | type_mark (formal_designator)

full_type_declaration ::=
    type identifier **is** type_definition ;

function_call ::=
    *function*_name [(actual_parameter_part)]

-- VHDL'87
generate_statement ::=
    *generate*_label : generation_scheme **generate**
    {concurrent_statement }
    end generate [*generate*_label];

-- VHDL'93
generate_statement ::=
    *generate*_label :
        generation_scheme **generate**
        [{ block_declarative_item }
        begin]
        { concurrent_statement }
        end generate [generate_label] ;

generation_scheme ::=

for *generate*_parameter_specification
| **if** condition

generic_clause ::=
generic (generic_list) ;

generic_list ::= *generic*_interface_list

generic_map_aspect ::=
generic map (*generic*_association_list)

graphic_character ::=
basic_graphic_character | lower_case_letter |
other_special_character

group_constituent ::= name | character_literal

group_constituent_list ::= group_constituent
{ , group_constituent }

group_declaration ::=
group identifier : *group_template*_name
(group_constituent_list) ;

group_template_declaration ::=
group identifier **is** (entity_class_entry_list) ;

guarded_signal_specification ::=
*guarded*_signal_list : type_mark

-- VHDL'87
identifier ::=
letter { [underline] letter_or_digit }

-- VHDL'93
identifier ::=
basic_identifier | extended_identifier

identifier_list ::= identifier { , identifier }

if_statement ::=
[if_label :]

if condition **then**
sequence_of_statements
{ **elsif** condition **then**
sequence_of_statements }
[**else**
sequence_of_statements]
end if [if_label] ;

incomplete_type_declaration ::= **type** identifier ;

index_constraint ::= (discrete_range
{ , discrete_range })

index_specification ::=
discrete_range
| *static*_expression

index_subtype_definition ::= type_mark **range** <>

indexed_name ::= prefix (expression { , expression })

-- VHDL'93
instantiated_unit ::=
[**component**] *component*_name
| **entity** *entity*_name [(*architecture*_identifier)]
| **configuration** *configuration*_name

instantiation_list ::=
*instantiation*_label { , instantiation_label }
| **others**
| **all**

integer ::= digit { [underline] digit }

integer_type_definition ::= range_constraint

interface_constant_declaration ::=
[**constant**] identifier_list : [**in**]
subtype_indication [:= *static*_expression]

interface_declaration ::=
interface_constant_declaration
| interface_signal_declaration
| interface_variable_declaration
| interface_file_declaration

interface_element ::= interface_declaration

-- VHDL'93
interface_file_declaration ::=
file identifier_list : subtype_indication

interface_list ::=
interface_element { ; interface_element }

interface_signal_declaration ::=
[**signal**] identifier_list : [mode]
subtype_indication [**bus**] [:= *static*_expression]

interface_variable_declaration ::=
[**variable**] identifier_list : [mode]
subtype_indication [:= *static*_expression]

iteration_scheme ::=
while condition
| **for** *loop*_parameter_specification

label ::= identifier

letter ::= upper_case_letter | lower_case_letter

letter_or_digit ::= letter | digit

library_clause ::= **library** logical_name_list ;

library_unit ::=
primary_unit
| secondary_unit

literal ::=
numeric_literal
| enumeration_literal
| string_literal
| bit_string_literal
| **null**

logical_name ::= identifier

logical_name_list ::= logical_name { , logical_name }

logical_operator ::= **and | or | nand | nor | xor | xnor**

loop_statement ::= [*loop*_label :]
[iteration_scheme] **loop**
sequence_of_statements
end loop [*loop*_label] ;

miscellaneous_operator ::= ** | **abs** | **not**

mode ::= **in** | **out** | **inout** | **buffer** | **linkage**

multiplying_operator ::= * | / | **mod** | **rem**

name ::=
simple_name
| operator_symbol
| selected_name
| indexed_name
| slice_name
| attribute_name

next_statement ::=
[label :] **next** [*loop*_label] [**when** condition] ;

null_statement ::= [label :] **null** ;

numeric_literal ::=
abstract_literal
| physical_literal

object_declaration ::=
constant_declaration
| signal_declaration
| variable_declaration
| file_declaration

operator_symbol ::= string_literal

-- VHDL'87
options ::=
[**guarded**] [**transport**]

-- VHDL'93
options ::= [**guarded**] [delay_mechanism]

package_body ::=
package body *package*_simple_name **is**
package_body_declarative_part
end [**package body**] [*package*_simple_name] ;

package_body_declarative_item ::=
subprogram_declaration
| subprogram_body
| type_declaration
| subtype_declaration
| constant_declaration
| shared_variable_declaration
| file_declaration
| alias_declaration
| use_clause
| group_template_declaration
| group_declaration

package_body_declarative_part ::=
{ package_body_declarative_item }

package_declaration ::=
package identifier **is**
package_declarative_part
end [**package**] [*package*_simple_name] ;

package_declarative_item ::=
subprogram_declaration
| type_declaration
| subtype_declaration
| constant_declaration
| signal_declaration
| shared_variable_declaration
| file_declaration
| alias_declaration
| component_declaration
| attribute_declaration
| attribute_specification
| disconnection_specification
| use_clause

| group_template_declaration

| group_declaration

package_declarative_part ::=
{ package_declarative_item }

parameter_specification ::=
identifier **in** discrete_range

physical_literal ::= [abstract_literal] unit_name

physical_type_definition ::=
range_constraint
units
base_unit_declaration
{ secondary_unit_declaration }
end units [*physical_type*_simple_name]

port_clause ::=
port (port_list) ;

port_list ::= *port*_interface_list

port_map_aspect ::=
port map (*port*_association_list)

prefix ::=
name
| **function**_call

primary ::=
name
| literal
| aggregate
| **function**_call
| qualified_expression
| type_conversion
| allocator
| (expression)

primary_unit ::=
entity_declaration
| configuration_declaration
| package_declaration

procedure_call ::= *procedure*_name
[(actual_parameter_part)]

procedure_call_statement ::=
[label :] procedure_call ;

process_declarative_item ::=
subprogram_declaration
| subprogram_body
| type_declaration
| subtype_declaration
| constant_declaration
| variable_declaration
| file_declaration
| alias_declaration
| attribute_declaration
| attribute_specification
| use_clause
| group_template_declaration

| group_declaration

process_declarative_part ::=
{ process_declarative_item }

process_statement ::=
[*process*_label :]
[**postponed**] **process** [(sensitivity_list)] [**is**]
process_declarative_part
begin
process_statement_part
end [**postponed**] **process** [*process*_label] ;

process_statement_part ::=
{ sequential_statement }

qualified_expression ::=
type_mark ' (expression)
| type_mark ' aggregate

range ::=
*range*_attribute_name
| simple_expression direction simple_expression

range_constraint ::= **range** range

record_type_definition ::=
record
element_declaration
{ element_declaration }
end record [record_type_simple_name]

relation ::=
shift_expression [relational_operator
shift_expression]

relational_operator ::=
= | /= | < | <= | > | >=

return_statement ::=
[label :] **return** [expression] ;

report_statement ::=
[label :]
report expression

[**severity** expression] ;

scalar_type_definition ::=
enumeration_type_definition |
integer_type_definition |
floating_type_definition |
physical_type_definition

secondary_unit ::=
architecture_body
| package_body

secondary_unit_declaration ::=
identifier = physical_literal ;

selected_name ::= prefix . suffix

selected_signal_assignment ::=
with expression **select**
target <= options selected_waveforms ;

selected_waveforms ::=
{ waveform **when** choices , }
waveform **when** choices

sensitivity_clause ::= **on** sensitivity_list

sensitivity_list ::= *signal*_name { , *signal*_name }

sequence_of_statements ::=
{ sequential_statement }

sequential_statement ::=
wait_statement
| assertion_statement
| report_statement
| signal_assignment_statement
| variable_assignment_statement
| procedure_call_statement
| if_statement
| case_statement
| loop_statement
| next_statement
| exit_statement
| **return**_statement
| null_statement

-- VHDL'93
shift_expression ::=
simple_expression
[shift_operator simple_expression]

-- VHDL'93
shift_operator ::= **sll** | **srl** | **sla** | **sra** | **rol** | **ror**

sign ::= + | -

-- VHDL'87
signal_assignment_statement ::=
target <= [**transport**] waveform ;

-- VHDL'93
signal_assignment_statement ::=
[label :] target <= [delay_mechanism]
waveform ;

signal_declaration ::=
signal identifier_list : subtype_indication
[signal_kind] [:= expression] ;

signal_kind ::= **register** | **bus**
signal_list ::=
*signal*_name { , *signal*_name }
| **others**
| **all**
-- VHDL'93
signature ::= [[type_mark { , type_mark }]
[**return** type_mark]]

simple_expression ::=
[sign] term { adding_operator term }

simple_name ::= identifier

slice_name ::= prefix (discrete_range)

string_literal ::= " { graphic_character } "

subprogram_body ::=
subprogram_specification **is**
subprogram_declarative_part
begin
subprogram_statement_part
end [subprogram_kind] [designator] ;

subprogram_declaration ::=
subprogram_specification ;

subprogram_declarative_item ::=
subprogram_declaration
| subprogram_body
| type_declaration
| subtype_declaration
| constant_declaration
| variable_declaration
| file_declaration
| alias_declaration
| attribute_declaration
| attribute_specification
| use_clause
| group_template_declaration

| group_declaration

subprogram_declarative_part ::=
{ subprogram_declarative_item }

subprogram_kind ::= **procedure** | **function**

subprogram_specification ::=
procedure designator [(formal_parameter_list)]
| [**pure | impure**] **function** designator
[(formal_parameter_list)]
return type_mark

subprogram_statement_part ::=
{ sequential_statement }

subtype_declaration ::=
subtype identifier **is** subtype_indication ;

subtype_indication ::=
[*resolution_function*_name] type_mark
[constraint]

suffix ::=
simple_name
| character_literal
| operator_symbol
| **all**

target ::=
name
| aggregate

term ::=
factor { multiplying_operator factor }

timeout_clause ::= **for** *time*_expression

type_conversion ::= type_mark (expression)

type_declaration ::=
full_type_declaration
| incomplete_type_declaration

type_definition ::=
scalar_type_definition
| composite_type_definition
| access_type_definition
| file_type_definition

type_mark ::=
*type*_name
| *subtype*_name

unconstrained_array_definition ::=
array (index_subtype_definition
{ , index_subtype_definition })
of *element*_subtype_indication

use_clause ::=
use selected_name { , selected_name } ;

variable_assignment_statement ::=
[label :] target := expression ;

variable_declaration ::=
[**shared**] variable identifier_list :
subtype_indication [:= expression] ;

wait_statement ::=
[label :] **wait** [sensitivity_clause]
[condition_clause] [timeout_clause] ;

waveform ::=
waveform_element { , waveform_element }
| unaffected

waveform_element ::=
value_expression [**after** *time*_expression]
| **null** [**after** *time*_expression]

APPENDIX B : PACKAGE STANDARD

```
-- This is Package STANDARD as defined in the VHDL 1992 Language Reference Manual.
--  Reprinted by permission from Model Technology Inc.
--   NOTE: VCOM and VSIM will not work properly if these declarations
--         are modified.

-- Version information: @(#)standard.vhd

package standard is
    type boolean is (false,true);
    type bit is ('0', '1');
    type character is (
              nul, soh, stx, etx, eot, enq, ack, bel,
              bs,  ht,  lf,  vt,  ff,  cr,  so,  si,
              dle, dc1, dc2, dc3, dc4, nak, syn, etb,
              can, em,  sub, esc, fsp, gsp, rsp, usp,

              ' ', '!', '"', '#', '$', '%', '&', ''',
              '(', ')', '*', '+', ',', '-', '.', '/',
              '0', '1', '2', '3', '4', '5', '6', '7',
              '8', '9', ':', ';', '<', '=', '>', '?',

              '@', 'A', 'B', 'C', 'D', 'E', 'F', 'G',
              'H', 'I', 'J', 'K', 'L', 'M', 'N', 'O',
              'P', 'Q', 'R', 'S', 'T', 'U', 'V', 'W',
              'X', 'Y', 'Z', '[', '\', ']', '^', '_',

              '`', 'a', 'b', 'c', 'd', 'e', 'f', 'g',
              'h', 'i', 'j', 'k', 'l', 'm', 'n', 'o',
              'p', 'q', 'r', 's', 't', 'u', 'v', 'w',
              'x', 'y', 'z', '{', '|', '}', '~', del,

              c128, c129, c130, c131, c132, c133, c134, c135,
              c136, c137, c138, c139, c140, c141, c142, c143,
              c144, c145, c146, c147, c148, c149, c150, c151,
              c152, c153, c154, c155, c156, c157, c158, c159,

              -- the character code for 160 is there (NBSP),
              -- but prints as no char
```

```
                    ' ', '¡', '¢', '£', '¤', '¥', '¦', '§',
                    '¨', '©', 'ª', '«', '¬', '-', '®', '¯',
                    '°', '±', '²', '³', '´', 'µ', '¶', '·',
                    '¸', '¹', 'º', '»', '¼', '½', '¾', '¿',

                    'À', 'Á', 'Â', 'Ã', 'Ä', 'Å', 'Æ', 'Ç',
                    'È', 'É', 'Ê', 'Ë', 'Ì', 'Í', 'Î', 'Ï',
                    'Ð', 'Ñ', 'Ò', 'Ó', 'Ô', 'Õ', 'Ö', '×',
                    'Ø', 'Ù', 'Ú', 'Û', 'Ü', 'Ý', 'Þ', 'ß',

                    'à', 'á', 'â', 'ã', 'ä', 'å', 'æ', 'ç',
                    'è', 'é', 'ê', 'ë', 'ì', 'í', 'î', 'ï',
                    'ð', 'ñ', 'ò', 'ó', 'ô', 'õ', 'ö', '÷',
                    'ø', 'ù', 'ú', 'û', 'ü', 'ý', 'þ', 'ÿ' );

    type severity_level is (note, warning, error, failure);
    type integer is range -2147483647 to 2147483647;
    type real is range -1.0E308 to 1.0E308;
    type time is range -2147483647 to 2147483647
             units
                fs;
                ps = 1000 fs;
                ns = 1000 ps;
                us = 1000 ns;
                ms = 1000 us;
                sec = 1000 ms;
                min = 60 sec;
                hr = 60 min;
             end units;
    subtype delay_length is time range 0 fs to time'high;
    impure function now return delay_length;
    subtype natural is integer range 0 to integer'high;
    subtype positive is integer range 1 to integer'high;
    type string is array (positive range <>) of character;
    type bit_vector is array (natural range <>) of bit;
    type file_open_kind is (
             read_mode,
             write_mode,
             append_mode);
    type file_open_status is (
             open_ok,
             status_error,
             name_error,
             mode_error);
    attribute foreign : string;
end standard;
```

APPENDIX C : PACKAGE TEXTIO

```
--------------------------------------------------------------------------
-- Package TEXTIO as defined in Chapter 14 of the IEEE Standard VHDL
--   Language Reference Manual (IEEE Std. 1076-1987), as modified
--   by the Issues Screening and Analysis Committee (ISAC), a subcommittee
--   of the VHDL Analysis and Standardization Group (VASG) on
--   10 November, 1988.  See "The Sense of the VASG", October, 1989.
-- Reprinted by permission from Model Technology Inc.
--------------------------------------------------------------------------
-- Version information: %W% %G%
--------------------------------------------------------------------------

package TEXTIO is
    type LINE is access string;
    type TEXT is file of string;
    type SIDE is (right, left);
    subtype WIDTH is natural;

    -- changed for vhdl92 syntax:
    file input : TEXT open read_mode is "STD_INPUT";
    file output : TEXT open write_mode is "STD_OUTPUT";

    -- changed for vhdl92 syntax (and now a built-in):
    procedure READLINE(file f: TEXT; L: out LINE);

    procedure READ(L:inout LINE; VALUE: out bit; GOOD : out BOOLEAN);
    procedure READ(L:inout LINE; VALUE: out bit);

    procedure READ(L:inout LINE; VALUE: out bit_vector; GOOD : out BOOLEAN);
    procedure READ(L:inout LINE; VALUE: out bit_vector);

    procedure READ(L:inout LINE; VALUE: out BOOLEAN; GOOD : out BOOLEAN);
    procedure READ(L:inout LINE; VALUE: out BOOLEAN);

    procedure READ(L:inout LINE; VALUE: out character; GOOD : out BOOLEAN);
    procedure READ(L:inout LINE; VALUE: out character);

    procedure READ(L:inout LINE; VALUE: out integer; GOOD : out BOOLEAN);
```

```
    procedure READ(L:inout LINE; VALUE: out integer);

    procedure READ(L:inout LINE; VALUE: out real; GOOD : out BOOLEAN);
    procedure READ(L:inout LINE; VALUE: out real);

    procedure READ(L:inout LINE; VALUE: out string; GOOD : out BOOLEAN);
    procedure READ(L:inout LINE; VALUE: out string);

    procedure READ(L:inout LINE; VALUE: out time; GOOD : out BOOLEAN);
    procedure READ(L:inout LINE; VALUE: out time);

    -- changed for vhdl92 syntax (and now a built-in):
    procedure WRITELINE(file f : TEXT; L : inout LINE);

    procedure WRITE(L : inout LINE; VALUE : in bit;
          JUSTIFIED: in SIDE := right;
          FIELD: in WIDTH := 0);

    procedure WRITE(L : inout LINE; VALUE : in bit_vector;
          JUSTIFIED: in SIDE := right;
          FIELD: in WIDTH := 0);

    procedure WRITE(L : inout LINE; VALUE : in BOOLEAN;
          JUSTIFIED: in SIDE := right;
          FIELD: in WIDTH := 0);

    procedure WRITE(L : inout LINE; VALUE : in character;
          JUSTIFIED: in SIDE := right;
          FIELD: in WIDTH := 0);

    procedure WRITE(L : inout LINE; VALUE : in integer;
          JUSTIFIED: in SIDE := right;
          FIELD: in WIDTH := 0);

    procedure WRITE(L : inout LINE; VALUE : in real;
          JUSTIFIED: in SIDE := right;
          FIELD: in WIDTH := 0;
          DIGITS: in NATURAL := 0);

    procedure WRITE(L : inout LINE; VALUE : in string;
          JUSTIFIED: in SIDE := right;
          FIELD: in WIDTH := 0);

    procedure WRITE(L : inout LINE; VALUE : in time;
          JUSTIFIED: in SIDE := right;
          FIELD: in WIDTH := 0;
          UNIT: in TIME := ns);

    -- is implicit built-in:
    -- function ENDFILE(file F : TEXT) return boolean;

    -- function ENDLINE(variable L : in LINE) return BOOLEAN;
    --
    -- Function ENDLINE as declared cannot be legal VHDL, and
    --   the entire function was deleted from the definition
    --   by the Issues Screening and Analysis Committee (ISAC),
    --   a subcommittee of the VHDL Analysis and Standardization
    --   Group (VASG) on 10 November, 1988.  See "The Sense of
    --   the VASG", October, 1989, VHDL Issue Number 0032.
end;
--*******************************************************
--** Copyright (c) Model Technology Incorporated 1991  **
--**              All Rights Reserved                  **
--*******************************************************
```

APPENDIX D : STD_LOGIC_TEXTIO

```
-----------------------------------------------------------------------
--
-- Copyright (c) 1990, 1991, 1992 by Synopsys, Inc.  All rights reserved.
--
-- This source file may be used and distributed without restriction
-- provided that this copyright statement is not removed from the file
-- and that any derivative work contains this copyright notice.
--
--  Package name: STD_LOGIC_TEXTIO
--
--  Purpose: This package overloads the standard TEXTIO procedures
--               READ and WRITE.
--
--  Author: CRC, TS
--
-----------------------------------------------------------------------

use STD.textio.all;
library IEEE;
use IEEE.std_logic_1164.all;

package STD_LOGIC_TEXTIO is
--synopsys synthesis_off
    -- Read and Write procedures for STD_ULOGIC and STD_ULOGIC_VECTOR
    procedure READ(L:inout LINE; VALUE:out STD_ULOGIC);
    procedure READ(L:inout LINE; VALUE:out STD_ULOGIC; GOOD: out BOOLEAN);
    procedure READ(L:inout LINE; VALUE:out STD_ULOGIC_VECTOR);
    procedure READ(L:inout LINE; VALUE:out STD_ULOGIC_VECTOR;
                GOOD: out BOOLEAN);
    procedure WRITE(L:inout LINE; VALUE:in STD_ULOGIC;
                JUSTIFIED:in SIDE := RIGHT; FIELD:in WIDTH := 0);
    procedure WRITE(L:inout LINE; VALUE:in STD_ULOGIC_VECTOR;
                JUSTIFIED:in SIDE := RIGHT; FIELD:in WIDTH := 0);

    -- Read and Write procedures for STD_LOGIC_VECTOR
    procedure READ(L:inout LINE; VALUE:out STD_LOGIC_VECTOR);
    procedure READ(L:inout LINE; VALUE:out STD_LOGIC_VECTOR; GOOD: out BOOLEAN);
    procedure WRITE(L:inout LINE; VALUE:in STD_LOGIC_VECTOR;
```

```
                JUSTIFIED:in SIDE := RIGHT; FIELD:in WIDTH := 0);

    --
    -- Read and Write procedures for Hex and Octal values.
    -- The values appear in the file as a series of characters
    -- between 0-F (Hex), or 0-7 (Octal) respectively.
    --

    -- Hex
    procedure HREAD(L:inout LINE; VALUE:out STD_ULOGIC_VECTOR);
    procedure HREAD(L:inout LINE; VALUE:out STD_ULOGIC_VECTOR;
            GOOD: out BOOLEAN);
    procedure HWRITE(L:inout LINE; VALUE:in STD_ULOGIC_VECTOR;
                JUSTIFIED:in SIDE := RIGHT; FIELD:in WIDTH := 0);
    procedure HREAD(L:inout LINE; VALUE:out STD_LOGIC_VECTOR);
    procedure HREAD(L:inout LINE; VALUE:out STD_LOGIC_VECTOR;
                GOOD: out BOOLEAN);
    procedure HWRITE(L:inout LINE; VALUE:in STD_LOGIC_VECTOR;
                JUSTIFIED:in SIDE := RIGHT; FIELD:in WIDTH := 0);

    -- Octal
    procedure OREAD(L:inout LINE; VALUE:out STD_ULOGIC_VECTOR);
    procedure OREAD(L:inout LINE; VALUE:out STD_ULOGIC_VECTOR;
                GOOD: out BOOLEAN);
    procedure OWRITE(L:inout LINE; VALUE:in STD_ULOGIC_VECTOR;
                JUSTIFIED:in SIDE := RIGHT; FIELD:in WIDTH := 0);
    procedure OREAD(L:inout LINE; VALUE:out STD_LOGIC_VECTOR);
    procedure OREAD(L:inout LINE; VALUE:out STD_LOGIC_VECTOR;
                GOOD: out BOOLEAN);
    procedure OWRITE(L:inout LINE; VALUE:in STD_LOGIC_VECTOR;
                JUSTIFIED:in SIDE := RIGHT; FIELD:in WIDTH := 0);

--synopsys synthesis_on
end STD_LOGIC_TEXTIO;
```

APPENDIX E :PACKAGE STD_LOGIC_1164

```
-- --------------------------------------------------------------------
--
--    Title       :  std_logic_1164 multi-value logic system
--    Library     :  This package shall be compiled into a library
--                :  symbolically named IEEE.
--                :
--    Developers:    IEEE model standards group (par 1164)
--    Purpose     :  This packages defines a standard for designers
--                :  to use in describing the interconnection data types
--                :  used in vhdl modeling.
--                :
--    Limitation:    The logic system defined in this package may
--                :  be insufficient for modeling switched transistors,
--                :  since such a requirement is out of the scope of this
--                :  effort. Furthermore, mathematics, primitives,
--                :  timing standards, etc. are considered orthogonal
--                :  issues as it relates to this package and are therefore
--                :  beyond the scope of this effort.
--                :
--    Note        :  No declarations or definitions shall be included in,
--                :  or excluded from this package. The "package declaration"
--                :  defines the types, subtypes and declarations of
--                :  std_logic_1164. The std_logic_1164 package body shall be
--                :  considered the formal definition of the semantics of
--                :  this package. Tool developers may choose to implement
--                :  the package body in the most efficient manner available
--                :  to them.
--                :
-- --------------------------------------------------------------------
--    modification history :
-- --------------------------------------------------------------------
--   version | mod. date:|
--    v4.200 | 01/02/92  |
-- --------------------------------------------------------------------
```

```
PACKAGE std_logic_1164 IS
    -- logic state system  (unresolved)
    -------------------------------------------------------------------
    TYPE std_ulogic IS ( 'U',  -- Uninitialized
                         'X',  -- Forcing  Unknown
                         '0',  -- Forcing  0
                         '1',  -- Forcing  1
                         'Z',  -- High Impedance
                         'W',  -- Weak     Unknown
                         'L',  -- Weak     0
                         'H',  -- Weak     1
                         '-'   -- Don't care
                       );
    -------------------------------------------------------------------
    -- unconstrained array of std_ulogic for use with the resolution function
    -------------------------------------------------------------------
    TYPE std_ulogic_vector IS ARRAY ( NATURAL RANGE <> ) OF std_ulogic;
    -------------------------------------------------------------------
    -- resolution function
    -------------------------------------------------------------------
    FUNCTION resolved ( s : std_ulogic_vector ) RETURN std_ulogic;
    -------------------------------------------------------------------
    -- *** industry standard logic type ***
    -------------------------------------------------------------------
    SUBTYPE std_logic IS resolved std_ulogic;

    -------------------------------------------------------------------
    -- unconstrained array of std_logic for use in declaring signal arrays
    -------------------------------------------------------------------
    TYPE std_logic_vector IS ARRAY ( NATURAL RANGE <>) OF std_logic;

    -------------------------------------------------------------------
    -- common subtypes
    -------------------------------------------------------------------
    SUBTYPE X01     IS resolved std_ulogic RANGE 'X' TO '1'; -- ('X','0','1')
    SUBTYPE X01Z    IS resolved std_ulogic RANGE 'X' TO 'Z'; -- ('X','0','1','Z')
    SUBTYPE UX01    IS resolved std_ulogic RANGE 'U' TO '1'; -- ('U','X','0','1')
    SUBTYPE UX01Z   IS resolved std_ulogic RANGE 'U' TO 'Z'; -- ('U','X','0','1','Z')
    -------------------------------------------------------------------
    -- overloaded logical operators
    -------------------------------------------------------------------

    FUNCTION "and"  ( l : std_ulogic; r : std_ulogic ) RETURN UX01;
    FUNCTION "nand" ( l : std_ulogic; r : std_ulogic ) RETURN UX01;
    FUNCTION "or"   ( l : std_ulogic; r : std_ulogic ) RETURN UX01;
    FUNCTION "nor"  ( l : std_ulogic; r : std_ulogic ) RETURN UX01;
    FUNCTION "xor"  ( l : std_ulogic; r : std_ulogic ) RETURN UX01;
--  function "xnor" ( l : std_ulogic; r : std_ulogic ) return ux01;
    FUNCTION "not"  ( l : std_ulogic                 ) RETURN UX01;
    -------------------------------------------------------------------
    -- vectorized overloaded logical operators
    -------------------------------------------------------------------
    FUNCTION "and"  ( l, r : std_logic_vector  ) RETURN std_logic_vector;
    FUNCTION "and"  ( l, r : std_ulogic_vector ) RETURN std_ulogic_vector;
    FUNCTION "nand" ( l, r : std_logic_vector  ) RETURN std_logic_vector;
    FUNCTION "nand" ( l, r : std_ulogic_vector ) RETURN std_ulogic_vector;
    FUNCTION "or"   ( l, r : std_logic_vector  ) RETURN std_logic_vector;
    FUNCTION "or"   ( l, r : std_ulogic_vector ) RETURN std_ulogic_vector;
    FUNCTION "nor"  ( l, r : std_logic_vector  ) RETURN std_logic_vector;
    FUNCTION "nor"  ( l, r : std_ulogic_vector ) RETURN std_ulogic_vector;
    FUNCTION "xor"  ( l, r : std_logic_vector  ) RETURN std_logic_vector;
    FUNCTION "xor"  ( l, r : std_ulogic_vector ) RETURN std_ulogic_vector;
```

```
--  ---------------------------------------------------------------------
--  Note : The declaration and implementation of the "xnor" function is
--  specifically commented until at which time the VHDL language has been
--  officially adopted as containing such a function. At such a point,
--  the following comments may be removed along with this notice without
--  further "official" ballotting of this std_logic_1164 package. It is
--  the intent of this effort to provide such a function once it becomes
--  available in the VHDL standard.
--  ---------------------------------------------------------------------
--  function "xnor" ( l, r : std_logic_vector  ) return std_logic_vector;
--  function "xnor" ( l, r : std_ulogic_vector ) return std_ulogic_vector;

    FUNCTION "not"  ( l : std_logic_vector  ) RETURN std_logic_vector;
    FUNCTION "not"  ( l : std_ulogic_vector ) RETURN std_ulogic_vector;

    -------------------------------------------------------------------
    -- conversion functions
    -------------------------------------------------------------------
    FUNCTION To_bit       ( s : std_ulogic;        xmap : BIT := '0') RETURN BIT;
    FUNCTION To_bitvector ( s : std_logic_vector ; xmap : BIT := '0')
        RETURN BIT_VECTOR;
    FUNCTION To_bitvector ( s : std_ulogic_vector; xmap : BIT := '0')
        RETURN BIT_VECTOR;

    FUNCTION To_StdULogic       ( b : BIT        ) RETURN std_ulogic;
    FUNCTION To_StdLogicVector  ( b : BIT_VECTOR ) RETURN std_logic_vector;
    FUNCTION To_StdLogicVector  ( s : std_ulogic_vector) RETURN std_logic_vector;
    FUNCTION To_StdULogicVector ( b : BIT_VECTOR ) RETURN std_ulogic_vector;
    FUNCTION To_StdULogicVector( s : std_logic_vector) RETURN std_ulogic_vector;

    -------------------------------------------------------------------
    -- strength strippers and type convertors
    -------------------------------------------------------------------
    FUNCTION To_X01  ( s : std_logic_vector  ) RETURN  std_logic_vector;
    FUNCTION To_X01  ( s : std_ulogic_vector ) RETURN  std_ulogic_vector;
    FUNCTION To_X01  ( s : std_ulogic        ) RETURN  X01;
    FUNCTION To_X01  ( b : BIT_VECTOR        ) RETURN  std_logic_vector;
    FUNCTION To_X01  ( b : BIT_VECTOR        ) RETURN  std_ulogic_vector;
    FUNCTION To_X01  ( b : BIT               ) RETURN  X01;

    FUNCTION To_X01Z ( s : std_logic_vector  ) RETURN  std_logic_vector;
    FUNCTION To_X01Z ( s : std_ulogic_vector ) RETURN  std_ulogic_vector;
    FUNCTION To_X01Z ( s : std_ulogic        ) RETURN  X01Z;
    FUNCTION To_X01Z ( b : BIT_VECTOR        ) RETURN  std_logic_vector;
    FUNCTION To_X01Z ( b : BIT_VECTOR        ) RETURN  std_ulogic_vector;
    FUNCTION To_X01Z ( b : BIT               ) RETURN  X01Z;

    FUNCTION To_UX01  ( s : std_logic_vector  ) RETURN  std_logic_vector;
    FUNCTION To_UX01  ( s : std_ulogic_vector ) RETURN  std_ulogic_vector;
    FUNCTION To_UX01  ( s : std_ulogic        ) RETURN  UX01;
    FUNCTION To_UX01  ( b : BIT_VECTOR        ) RETURN  std_logic_vector;
    FUNCTION To_UX01  ( b : BIT_VECTOR        ) RETURN  std_ulogic_vector;
    FUNCTION To_UX01  ( b : BIT               ) RETURN  UX01;
    -------------------------------------------------------------------
    -- edge detection
    -------------------------------------------------------------------
    FUNCTION rising_edge  (SIGNAL s : std_ulogic) RETURN BOOLEAN;
    FUNCTION falling_edge (SIGNAL s : std_ulogic) RETURN BOOLEAN;
    -------------------------------------------------------------------
    -- object contains an unknown
    -------------------------------------------------------------------
    FUNCTION Is_X ( s : std_ulogic_vector ) RETURN  BOOLEAN;
    FUNCTION Is_X ( s : std_logic_vector  ) RETURN  BOOLEAN;
    FUNCTION Is_X ( s : std_ulogic        ) RETURN  BOOLEAN;
END std_logic_1164;
```

APPENDIX F : NUMERIC_STD

```
-- -------------------------------------------------------------------
--
-- Copyright 1995 by IEEE. All rights reserved.
--
-- This source file is considered by the IEEE to be an essential part of the use
-- of the standard 1076.3 and as such may be distributed without change, except
-- as permitted by the standard. This source file may not be sold or distributed
-- for profit. This package may be modified to include additional data required
-- by tools, but must in no way change the external interfaces or simulation
-- behaviour of the description. It is permissible to add comments and/or
-- attributes to the package declarations, but not to change or delete any
-- original lines of the approved package declaration. The package body may be
-- changed only in accordance with the terms of clauses 7.1 and 7.2 of the
-- standard.
--
-- Title      : Standard VHDL Synthesis Package (1076.3, NUMERIC_STD)
--
-- Library    : This package shall be compiled into a library symbolically
--            : named IEEE.
--
-- Developers : IEEE DASC Synthesis Working Group, PAR 1076.3
--
-- Purpose    : This package defines numeric types and arithmetic functions
--            : for use with synthesis tools. Two numeric types are defined:
--            : -- > UNSIGNED: represents UNSIGNED number in vector form
--            : -- > SIGNED: represents a SIGNED number in vector form
--            : The base element type is type STD_LOGIC.
--            : The leftmost bit is treated as the most significant bit.
--            : Signed vectors are represented in two's complement form.
--            : This package contains overloaded arithmetic operators on
--            : the SIGNED and UNSIGNED types. The package also contains
--            : useful type conversions functions.
--            :
--            : If any argument to a function is a null array, a null array is
--            : returned (exceptions, if any, are noted individually).
--
-- Limitation :
--
-- Note       : No declarations or definitions shall be included in,
--            : or excluded from this package. The "package declaration"
--            : defines the types, subtypes and declarations of
```

```
--             : NUMERIC_STD. The NUMERIC_STD package body shall be
--             : considered the formal definition of the semantics of
--             : this package. Tool developers may choose to implement
--             : the package body in the most efficient manner available
--             : to them.
--
-- ----------------------------------------------------------------
--   modification history :
-- ----------------------------------------------------------------
--   Version:  2.4
--   Date   :  12 April 1995
-- -----------------------------------------------------------------------
library IEEE;
use IEEE.STD_LOGIC_1164.all;

package NUMERIC_STD is
  constant CopyRightNotice: STRING
      := "Copyright 1995 IEEE. All rights reserved.";

  attribute builtin_subprogram: string;

  --============================================================================
  -- Numeric array type definitions
  --============================================================================

  type UNSIGNED is array (NATURAL range <>) of STD_LOGIC;
  type SIGNED is array (NATURAL range <>) of STD_LOGIC;

  --============================================================================
  -- Arithmetic Operators:
  --============================================================================
  function "abs" (ARG: SIGNED) return SIGNED;
  -- Result subtype: SIGNED(ARG'LENGTH-1 downto 0).
  -- Result: Returns the absolute value of a SIGNED vector ARG.

  function "-" (ARG: SIGNED) return SIGNED;
  -- Result subtype: SIGNED(ARG'LENGTH-1 downto 0).
  -- Result: Returns the value of the unary minus operation on a
  --         SIGNED vector ARG.

  --============================================================================
  function "+" (L, R: UNSIGNED) return UNSIGNED;
  -- Result subtype: UNSIGNED(MAX(L'LENGTH, R'LENGTH)-1 downto 0).
  -- Result: Adds two UNSIGNED vectors that may be of different lengths.

  function "+" (L, R: SIGNED) return SIGNED;
  -- Result subtype: SIGNED(MAX(L'LENGTH, R'LENGTH)-1 downto 0).
  -- Result: Adds two SIGNED vectors that may be of different lengths.

  function "+" (L: UNSIGNED; R: NATURAL) return UNSIGNED;
  -- Result subtype: UNSIGNED(L'LENGTH-1 downto 0).
  -- Result: Adds an UNSIGNED vector, L, with a non-negative INTEGER, R.

  function "+" (L: NATURAL; R: UNSIGNED) return UNSIGNED;
  -- Result subtype: UNSIGNED(R'LENGTH-1 downto 0).
  -- Result: Adds a non-negative INTEGER, L, with an UNSIGNED vector, R.

  function "+" (L: INTEGER; R: SIGNED) return SIGNED;
  -- Result subtype: SIGNED(R'LENGTH-1 downto 0).
  -- Result: Adds an INTEGER, L(may be positive or negative), to a SIGNED
  --         vector, R.
```

```
function "+" (L: SIGNED; R: INTEGER) return SIGNED;
-- Result subtype: SIGNED(L'LENGTH-1 downto 0).
-- Result: Adds a SIGNED vector, L, to an INTEGER, R.

--=========================================================================
function "-" (L, R: UNSIGNED) return UNSIGNED;
-- Result subtype: UNSIGNED(MAX(L'LENGTH, R'LENGTH)-1 downto 0).
-- Result: Subtracts two UNSIGNED vectors that may be of different lengths.

function "-" (L, R: SIGNED) return SIGNED;
-- Result subtype: SIGNED(MAX(L'LENGTH, R'LENGTH)-1 downto 0).
-- Result: Subtracts a SIGNED vector, R, from another SIGNED vector, L,
--         that may possibly be of different lengths.

function "-" (L: UNSIGNED;R: NATURAL) return UNSIGNED;
-- Result subtype: UNSIGNED(L'LENGTH-1 downto 0).
-- Result: Subtracts a non-negative INTEGER, R, from an UNSIGNED vector, L.

function "-" (L: NATURAL; R: UNSIGNED) return UNSIGNED;
-- Result subtype: UNSIGNED(R'LENGTH-1 downto 0).
-- Result: Subtracts an UNSIGNED vector, R, from a non-negative INTEGER, L.

function "-" (L: SIGNED; R: INTEGER) return SIGNED;
-- Result subtype: SIGNED(L'LENGTH-1 downto 0).
-- Result: Subtracts an INTEGER, R, from a SIGNED vector, L.

function "-" (L: INTEGER; R: SIGNED) return SIGNED;
-- Result subtype: SIGNED(R'LENGTH-1 downto 0).
-- Result: Subtracts a SIGNED vector, R, from an INTEGER, L.

--=========================================================================
function "*" (L, R: UNSIGNED) return UNSIGNED;
-- Result subtype: UNSIGNED((L'LENGTH+R'LENGTH-1) downto 0).
-- Result: Performs the multiplication operation on two UNSIGNED vectors
--         that may possibly be of different lengths.

function "*" (L, R: SIGNED) return SIGNED;
-- Result subtype: SIGNED((L'LENGTH+R'LENGTH-1) downto 0)
-- Result: Multiplies two SIGNED vectors that may possibly be of
--         different lengths.

function "*" (L: UNSIGNED; R: NATURAL) return UNSIGNED;
-- Result subtype: UNSIGNED((L'LENGTH+L'LENGTH-1) downto 0).
-- Result: Multiplies an UNSIGNED vector, L, with a non-negative
--         INTEGER, R. R is converted to an UNSIGNED vector of
--         SIZE L'LENGTH before multiplication.

function "*" (L: NATURAL; R: UNSIGNED) return UNSIGNED;
-- Result subtype: UNSIGNED((R'LENGTH+R'LENGTH-1) downto 0).
-- Result: Multiplies an UNSIGNED vector, R, with a non-negative
--         INTEGER, L. L is converted to an UNSIGNED vector of
--         SIZE R'LENGTH before multiplication.

function "*" (L: SIGNED; R: INTEGER) return SIGNED;
-- Result subtype: SIGNED((L'LENGTH+L'LENGTH-1) downto 0)
-- Result: Multiplies a SIGNED vector, L, with an INTEGER, R. R is
--         converted to a SIGNED vector of SIZE L'LENGTH before
--         multiplication.

function "*" (L: INTEGER; R: SIGNED) return SIGNED;
-- Result subtype: SIGNED((R'LENGTH+R'LENGTH-1) downto 0)
-- Result: Multiplies a SIGNED vector, R, with an INTEGER, L. L is
--         converted to a SIGNED vector of SIZE R'LENGTH before
--         multiplication.

--=========================================================================
--
-- NOTE: If second argument is zero for "/" operator, a severity level
--       of ERROR is issued.
```

```
function "/" (L, R: UNSIGNED) return UNSIGNED;
  attribute builtin_subprogram of
      "/"[UNSIGNED, UNSIGNED return UNSIGNED]: function is "numstd_div_uuu";
-- Result subtype: UNSIGNED(L'LENGTH-1 downto 0)
-- Result: Divides an UNSIGNED vector, L, by another UNSIGNED vector, R.

function "/" (L, R: SIGNED) return SIGNED;
-- Result subtype: SIGNED(L'LENGTH-1 downto 0)
-- Result: Divides an SIGNED vector, L, by another SIGNED vector, R.

function "/" (L: UNSIGNED; R: NATURAL) return UNSIGNED;
-- Result subtype: UNSIGNED(L'LENGTH-1 downto 0)
-- Result: Divides an UNSIGNED vector, L, by a non-negative INTEGER, R.
--         If NO_OF_BITS(R) > L'LENGTH, result is truncated to L'LENGTH.

function "/" (L: NATURAL; R: UNSIGNED) return UNSIGNED;
-- Result subtype: UNSIGNED(R'LENGTH-1 downto 0)
-- Result: Divides a non-negative INTEGER, L, by an UNSIGNED vector, R.
--         If NO_OF_BITS(L) > R'LENGTH, result is truncated to R'LENGTH.

function "/" (L: SIGNED; R: INTEGER) return SIGNED;
-- Result subtype: SIGNED(L'LENGTH-1 downto 0)
-- Result: Divides a SIGNED vector, L, by an INTEGER, R.
--         If NO_OF_BITS(R) > L'LENGTH, result is truncated to L'LENGTH.

function "/" (L: INTEGER; R: SIGNED) return SIGNED;
-- Result subtype: SIGNED(R'LENGTH-1 downto 0)
-- Result: Divides an INTEGER, L, by a SIGNED vector, R.
--         If NO_OF_BITS(L) > R'LENGTH, result is truncated to R'LENGTH.

--============================================================================
--
-- NOTE: If second argument is zero for "rem" operator, a severity level
--       of ERROR is issued.

function "rem" (L, R: UNSIGNED) return UNSIGNED;
-- Result subtype: UNSIGNED(R'LENGTH-1 downto 0)
-- Result: Computes "L rem R" where L and R are UNSIGNED vectors.

function "rem" (L, R: SIGNED) return SIGNED;
-- Result subtype: SIGNED(R'LENGTH-1 downto 0)
-- Result: Computes "L rem R" where L and R are SIGNED vectors.

function "rem" (L: UNSIGNED; R: NATURAL) return UNSIGNED;
-- Result subtype: UNSIGNED(L'LENGTH-1 downto 0)
-- Result: Computes "L rem R" where L is an UNSIGNED vector and R is a
--         non-negative INTEGER.
--         If NO_OF_BITS(R) > L'LENGTH, result is truncated to L'LENGTH.

function "rem" (L: NATURAL; R: UNSIGNED) return UNSIGNED;
-- Result subtype: UNSIGNED(R'LENGTH-1 downto 0)
-- Result: Computes "L rem R" where R is an UNSIGNED vector and L is a
--         non-negative INTEGER.
--         If NO_OF_BITS(L) > R'LENGTH, result is truncated to R'LENGTH.

function "rem" (L: SIGNED; R: INTEGER) return SIGNED;
-- Result subtype: SIGNED(L'LENGTH-1 downto 0)
-- Result: Computes "L rem R" where L is SIGNED vector and R is an INTEGER.
--         If NO_OF_BITS(R) > L'LENGTH, result is truncated to L'LENGTH.

function "rem" (L: INTEGER; R: SIGNED) return SIGNED;
-- Result subtype: SIGNED(R'LENGTH-1 downto 0)
-- Result: Computes "L rem R" where R is SIGNED vector and L is an INTEGER.
--         If NO_OF_BITS(L) > R'LENGTH, result is truncated to R'LENGTH.

--============================================================================
--
-- NOTE: If second argument is zero for "mod" operator, a severity level
```

```
--       of ERROR is issued.

function "mod" (L, R: UNSIGNED) return UNSIGNED;
-- Result subtype: UNSIGNED(R'LENGTH-1 downto 0)
-- Result: Computes "L mod R" where L and R are UNSIGNED vectors.

function "mod" (L, R: SIGNED) return SIGNED;
-- Result subtype: SIGNED(R'LENGTH-1 downto 0)
-- Result: Computes "L mod R" where L and R are SIGNED vectors.

function "mod" (L: UNSIGNED; R: NATURAL) return UNSIGNED;
-- Result subtype: UNSIGNED(L'LENGTH-1 downto 0)
-- Result: Computes "L mod R" where L is an UNSIGNED vector and R
--          is a non-negative INTEGER.
--          If NO_OF_BITS(R) > L'LENGTH, result is truncated to L'LENGTH.

function "mod" (L: NATURAL; R: UNSIGNED) return UNSIGNED;
-- Result subtype: UNSIGNED(R'LENGTH-1 downto 0)
-- Result: Computes "L mod R" where R is an UNSIGNED vector and L
--          is a non-negative INTEGER.
--          If NO_OF_BITS(L) > R'LENGTH, result is truncated to R'LENGTH.

function "mod" (L: SIGNED; R: INTEGER) return SIGNED;
-- Result subtype: SIGNED(L'LENGTH-1 downto 0)
-- Result: Computes "L mod R" where L is a SIGNED vector and
--          R is an INTEGER.
--          If NO_OF_BITS(R) > L'LENGTH, result is truncated to L'LENGTH.

function "mod" (L: INTEGER; R: SIGNED) return SIGNED;
-- Result subtype: SIGNED(R'LENGTH-1 downto 0)
-- Result: Computes "L mod R" where L is an INTEGER and
--          R is a SIGNED vector.
--          If NO_OF_BITS(L) > R'LENGTH, result is truncated to R'LENGTH.

--============================================================================
-- Comparison Operators
--============================================================================
function ">" (L, R: UNSIGNED) return BOOLEAN;
-- Result: Computes "L > R" where L and R are UNSIGNED vectors possibly
--          of different lengths.

function ">" (L, R: SIGNED) return BOOLEAN;
-- Result: Computes "L > R" where L and R are SIGNED vectors possibly
--          of different lengths.

function ">" (L: NATURAL; R: UNSIGNED) return BOOLEAN;
-- Result: Computes "L > R" where L is a non-negative INTEGER and
--          R is an UNSIGNED vector.

function ">" (L: INTEGER; R: SIGNED) return BOOLEAN;
-- Result: Computes "L > R" where L is a INTEGER and
--          R is a SIGNED vector.

function ">" (L: UNSIGNED; R: NATURAL) return BOOLEAN;
-- Result: Computes "L > R" where L is an UNSIGNED vector and
--          R is a non-negative INTEGER.

function ">" (L: SIGNED; R: INTEGER) return BOOLEAN;
-- Result: Computes "L > R" where L is a SIGNED vector and
--          R is a INTEGER.

--============================================================================
function "<" (L, R: UNSIGNED) return BOOLEAN;
-- Result: Computes "L < R" where L and R are UNSIGNED vectors possibly
--          of different lengths.

function "<" (L, R: SIGNED) return BOOLEAN;
-- Result: Computes "L < R" where L and R are SIGNED vectors possibly
--          of different lengths.
```

```
function "<" (L: NATURAL; R: UNSIGNED) return BOOLEAN;
-- Result: Computes "L < R" where L is a non-negative INTEGER and
--         R is an UNSIGNED vector.

function "<" (L: INTEGER; R: SIGNED) return BOOLEAN;
-- Result: Computes "L < R" where L is an INTEGER and
--         R is a SIGNED vector.

function "<" (L: UNSIGNED; R: NATURAL) return BOOLEAN;
-- Result: Computes "L < R" where L is an UNSIGNED vector and
--         R is a non-negative INTEGER.

function "<" (L: SIGNED; R: INTEGER) return BOOLEAN;
-- Result: Computes "L < R" where L is a SIGNED vector and
--         R is an INTEGER.

--============================================================================
function "<=" (L, R: UNSIGNED) return BOOLEAN;
-- Result: Computes "L <= R" where L and R are UNSIGNED vectors possibly
--         of different lengths.

function "<=" (L, R: SIGNED) return BOOLEAN;
-- Result: Computes "L <= R" where L and R are SIGNED vectors possibly
--         of different lengths.

function "<=" (L: NATURAL; R: UNSIGNED) return BOOLEAN;
-- Result: Computes "L <= R" where L is a non-negative INTEGER and
--         R is an UNSIGNED vector.

function "<=" (L: INTEGER; R: SIGNED) return BOOLEAN;
-- Result: Computes "L <= R" where L is an INTEGER and
--         R is a SIGNED vector.

function "<=" (L: UNSIGNED; R: NATURAL) return BOOLEAN;
-- Result: Computes "L <= R" where L is an UNSIGNED vector and
--         R is a non-negative INTEGER.

function "<=" (L: SIGNED; R: INTEGER) return BOOLEAN;
-- Result: Computes "L <= R" where L is a SIGNED vector and
--         R is an INTEGER.

--============================================================================
function ">=" (L, R: UNSIGNED) return BOOLEAN;
-- Result: Computes "L >= R" where L and R are UNSIGNED vectors possibly
--         of different lengths.

function ">=" (L, R: SIGNED) return BOOLEAN;
-- Result: Computes "L >= R" where L and R are SIGNED vectors possibly
--         of different lengths.
```

```
function ">=" (L: NATURAL; R: UNSIGNED) return BOOLEAN;
-- Result subtype: BOOLEAN
-- Result: Computes "L >= R" where L is a non-negative INTEGER and
--         R is an UNSIGNED vector.

function ">=" (L: INTEGER; R: SIGNED) return BOOLEAN;
-- Result: Computes "L >= R" where L is an INTEGER and
--         R is a SIGNED vector.

function ">=" (L: UNSIGNED; R: NATURAL) return BOOLEAN;
-- Result: Computes "L >= R" where L is an UNSIGNED vector and
--         R is a non-negative INTEGER.

function ">=" (L: SIGNED; R: INTEGER) return BOOLEAN;
-- Result: Computes "L >= R" where L is a SIGNED vector and
--         R is an INTEGER.
--=============================================================================
function "=" (L, R: UNSIGNED) return BOOLEAN;
-- Result: Computes "L = R" where L and R are UNSIGNED vectors possibly
--         of different lengths.

function "=" (L, R: SIGNED) return BOOLEAN;
-- Result: Computes "L = R" where L and R are SIGNED vectors possibly
--         of different lengths.

function "=" (L: NATURAL; R: UNSIGNED) return BOOLEAN;
-- Result: Computes "L = R" where L is a non-negative INTEGER and
--         R is an UNSIGNED vector.

function "=" (L: INTEGER; R: SIGNED) return BOOLEAN;
-- Result: Computes "L = R" where L is an INTEGER and
--         R is a SIGNED vector.

function "=" (L: UNSIGNED; R: NATURAL) return BOOLEAN;
-- Result: Computes "L = R" where L is an UNSIGNED vector and
--         R is a non-negative INTEGER.

function "=" (L: SIGNED; R: INTEGER) return BOOLEAN;
-- Result: Computes "L = R" where L is a SIGNED vector and
--         R is an INTEGER.

--=============================================================================
function "/=" (L, R: UNSIGNED) return BOOLEAN;
-- Result: Computes "L /= R" where L and R are UNSIGNED vectors possibly
--         of different lengths.

function "/=" (L, R: SIGNED) return BOOLEAN;
-- Result: Computes "L /= R" where L and R are SIGNED vectors possibly
--         of different lengths.

function "/=" (L: NATURAL; R: UNSIGNED) return BOOLEAN;
-- Result: Computes "L /= R" where L is a non-negative INTEGER and
--         R is an UNSIGNED vector.

function "/=" (L: INTEGER; R: SIGNED) return BOOLEAN;
-- Result: Computes "L /= R" where L is an INTEGER and
--         R is a SIGNED vector.

function "/=" (L: UNSIGNED; R: NATURAL) return BOOLEAN;
-- Result: Computes "L /= R" where L is an UNSIGNED vector and
--         R is a non-negative INTEGER.

function "/=" (L: SIGNED; R: INTEGER) return BOOLEAN;
-- Result: Computes "L /= R" where L is a SIGNED vector and
--         R is an INTEGER.

--=============================================================================
-- Shift and Rotate Functions
--=============================================================================
```

```
function SHIFT_LEFT (ARG: UNSIGNED; COUNT: NATURAL) return UNSIGNED;
-- Result subtype: UNSIGNED(ARG'LENGTH-1 downto 0)
-- Result: Performs a shift-left on an UNSIGNED vector COUNT times.
--         The vacated positions are filled with '0'.
--         The COUNT leftmost elements are lost.

function SHIFT_RIGHT (ARG: UNSIGNED; COUNT: NATURAL) return UNSIGNED;
-- Result subtype: UNSIGNED(ARG'LENGTH-1 downto 0)
-- Result: Performs a shift-right on an UNSIGNED vector COUNT times.
--         The vacated positions are filled with '0'.
--         The COUNT rightmost elements are lost.

function SHIFT_LEFT (ARG: SIGNED; COUNT: NATURAL) return SIGNED;
-- Result subtype: SIGNED(ARG'LENGTH-1 downto 0)
-- Result: Performs a shift-left on a SIGNED vector COUNT times.
--         The vacated positions are filled with '0'.
--         The COUNT leftmost elements are lost.

function SHIFT_RIGHT (ARG: SIGNED; COUNT: NATURAL) return SIGNED;
-- Result subtype: SIGNED(ARG'LENGTH-1 downto 0)
-- Result: Performs a shift-right on a SIGNED vector COUNT times.
--         The vacated positions are filled with the leftmost
--         element, ARG'LEFT. The COUNT rightmost elements are lost.

--============================================================================
function ROTATE_LEFT (ARG: UNSIGNED; COUNT: NATURAL) return UNSIGNED;
-- Result subtype: UNSIGNED(ARG'LENGTH-1 downto 0)
-- Result: Performs a rotate-left of an UNSIGNED vector COUNT times.

function ROTATE_RIGHT (ARG: UNSIGNED; COUNT: NATURAL) return UNSIGNED;
-- Result subtype: UNSIGNED(ARG'LENGTH-1 downto 0)
-- Result: Performs a rotate-right of an UNSIGNED vector COUNT times.

function ROTATE_LEFT (ARG: SIGNED; COUNT: NATURAL) return SIGNED;
-- Result subtype: SIGNED(ARG'LENGTH-1 downto 0)
-- Result: Performs a logical rotate-left of a SIGNED
--         vector COUNT times.

function ROTATE_RIGHT (ARG: SIGNED; COUNT: NATURAL) return SIGNED;
-- Result subtype: SIGNED(ARG'LENGTH-1 downto 0)
-- Result: Performs a logical rotate-right of a SIGNED
--         vector COUNT times.

--============================================================================

--============================================================================
------------------------------------------------------------------------------
--   Note : Function S.9 is not compatible with VHDL 1076-1987. Comment
--   out the function (declaration and body) for VHDL 1076-1987 compatibility.
------------------------------------------------------------------------------
function "sll" (ARG: UNSIGNED; COUNT: INTEGER) return UNSIGNED;
-- Result subtype: UNSIGNED(ARG'LENGTH-1 downto 0)
-- Result: SHIFT_LEFT(ARG, COUNT)

------------------------------------------------------------------------------
-- Note : Function S.10 is not compatible with VHDL 1076-1987. Comment
--   out the function (declaration and body) for VHDL 1076-1987 compatibility.
------------------------------------------------------------------------------
function "sll" (ARG: SIGNED; COUNT: INTEGER) return SIGNED;
-- Result subtype: SIGNED(ARG'LENGTH-1 downto 0)
-- Result: SHIFT_LEFT(ARG, COUNT)
------------------------------------------------------------------------------
--   Note : Function S.11 is not compatible with VHDL 1076-1987. Comment
--   out the function (declaration and body) for VHDL 1076-1987 compatibility.
------------------------------------------------------------------------------
function "srl" (ARG: UNSIGNED; COUNT: INTEGER) return UNSIGNED;
-- Result subtype: UNSIGNED(ARG'LENGTH-1 downto 0)
-- Result: SHIFT_RIGHT(ARG, COUNT)
```

```
--   Note : Function S.12 is not compatible with VHDL 1076-1987. Comment
--   out the function (declaration and body) for VHDL 1076-1987 compatibility.
------------------------------------------------------------------------------
function "srl" (ARG: SIGNED; COUNT: INTEGER) return SIGNED;
-- Result subtype: SIGNED(ARG'LENGTH-1 downto 0)
-- Result: SIGNED(SHIFT_RIGHT(UNSIGNED(ARG), COUNT))

------------------------------------------------------------------------------
--   Note : Function S.13 is not compatible with VHDL 1076-1987. Comment
-- out the function (declaration and body) for VHDL 1076-1987 compatibility.
------------------------------------------------------------------------------
function "rol" (ARG: UNSIGNED; COUNT: INTEGER) return UNSIGNED;
-- Result subtype: UNSIGNED(ARG'LENGTH-1 downto 0)
-- Result: ROTATE_LEFT(ARG, COUNT)

------------------------------------------------------------------------------
--   Note : Function S.14 is not compatible with VHDL 1076-1987. Comment
--   out the function (declaration and body) for VHDL 1076-1987 compatibility.
------------------------------------------------------------------------------
function "rol" (ARG: SIGNED; COUNT: INTEGER) return SIGNED;
-- Result subtype: SIGNED(ARG'LENGTH-1 downto 0)
-- Result: ROTATE_LEFT(ARG, COUNT)

------------------------------------------------------------------------------
-- Note : Function S.15 is not compatible with VHDL 1076-1987. Comment
--   out the function (declaration and body) for VHDL 1076-1987 compatibility.
------------------------------------------------------------------------------
function "ror" (ARG: UNSIGNED; COUNT: INTEGER) return UNSIGNED;
-- Result subtype: UNSIGNED(ARG'LENGTH-1 downto 0)
-- Result: ROTATE_RIGHT(ARG, COUNT)

------------------------------------------------------------------------------
--   Note : Function S.16 is not compatible with VHDL 1076-1987. Comment
--   out the function (declaration and body) for VHDL 1076-1987 compatibility.
------------------------------------------------------------------------------
function "ror" (ARG: SIGNED; COUNT: INTEGER) return SIGNED;
-- Result subtype: SIGNED(ARG'LENGTH-1 downto 0)
-- Result: ROTATE_RIGHT(ARG, COUNT)

--============================================================================
--   RESIZE Functions
--============================================================================
function RESIZE (ARG: SIGNED; NEW_SIZE: NATURAL) return SIGNED;
-- Result subtype: SIGNED(NEW_SIZE-1 downto 0)
-- Result: Resizes the SIGNED vector ARG to the specified size.
--         To create a larger vector, the new [leftmost] bit positions
--         are filled with the sign bit (ARG'LEFT). When truncating,
--         the sign bit is retained along with the rightmost part.

function RESIZE (ARG: UNSIGNED; NEW_SIZE: NATURAL) return UNSIGNED;
-- Result subtype: UNSIGNED(NEW_SIZE-1 downto 0)
-- Result: Resizes the SIGNED vector ARG to the specified size.
--         To create a larger vector, the new [leftmost] bit positions
--         are filled with '0'. When truncating, the leftmost bits
--         are dropped.

--============================================================================
-- Conversion Functions
--============================================================================

function TO_INTEGER (ARG: UNSIGNED) return NATURAL;
-- Result subtype: NATURAL. Value cannot be negative since parameter is an
--           UNSIGNED vector.
-- Result: Converts the UNSIGNED vector to an INTEGER.

function TO_INTEGER (ARG: SIGNED) return INTEGER;
-- Result subtype: INTEGER
-- Result: Converts a SIGNED vector to an INTEGER.
```

```
function TO_UNSIGNED (ARG, SIZE: NATURAL) return UNSIGNED;
-- Result subtype: UNSIGNED(SIZE-1 downto 0)
-- Result: Converts a non-negative INTEGER to an UNSIGNED vector with
--         the specified SIZE.

function TO_SIGNED (ARG: INTEGER; SIZE: NATURAL) return SIGNED;
-- Result subtype: SIGNED(SIZE-1 downto 0)
-- Result: Converts an INTEGER to a SIGNED vector of the specified SIZE.

--============================================================================
-- Logical Operators
--============================================================================
function "not" (L: UNSIGNED) return UNSIGNED;
-- Result subtype: UNSIGNED(L'LENGTH-1 downto 0)
-- Result: Termwise inversion

function "and" (L, R: UNSIGNED) return UNSIGNED;
-- Result subtype: UNSIGNED(L'LENGTH-1 downto 0)
-- Result: Vector AND operation

function "or" (L, R: UNSIGNED) return UNSIGNED;
-- Result subtype: UNSIGNED(L'LENGTH-1 downto 0)
-- Result: Vector OR operation

function "nand" (L, R: UNSIGNED) return UNSIGNED;
-- Result subtype: UNSIGNED(L'LENGTH-1 downto 0)
-- Result: Vector NAND operation

function "nor" (L, R: UNSIGNED) return UNSIGNED;
-- Result subtype: UNSIGNED(L'LENGTH-1 downto 0)
-- Result: Vector NOR operation

function "xor" (L, R: UNSIGNED) return UNSIGNED;
-- Result subtype: UNSIGNED(L'LENGTH-1 downto 0)
-- Result: Vector XOR operation

-- ---------------------------------------------------------------------------
-- Note : Function L.7 is not compatible with VHDL 1076-1987. Comment
-- out the function (declaration and body) for VHDL 1076-1987 compatibility.
-- ---------------------------------------------------------------------------
function "xnor" (L, R: UNSIGNED) return UNSIGNED;
-- Result subtype: UNSIGNED(L'LENGTH-1 downto 0)
-- Result: Vector XNOR operation

function "not" (L: SIGNED) return SIGNED;
-- Result subtype: SIGNED(L'LENGTH-1 downto 0)
-- Result: Termwise inversion

function "and" (L, R: SIGNED) return SIGNED;
-- Result subtype: SIGNED(L'LENGTH-1 downto 0)
-- Result: Vector AND operation

function "or" (L, R: SIGNED) return SIGNED;
-- Result subtype: SIGNED(L'LENGTH-1 downto 0)
-- Result: Vector OR operation

function "nand" (L, R: SIGNED) return SIGNED;
-- Result subtype: SIGNED(L'LENGTH-1 downto 0)
-- Result: Vector NAND operation

function "nor" (L, R: SIGNED) return SIGNED;
-- Result subtype: SIGNED(L'LENGTH-1 downto 0)
-- Result: Vector NOR operation

function "xor" (L, R: SIGNED) return SIGNED;
-- Result subtype: SIGNED(L'LENGTH-1 downto 0)
-- Result: Vector XOR operation
```

```
  -- Note : Function L.14 is not compatible with VHDL 1076-1987. Comment
  -- out the function (declaration and body) for VHDL 1076-1987 compatibility.
  -- -------------------------------------------------------------------------
  function "xnor" (L, R: SIGNED) return SIGNED;
  -- Result subtype: SIGNED(L'LENGTH-1 downto 0)
  -- Result: Vector XNOR operation

  --=========================================================================
  -- Match Functions
  --=========================================================================
  function STD_MATCH (L, R: STD_ULOGIC) return BOOLEAN;
  -- Result: terms compared per STD_LOGIC_1164 intent

  function STD_MATCH (L, R: UNSIGNED) return BOOLEAN;
  -- Result: terms compared per STD_LOGIC_1164 intent

  function STD_MATCH (L, R: SIGNED) return BOOLEAN;
  -- Result: terms compared per STD_LOGIC_1164 intent

  function STD_MATCH (L, R: STD_LOGIC_VECTOR) return BOOLEAN;
  -- Result: terms compared per STD_LOGIC_1164 intent

  function STD_MATCH (L, R: STD_ULOGIC_VECTOR) return BOOLEAN;
  -- Result: terms compared per STD_LOGIC_1164 intent

  --=========================================================================
  -- Translation Functions
  --=========================================================================
  function TO_01 (S: UNSIGNED; XMAP: STD_LOGIC := '0') return UNSIGNED;
  -- Result subtype: UNSIGNED(S'RANGE)
  -- Result: Termwise, 'H' is translated to '1', and 'L' is translated
  --         to '0'. If a value other than '0'|'1'|'H'|'L' is found,
  --         the array is set to (others => XMAP), and a warning is
  --         issued.

  function TO_01 (S: SIGNED; XMAP: STD_LOGIC := '0') return SIGNED;
  -- Result subtype: SIGNED(S'RANGE)
  -- Result: Termwise, 'H' is translated to '1', and 'L' is translated
  --         to '0'. If a value other than '0'|'1'|'H'|'L' is found,
  --         the array is set to (others => XMAP), and a warning is
  --         issued.

end NUMERIC_STD;
```

APPENDIX G : STD_LOGIC_UNSIGNED

```
--------------------------------------------------------------------------
--                                                                      --
-- Copyright (c) 1990, 1991, 1992 by Synopsys, Inc.                     --
--                                             All rights reserved.     --
--                                                                      --
-- This source file may be used and distributed without restriction     --
-- provided that this copyright statement is not removed from the file  --
-- and that any derivative work contains this copyright notice.         --
--                                                                      --
--  Package name: STD_LOGIC_UNSIGNED                                  --
--                                                      --
--                                              --
--      Date:           09/11/92 KN                   --
--               10/08/92      AMT                    --
--                                              --
--  Purpose:                                --
--   A set of unsigned arithemtic, conversion,                        --
--          and comparision functions for STD_LOGIC_VECTOR.             --
--                                              --
-- Note:  comparision of same length discrete arrays is defined     --
--              by the LRM.  This package will "overload" those     --
--              definitions                           --
--                                              --
--------------------------------------------------------------------------

library IEEE;
use IEEE.std_logic_1164.all;
use IEEE.std_logic_arith.all;

package STD_LOGIC_UNSIGNED is

    function "+"(L: STD_LOGIC_VECTOR; R: STD_LOGIC_VECTOR)
       return STD_LOGIC_VECTOR;
    function "+"(L: STD_LOGIC_VECTOR; R: INTEGER) return STD_LOGIC_VECTOR;
    function "+"(L: INTEGER; R: STD_LOGIC_VECTOR) return STD_LOGIC_VECTOR;
    function "+"(L: STD_LOGIC_VECTOR; R: STD_LOGIC) return STD_LOGIC_VECTOR;
    function "+"(L: STD_LOGIC; R: STD_LOGIC_VECTOR) return STD_LOGIC_VECTOR;

    function "-"(L: STD_LOGIC_VECTOR; R: STD_LOGIC_VECTOR)
```

```
      return STD_LOGIC_VECTOR;
   function "-"(L: STD_LOGIC_VECTOR; R: INTEGER) return STD_LOGIC_VECTOR;
   function "-"(L: INTEGER; R: STD_LOGIC_VECTOR) return STD_LOGIC_VECTOR;
   function "-"(L: STD_LOGIC_VECTOR; R: STD_LOGIC) return STD_LOGIC_VECTOR;
   function "-"(L: STD_LOGIC; R: STD_LOGIC_VECTOR) return STD_LOGIC_VECTOR;

   function "+"(L: STD_LOGIC_VECTOR) return STD_LOGIC_VECTOR;

   function "*"(L: STD_LOGIC_VECTOR; R: STD_LOGIC_VECTOR)
      return STD_LOGIC_VECTOR;

   function "<"(L: STD_LOGIC_VECTOR; R: STD_LOGIC_VECTOR) return BOOLEAN;
   function "<"(L: STD_LOGIC_VECTOR; R: INTEGER) return BOOLEAN;
   function "<"(L: INTEGER; R: STD_LOGIC_VECTOR) return BOOLEAN;

   function "<="(L: STD_LOGIC_VECTOR; R: STD_LOGIC_VECTOR) return BOOLEAN;
   function "<="(L: STD_LOGIC_VECTOR; R: INTEGER) return BOOLEAN;
   function "<="(L: INTEGER; R: STD_LOGIC_VECTOR) return BOOLEAN;

   function ">"(L: STD_LOGIC_VECTOR; R: STD_LOGIC_VECTOR) return BOOLEAN;
   function ">"(L: STD_LOGIC_VECTOR; R: INTEGER) return BOOLEAN;
   function ">"(L: INTEGER; R: STD_LOGIC_VECTOR) return BOOLEAN;

   function ">="(L: STD_LOGIC_VECTOR; R: STD_LOGIC_VECTOR) return BOOLEAN;
   function ">="(L: STD_LOGIC_VECTOR; R: INTEGER) return BOOLEAN;
   function ">="(L: INTEGER; R: STD_LOGIC_VECTOR) return BOOLEAN;

   function "="(L: STD_LOGIC_VECTOR; R: STD_LOGIC_VECTOR) return BOOLEAN;
   function "="(L: STD_LOGIC_VECTOR; R: INTEGER) return BOOLEAN;
   function "="(L: INTEGER; R: STD_LOGIC_VECTOR) return BOOLEAN;

   function "/="(L: STD_LOGIC_VECTOR; R: STD_LOGIC_VECTOR) return BOOLEAN;
   function "/="(L: STD_LOGIC_VECTOR; R: INTEGER) return BOOLEAN;
   function "/="(L: INTEGER; R: STD_LOGIC_VECTOR) return BOOLEAN;
   function SHL(ARG:STD_LOGIC_VECTOR;COUNT: STD_LOGIC_VECTOR)
     return STD_LOGIC_VECTOR;
   function SHR(ARG:STD_LOGIC_VECTOR;COUNT: STD_LOGIC_VECTOR)
     return STD_LOGIC_VECTOR;

   function CONV_INTEGER(ARG: STD_LOGIC_VECTOR) return INTEGER;

-- remove this since it is already in std_logic_arith
--    function CONV_STD_LOGIC_VECTOR(ARG: INTEGER; SIZE: INTEGER)
--      return STD_LOGIC_VECTOR;
end STD_LOGIC_UNSIGNED;
```

APPENDIX H : STD_LOGIC_SIGNED

```
--------------------------------------------------------------------------
--                                                                      --
-- Copyright (c) 1990, 1991, 1992 by Synopsys, Inc.                     --
--                                             All rights reserved.     --
--                                                                      --
-- This source file may be used and distributed without restriction     --
-- provided that this copyright statement is not removed from the file  --
-- and that any derivative work contains this copyright notice.         --
--                                                                      --
--  Package name: STD_LOGIC_SIGNED                                  --
--                                                         --
--                                                --
--      Date:        09/11/91 KN                                        --
--                   10/08/92 AMT change std_ulogic to signed std_logic --
--                 10/28/92 AMT added signed functions, -, ABS      --
--                                                --
--  Purpose:                                  --
--   A set of signed arithemtic, conversion,                        --
--          and comparision functions for STD_LOGIC_VECTOR.             --
--                                                --
--  Note:      Comparision of same length std_logic_vector is defined  --
--             in the LRM.  The interpretation is for unsigned vectors --
--             This package will "overload" that definition.       --
--                                                --
--------------------------------------------------------------------------

library IEEE;
use IEEE.std_logic_1164.all;
use IEEE.std_logic_arith.all;

package STD_LOGIC_SIGNED is

    function "+"(L: STD_LOGIC_VECTOR; R: STD_LOGIC_VECTOR)
       return STD_LOGIC_VECTOR;
    function "+"(L: STD_LOGIC_VECTOR; R: INTEGER) return STD_LOGIC_VECTOR;
    function "+"(L: INTEGER; R: STD_LOGIC_VECTOR) return STD_LOGIC_VECTOR;
    function "+"(L: STD_LOGIC_VECTOR; R: STD_LOGIC) return STD_LOGIC_VECTOR;
    function "+"(L: STD_LOGIC; R: STD_LOGIC_VECTOR) return STD_LOGIC_VECTOR;
```

```
    function "-"(L: STD_LOGIC_VECTOR; R: STD_LOGIC_VECTOR)
       return STD_LOGIC_VECTOR;
    function "-"(L: STD_LOGIC_VECTOR; R: INTEGER) return STD_LOGIC_VECTOR;
    function "-"(L: INTEGER; R: STD_LOGIC_VECTOR) return STD_LOGIC_VECTOR;
    function "-"(L: STD_LOGIC_VECTOR; R: STD_LOGIC) return STD_LOGIC_VECTOR;
    function "-"(L: STD_LOGIC; R: STD_LOGIC_VECTOR) return STD_LOGIC_VECTOR;

    function "+"(L: STD_LOGIC_VECTOR) return STD_LOGIC_VECTOR;
    function "-"(L: STD_LOGIC_VECTOR) return STD_LOGIC_VECTOR;
    function "ABS"(L: STD_LOGIC_VECTOR) return STD_LOGIC_VECTOR;


    function "*"(L: STD_LOGIC_VECTOR; R: STD_LOGIC_VECTOR)
      return STD_LOGIC_VECTOR;

    function "<"(L: STD_LOGIC_VECTOR; R: STD_LOGIC_VECTOR) return BOOLEAN;
    function "<"(L: STD_LOGIC_VECTOR; R: INTEGER) return BOOLEAN;
    function "<"(L: INTEGER; R: STD_LOGIC_VECTOR) return BOOLEAN;

    function "<="(L: STD_LOGIC_VECTOR; R: STD_LOGIC_VECTOR) return BOOLEAN;
    function "<="(L: STD_LOGIC_VECTOR; R: INTEGER) return BOOLEAN;
    function "<="(L: INTEGER; R: STD_LOGIC_VECTOR) return BOOLEAN;

    function ">"(L: STD_LOGIC_VECTOR; R: STD_LOGIC_VECTOR) return BOOLEAN;
    function ">"(L: STD_LOGIC_VECTOR; R: INTEGER) return BOOLEAN;
    function ">"(L: INTEGER; R: STD_LOGIC_VECTOR) return BOOLEAN;

    function ">="(L: STD_LOGIC_VECTOR; R: STD_LOGIC_VECTOR) return BOOLEAN;
    function ">="(L: STD_LOGIC_VECTOR; R: INTEGER) return BOOLEAN;
    function ">="(L: INTEGER; R: STD_LOGIC_VECTOR) return BOOLEAN;

    function "="(L: STD_LOGIC_VECTOR; R: STD_LOGIC_VECTOR) return BOOLEAN;
    function "="(L: STD_LOGIC_VECTOR; R: INTEGER) return BOOLEAN;
    function "="(L: INTEGER; R: STD_LOGIC_VECTOR) return BOOLEAN;

    function "/="(L: STD_LOGIC_VECTOR; R: STD_LOGIC_VECTOR) return BOOLEAN;
    function "/="(L: STD_LOGIC_VECTOR; R: INTEGER) return BOOLEAN;
    function "/="(L: INTEGER; R: STD_LOGIC_VECTOR) return BOOLEAN;
    function SHL(ARG:STD_LOGIC_VECTOR;COUNT: STD_LOGIC_VECTOR)
      return STD_LOGIC_VECTOR;
    function SHR(ARG:STD_LOGIC_VECTOR;COUNT: STD_LOGIC_VECTOR)
      return STD_LOGIC_VECTOR;

    function CONV_INTEGER(ARG: STD_LOGIC_VECTOR) return INTEGER;

-- remove this since it is already in std_logic_arith
--    function CONV_STD_LOGIC_VECTOR(ARG: INTEGER; SIZE: INTEGER)
--      return STD_LOGIC_VECTOR;

end STD_LOGIC_SIGNED;
```

APPENDIX I : STD_LOGIC_ARITH

```
--------------------------------------------------------------------
--                                                                --
-- Copyright (c) 1990, 1991 by Synopsys, Inc.  All rights reserved.   --
--                                                                --
-- This source file may be used and distributed without restriction  --
-- provided that this copyright statement is not removed from the file --
-- and that any derivative work contains this copyright notice.      --
--                                                                --
--      Package name: STD_LOGIC_ARITH                             --
--                                                                --
--      Purpose:                                                  --
--       A set of arithemtic, conversion, and comparison functions   --
--       for SIGNED, UNSIGNED, SMALL_INT, INTEGER,                --
--       STD_ULOGIC, STD_LOGIC, and STD_LOGIC_VECTOR.             --
--                                                                --
--------------------------------------------------------------------
library IEEE;
  use IEEE.std_logic_1164.all;
package std_logic_arith is
    type UNSIGNED is array (NATURAL range <>) of STD_LOGIC;
    type SIGNED is array (NATURAL range <>) of STD_LOGIC;
    subtype SMALL_INT is INTEGER range 0 to 1;

    function "+"(L: UNSIGNED; R: UNSIGNED) return UNSIGNED;
    function "+"(L: SIGNED; R: SIGNED) return SIGNED;
    function "+"(L: UNSIGNED; R: SIGNED) return SIGNED;
    function "+"(L: SIGNED; R: UNSIGNED) return SIGNED;
    function "+"(L: UNSIGNED; R: INTEGER) return UNSIGNED;
    function "+"(L: INTEGER; R: UNSIGNED) return UNSIGNED;
    function "+"(L: SIGNED; R: INTEGER) return SIGNED;
    function "+"(L: INTEGER; R: SIGNED) return SIGNED;
    function "+"(L: UNSIGNED; R: STD_ULOGIC) return UNSIGNED;
    function "+"(L: STD_ULOGIC; R: UNSIGNED) return UNSIGNED;
    function "+"(L: SIGNED; R: STD_ULOGIC) return SIGNED;
    function "+"(L: STD_ULOGIC; R: SIGNED) return SIGNED;
    function "+"(L: UNSIGNED; R: UNSIGNED) return STD_LOGIC_VECTOR;
    function "+"(L: SIGNED; R: SIGNED) return STD_LOGIC_VECTOR;
    function "+"(L: UNSIGNED; R: SIGNED) return STD_LOGIC_VECTOR;
```

```
function "+"(L: SIGNED; R: UNSIGNED) return STD_LOGIC_VECTOR;
function "+"(L: UNSIGNED; R: INTEGER) return STD_LOGIC_VECTOR;
function "+"(L: INTEGER; R: UNSIGNED) return STD_LOGIC_VECTOR;
function "+"(L: SIGNED; R: INTEGER) return STD_LOGIC_VECTOR;
function "+"(L: INTEGER; R: SIGNED) return STD_LOGIC_VECTOR;
function "+"(L: UNSIGNED; R: STD_ULOGIC) return STD_LOGIC_VECTOR;
function "+"(L: STD_ULOGIC; R: UNSIGNED) return STD_LOGIC_VECTOR;
function "+"(L: SIGNED; R: STD_ULOGIC) return STD_LOGIC_VECTOR;
function "+"(L: STD_ULOGIC; R: SIGNED) return STD_LOGIC_VECTOR;

function "-"(L: UNSIGNED; R: UNSIGNED) return UNSIGNED;
function "-"(L: SIGNED; R: SIGNED) return SIGNED;
function "-"(L: UNSIGNED; R: SIGNED) return SIGNED;
function "-"(L: SIGNED; R: UNSIGNED) return SIGNED;
function "-"(L: UNSIGNED; R: INTEGER) return UNSIGNED;
function "-"(L: INTEGER; R: UNSIGNED) return UNSIGNED;
function "-"(L: SIGNED; R: INTEGER) return SIGNED;
function "-"(L: INTEGER; R: SIGNED) return SIGNED;
function "-"(L: UNSIGNED; R: STD_ULOGIC) return UNSIGNED;
function "-"(L: STD_ULOGIC; R: UNSIGNED) return UNSIGNED;
function "-"(L: SIGNED; R: STD_ULOGIC) return SIGNED;
function "-"(L: STD_ULOGIC; R: SIGNED) return SIGNED;

function "-"(L: UNSIGNED; R: UNSIGNED) return STD_LOGIC_VECTOR;
function "-"(L: SIGNED; R: SIGNED) return STD_LOGIC_VECTOR;
function "-"(L: UNSIGNED; R: SIGNED) return STD_LOGIC_VECTOR;
function "-"(L: SIGNED; R: UNSIGNED) return STD_LOGIC_VECTOR;
function "-"(L: UNSIGNED; R: INTEGER) return STD_LOGIC_VECTOR;
function "-"(L: INTEGER; R: UNSIGNED) return STD_LOGIC_VECTOR;
function "-"(L: SIGNED; R: INTEGER) return STD_LOGIC_VECTOR;
function "-"(L: INTEGER; R: SIGNED) return STD_LOGIC_VECTOR;
function "-"(L: UNSIGNED; R: STD_ULOGIC) return STD_LOGIC_VECTOR;
function "-"(L: STD_ULOGIC; R: UNSIGNED) return STD_LOGIC_VECTOR;
function "-"(L: SIGNED; R: STD_ULOGIC) return STD_LOGIC_VECTOR;
function "-"(L: STD_ULOGIC; R: SIGNED) return STD_LOGIC_VECTOR;

function "+"(L: UNSIGNED) return UNSIGNED;
function "+"(L: SIGNED) return SIGNED;
function "-"(L: SIGNED) return SIGNED;
function "ABS"(L: SIGNED) return SIGNED;

function "+"(L: UNSIGNED) return STD_LOGIC_VECTOR;
function "+"(L: SIGNED) return STD_LOGIC_VECTOR;
function "-"(L: SIGNED) return STD_LOGIC_VECTOR;
function "ABS"(L: SIGNED) return STD_LOGIC_VECTOR;

function "*"(L: UNSIGNED; R: UNSIGNED) return UNSIGNED;
function "*"(L: SIGNED; R: SIGNED) return SIGNED;
function "*"(L: SIGNED; R: UNSIGNED) return SIGNED;
function "*"(L: UNSIGNED; R: SIGNED) return SIGNED;

function "*"(L: UNSIGNED; R: UNSIGNED) return STD_LOGIC_VECTOR;
function "*"(L: SIGNED; R: SIGNED) return STD_LOGIC_VECTOR;
function "*"(L: SIGNED; R: UNSIGNED) return STD_LOGIC_VECTOR;
function "*"(L: UNSIGNED; R: SIGNED) return STD_LOGIC_VECTOR;

function "<"(L: UNSIGNED; R: UNSIGNED) return BOOLEAN;
function "<"(L: SIGNED; R: SIGNED) return BOOLEAN;
function "<"(L: UNSIGNED; R: SIGNED) return BOOLEAN;
function "<"(L: SIGNED; R: UNSIGNED) return BOOLEAN;
function "<"(L: UNSIGNED; R: INTEGER) return BOOLEAN;
function "<"(L: INTEGER; R: UNSIGNED) return BOOLEAN;
function "<"(L: SIGNED; R: INTEGER) return BOOLEAN;
function "<"(L: INTEGER; R: SIGNED) return BOOLEAN;

function "<="(L: UNSIGNED; R: UNSIGNED) return BOOLEAN;
function "<="(L: SIGNED; R: SIGNED) return BOOLEAN;
```

```
function "<="(L: UNSIGNED; R: SIGNED) return BOOLEAN;
function "<="(L: SIGNED; R: UNSIGNED) return BOOLEAN;
function "<="(L: UNSIGNED; R: INTEGER) return BOOLEAN;
function "<="(L: INTEGER; R: UNSIGNED) return BOOLEAN;
function "<="(L: SIGNED; R: INTEGER) return BOOLEAN;
function "<="(L: INTEGER; R: SIGNED) return BOOLEAN;

function ">"(L: UNSIGNED; R: UNSIGNED) return BOOLEAN;
function ">"(L: SIGNED; R: SIGNED) return BOOLEAN;
function ">"(L: UNSIGNED; R: SIGNED) return BOOLEAN;
function ">"(L: SIGNED; R: UNSIGNED) return BOOLEAN;
function ">"(L: UNSIGNED; R: INTEGER) return BOOLEAN;
function ">"(L: INTEGER; R: UNSIGNED) return BOOLEAN;
function ">"(L: SIGNED; R: INTEGER) return BOOLEAN;
function ">"(L: INTEGER; R: SIGNED) return BOOLEAN;

function ">="(L: UNSIGNED; R: UNSIGNED) return BOOLEAN;
function ">="(L: SIGNED; R: SIGNED) return BOOLEAN;
function ">="(L: UNSIGNED; R: SIGNED) return BOOLEAN;
function ">="(L: SIGNED; R: UNSIGNED) return BOOLEAN;
function ">="(L: UNSIGNED; R: INTEGER) return BOOLEAN;
function ">="(L: INTEGER; R: UNSIGNED) return BOOLEAN;
function ">="(L: SIGNED; R: INTEGER) return BOOLEAN;
function ">="(L: INTEGER; R: SIGNED) return BOOLEAN;

function "="(L: UNSIGNED; R: UNSIGNED) return BOOLEAN;
function "="(L: SIGNED; R: SIGNED) return BOOLEAN;
function "="(L: UNSIGNED; R: SIGNED) return BOOLEAN;
function "="(L: SIGNED; R: UNSIGNED) return BOOLEAN;
function "="(L: UNSIGNED; R: INTEGER) return BOOLEAN;
function "="(L: INTEGER; R: UNSIGNED) return BOOLEAN;
function "="(L: SIGNED; R: INTEGER) return BOOLEAN;
function "="(L: INTEGER; R: SIGNED) return BOOLEAN;

function "/="(L: UNSIGNED; R: UNSIGNED) return BOOLEAN;
function "/="(L: SIGNED; R: SIGNED) return BOOLEAN;
function "/="(L: UNSIGNED; R: SIGNED) return BOOLEAN;
function "/="(L: SIGNED; R: UNSIGNED) return BOOLEAN;
function "/="(L: UNSIGNED; R: INTEGER) return BOOLEAN;
function "/="(L: INTEGER; R: UNSIGNED) return BOOLEAN;
function "/="(L: SIGNED; R: INTEGER) return BOOLEAN;
function "/="(L: INTEGER; R: SIGNED) return BOOLEAN;

function SHL(ARG: UNSIGNED; COUNT: UNSIGNED) return UNSIGNED;
function SHL(ARG: SIGNED; COUNT: UNSIGNED) return SIGNED;
function SHR(ARG: UNSIGNED; COUNT: UNSIGNED) return UNSIGNED;
function SHR(ARG: SIGNED; COUNT: UNSIGNED) return SIGNED;

function CONV_INTEGER(ARG: INTEGER) return INTEGER;
function CONV_INTEGER(ARG: UNSIGNED) return INTEGER;
function CONV_INTEGER(ARG: SIGNED) return INTEGER;
function CONV_INTEGER(ARG: STD_ULOGIC) return SMALL_INT;

function CONV_UNSIGNED(ARG: INTEGER; SIZE: INTEGER) return UNSIGNED;
function CONV_UNSIGNED(ARG: UNSIGNED; SIZE: INTEGER) return UNSIGNED;
function CONV_UNSIGNED(ARG: SIGNED; SIZE: INTEGER) return UNSIGNED;
function CONV_UNSIGNED(ARG: STD_ULOGIC; SIZE: INTEGER) return UNSIGNED;

function CONV_SIGNED(ARG: INTEGER; SIZE: INTEGER) return SIGNED;
function CONV_SIGNED(ARG: UNSIGNED; SIZE: INTEGER) return SIGNED;
function CONV_SIGNED(ARG: SIGNED; SIZE: INTEGER) return SIGNED;
function CONV_SIGNED(ARG: STD_ULOGIC; SIZE: INTEGER) return SIGNED;
function CONV_STD_LOGIC_VECTOR(ARG: INTEGER; SIZE: INTEGER)
                                             return STD_LOGIC_VECTOR;
function CONV_STD_LOGIC_VECTOR(ARG: UNSIGNED; SIZE: INTEGER)
                                             return STD_LOGIC_VECTOR;
function CONV_STD_LOGIC_VECTOR(ARG: SIGNED; SIZE: INTEGER)
```

```
                                                    return STD_LOGIC_VECTOR;
    function CONV_STD_LOGIC_VECTOR(ARG: STD_ULOGIC; SIZE: INTEGER)
                                                    return STD_LOGIC_VECTOR;
    -- zero extend STD_LOGIC_VECTOR (ARG) to SIZE,
    -- SIZE < 0 is same as SIZE = 0
    -- returns STD_LOGIC_VECTOR(SIZE-1 downto 0)
    function EXT(ARG: STD_LOGIC_VECTOR; SIZE: INTEGER) return STD_LOGIC_VECTOR;

    -- sign extend STD_LOGIC_VECTOR (ARG) to SIZE,
    -- SIZE < 0 is same as SIZE = 0
    -- return STD_LOGIC_VECTOR(SIZE-1 downto 0)
    function SXT(ARG: STD_LOGIC_VECTOR; SIZE: INTEGER) return STD_LOGIC_VECTOR;

end Std_logic_arith;
```

APPENDIX J : STD_LOGIC_MISC

```
-----------------------------------------------------------------------
--
-- Copyright (c) 1990, 1991, 1992 by Synopsys, Inc.  All rights reserved.
--
-- This source file may be used and distributed without restriction
-- provided that this copyright statement is not removed from the file
-- and that any derivative work contains this copyright notice.
--
--  Package name: std_logic_misc
--
--  Purpose: This package defines supplemental types, subtypes,
--               constants, and functions for the Std_logic_1164 Package.
--
--  Author:  GWH
--
-----------------------------------------------------------------------
library IEEE;
use IEEE.STD_LOGIC_1164.all;
library SYNOPSYS;
use SYNOPSYS.attributes.all;
package std_logic_misc is
    -- output-strength types
    type STRENGTH is (strn_X01, strn_X0H, strn_XL1, strn_X0Z, strn_XZ1,
                      strn_WLH, strn_WLZ, strn_WZH, strn_W0H, strn_WL1);
--synopsys synthesis_off
    type MINOMAX is array (1 to 3) of TIME;
    -------------------------------------------------------------------
    --
    -- functions for mapping the STD_(U)LOGIC according to STRENGTH
    --
    -------------------------------------------------------------------
    function strength_map(input: STD_ULOGIC; strn: STRENGTH) return STD_LOGIC;

    function strength_map_z(input:STD_ULOGIC; strn:STRENGTH) return STD_LOGIC;
```

```
    -------------------------------------------------------------------
    --
    -- conversion functions for STD_ULOGIC_VECTOR and STD_LOGIC_VECTOR
    --
    -------------------------------------------------------------------

--synopsys synthesis_on
    function Drive (V: STD_ULOGIC_VECTOR) return STD_LOGIC_VECTOR;

    function Drive (V: STD_LOGIC_VECTOR) return STD_ULOGIC_VECTOR;
--synopsys synthesis_off

    attribute CLOSELY_RELATED_TCF of Drive: function is TRUE;

    -------------------------------------------------------------------
    --
    -- conversion functions for sensing various types
    -- (the second argument allows the user to specify the value to
    --  be returned when the network is undriven)
    --
    -------------------------------------------------------------------

    function Sense (V: STD_ULOGIC; vZ, vU, vDC: STD_ULOGIC) return STD_LOGIC;

    function Sense (V: STD_ULOGIC_VECTOR; vZ, vU, vDC: STD_ULOGIC)
                                 return STD_LOGIC_VECTOR;
    function Sense (V: STD_ULOGIC_VECTOR; vZ, vU, vDC: STD_ULOGIC)
                                 return STD_ULOGIC_VECTOR;

    function Sense (V: STD_LOGIC_VECTOR; vZ, vU, vDC: STD_ULOGIC)
                                 return STD_LOGIC_VECTOR;
    function Sense (V: STD_LOGIC_VECTOR; vZ, vU, vDC: STD_ULOGIC)
                                 return STD_ULOGIC_VECTOR;

--synopsys synthesis_on

    -------------------------------------------------------------------
    --
    --          Function: STD_LOGIC_VECTORtoBIT_VECTOR
STD_ULOGIC_VECTORtoBIT_VECTOR
    --
    --          Purpose: Conversion fun. from STD_(U)LOGIC_VECTOR to BIT_VECTOR
    --
    --          Mapping:    0, L --> 0
    --                  1, H --> 1
    --                  X, W --> vX if Xflag is TRUE
    --                  X, W --> 0  if Xflag is FALSE
    --                  Z --> vZ if Zflag is TRUE
    --                  Z --> 0  if Zflag is FALSE
    --                  U --> vU if Uflag is TRUE
    --                  U --> 0  if Uflag is FALSE
    --                  - --> vDC if DCflag is TRUE
    --                  - --> 0  if DCflag is FALSE
    --
    -------------------------------------------------------------------

    function STD_LOGIC_VECTORtoBIT_VECTOR (V: STD_LOGIC_VECTOR
--synopsys synthesis_off
              ; vX, vZ, vU, vDC: BIT := '0';
                Xflag, Zflag, Uflag, DCflag: BOOLEAN := FALSE
--synopsys synthesis_on
              ) return BIT_VECTOR;
```

```
    function STD_ULOGIC_VECTORtoBIT_VECTOR (V: STD_ULOGIC_VECTOR
--synopsys synthesis_off
              ; vX, vZ, vU, vDC: BIT := '0';
                Xflag, Zflag, Uflag, DCflag: BOOLEAN := FALSE
--synopsys synthesis_on
              ) return BIT_VECTOR;


    -------------------------------------------------------------------
    --
    --          Function: STD_ULOGICtoBIT
    --
    --          Purpose: Conversion function from STD_(U)LOGIC to BIT
    --
    --          Mapping:    0, L --> 0
    --                  1, H --> 1
    --                  X, W --> vX if Xflag is TRUE
    --                  X, W --> 0  if Xflag is FALSE
    --                  Z --> vZ if Zflag is TRUE
    --                  Z --> 0  if Zflag is FALSE
    --                  U --> vU if Uflag is TRUE
    --                  U --> 0  if Uflag is FALSE
    --                  - --> vDC if DCflag is TRUE
    --                  - --> 0  if DCflag is FALSE
    --
    -------------------------------------------------------------------

    function STD_ULOGICtoBIT (V: STD_ULOGIC
--synopsys synthesis_off
              ; vX, vZ, vU, vDC: BIT := '0';
                Xflag, Zflag, Uflag, DCflag: BOOLEAN := FALSE
--synopsys synthesis_on
              ) return BIT;

        -------------------------------------------------------------------
        function AND_REDUCE(ARG: STD_LOGIC_VECTOR) return UX01;
        function NAND_REDUCE(ARG: STD_LOGIC_VECTOR) return UX01;
        function OR_REDUCE(ARG: STD_LOGIC_VECTOR) return UX01;
        function NOR_REDUCE(ARG: STD_LOGIC_VECTOR) return UX01;
        function XOR_REDUCE(ARG: STD_LOGIC_VECTOR) return UX01;
        function XNOR_REDUCE(ARG: STD_LOGIC_VECTOR) return UX01;

        function AND_REDUCE(ARG: STD_ULOGIC_VECTOR) return UX01;
        function NAND_REDUCE(ARG: STD_ULOGIC_VECTOR) return UX01;
        function OR_REDUCE(ARG: STD_ULOGIC_VECTOR) return UX01;
        function NOR_REDUCE(ARG: STD_ULOGIC_VECTOR) return UX01;
        function XOR_REDUCE(ARG: STD_ULOGIC_VECTOR) return UX01;
        function XNOR_REDUCE(ARG: STD_ULOGIC_VECTOR) return UX01;

--synopsys synthesis_off

        function fun_BUF3S(Input, Enable: UX01; Strn: STRENGTH) return STD_LOGIC;
        function fun_BUF3SL(Input, Enable: UX01; Strn: STRENGTH) return STD_LOGIC;
        function fun_MUX2x1(Input0, Input1, Sel: UX01) return UX01;

        function fun_MAJ23(Input0, Input1, Input2: UX01) return UX01;
        function fun_WiredX(Input0, Input1: std_ulogic) return STD_LOGIC;

--synopsys synthesis_on

end;
```

APPENDIX K : VHDL PREDEFINED ATTRIBUTES

VHDL Attributes

Attribute	Prefix	Comments
T'base	Type	Base type of T. Must be prefix to another attribute
T'left	scalar	The left bound of T, result of type T
T'right	type/ST	The right bound of T, result of type T
T'high	scalar	The upper bound of T, result of type T
T'low	type/ST	The lower bound of T, result of type T
T'Ascending VHDL'93	scalar type/ST	TRUE if type T is ascending
T'image(X) VHDL'93	scalar type/ST	Function that converts scalar object X of type T into string
T'value(X) VHDL'93	scalar type/ST	Function that converts object X of type string into scalar of type T
T'pos(X)	discrete /PT/ST	Function that returns a universal integer representing the position number of parameter X of type T. First position = 0.
T'val(X)	discrete /PT/ST	Function that returns of base type T the value whose position is the universal integer value corresponding to X.
T'succ(X)	discrete /PT/ST	Function returning a value of type T whose value is the position number one greater than the one of the parameter. It is an error if X = T'high or if does not belong to the range T'low to T'high
T'pred(X)	discrete /PT/ST	Function returning a value of type T whose value is the position number one less than the one of the parameter. It is an error if X = T'low or if does not belong to the range T'low to T'high
T'leftof(X)	discrete /PT/ST	Function that returns the value that is to the left of parameter X of type T.

		Result type is of type T. Error if X = T'left
T'rightof(X)	discrete /PT/ST	Function that returns the value that is to the right of parameter X of type T. Result type is of type T. Error is X = T'right
A'left(N)	Array*	Function that returns the left bound of of the Nth index range of A. X is of type universal integer. Result type is of type of the left bound of the left index range of A. N = 1 if omitted.
A'right(A)	Array*	Same as A'left(N), except right bound is returned
A'high(N)	Array*	Function that returns the upper bound of the range of A. Result type is the type of the Nth index range of A. N = 1 if omitted.
A'low(N)	Array*	Same as A'high(N), e lower bound is returned.
A'range(N)	Array*	The range of A'left(N) to A'right(N)
A'reverse _range(N)		The range of A'right(N) to A'left(N)
A'length	Array*	returns 0 is array os null. Else, returns T'pos(A'high(N)) - T'pos(A'low(N) where T is the subtype of the Nth index of A.
A'Ascending	Array*	True if Nth index range of A is defined in an ascending range, else returns false.

PT = physical type
ST = Subtype
Array* = Any prefix that is appropriate for an array object, (e.g. type, variable, signal) or alias therof, or that denotes a constrained array subtype

Summary of the VHDL Signal Attributes

S'event	Function returning a **Boolean** that identifies if signal S has a new value assigned onto this signal (i.e. value is different that last value). **if** Clk'event **then** -- if Clk just changed in value then ... **wait until** Clk'event **and** Clk = '1'; -- rising edge of clock
S'active	Function returning a **Boolean** that identifies if signal S had a new assignment made onto it (whether the value of the assignment is the SAME or DIFFERENT. **if** Data'active **then** -- New assignment of Data
S'transaction	**Implicit signal** of type **bit** that is created for signal S when it S'transaction is used in the code. This implicit signal is NOT declared since it is implicitly defined. This signal toggles in value (between '0' and '1') when signal S had a new assignment made onto it (whether the value of the assignment is the SAME or DIFFERENT. The user should NOT rely on its VALUE. **wait on** ReceivedData'transaction; -- process is -- sensitive to ReceiveData changing value
S'delayed(T)	**Implicit signal** of the same **base type as S**. It represents the value of signal S delayed by a time Tn. Thus, the value of S'delayed(T) at time Tn is always equal to the value of S at time Tn -t. For example, the value of S'delayed(5 ns) at time 1000 ns is the value of S at time 995 ns. Note if time is omitted, it defaults to 0 ns. **wait on** Data'transaction; **case** BV2'(Data'Delayed & Data) **is** -- Data @ last delta time **when "X0" => ...** -- from X to 0 transition **when "10:"** => ... -- from 1 to 0 transition **when others => ... --** **end case;**
S'stable(T)	**Implicit signal** of **Boolean** type. This implicit signal is true when an event (change in value) has NOT occurred on signal S for T time units, and the value FALSE otherwise. If time is omitted, it defaults to 0 ns. **if** Data'stable(40 **ns**) **then** -- met set up time
S'quiet(T)	**Implicit signal** of **Boolean** type. This implicit signal is true when the signal has been quiet (i.e. no activity or signal assignment) for T time units, and the value FALSE otherwise. If time is omitted, it defaults to 0 ns. **if** Data'quiet(40 **ns**) **then** -- Really quiet, not even an assignment of -- the same value during the last T time units
S'last_event	The amount of **time** that has elapsed since the last event (change in value) occurred on signal S. If there was no previous event, it returns Time'high. **variable** : TsinceLastEvent : time; -- .. TsinceLastEvent := Data'last_event;
S'last_active	The amount of **time** that has elapsed since the last activity (assignment) occurred on signal S. If there was no previous event, it returns Time'high. **variable** : TsinceLastEvent : time; -- .. TsinceLastEvent := Data'last_active;
S'last_value	Function of the **base type of S** returning the previous value of S, immediately before the last change of S. **wait on** Data'**transaction**; **case** BV2'(Data'last_value & Data) **is** -- Data @ last value **when "X0" => ...** -- from X to 0 transition

	when "10" => ... -- from 1 to 0 transition **when others => ... --** **end case;**

INDEX

A

B

C

D

E

H

I

L

M

R

S

T

U

V

W

CD INCLUDES

- **All source code in book**

- **30-day ModelSim evaluation code**
  The most widely used VHDL simulator, power-packed for advanced ASIC, FPGA, and system design with an intuitive interface ideal for the first time VHDL user
  (See www.model.com for more information)

- **20-day Synplify evaluation code**
  The most powerful VHDL and Verilog Synthesis Solution for FPGA and CPLD Designers, industry's best quality of results, easy to learn and use, unmatched productivity, with unique code analysis and debugging.
  (see http://www.synplicity.com for more information)

- **GNU toolset with**
  - **EMACS, a very powerful language sensitive editor with many modes including VHDL, Verilog, C, Ada, and Java.**
  - **Tshell and GNU help files *(in Windows Help format)***

- **NotGnu, a compact EMACS editor**

- **VHDL and Verilog reference cards**

- **VHDL Syntax in HTML**